Memoirs
of the
American Mathematical Society

Number 1299

C-Projective Geometry

David M. J. Calderbank
Michael G. Eastwood
Vladimir S. Matveev
Katharina Neusser

September 2020 • Volume 267 • Number 1299 (third of 7 numbers)

Library of Congress Cataloging-in-Publication Data

Cataloging-in-Publication Data has been applied for by the AMS.
See http://www.loc.gov/publish/cip/.
DOI: https://doi.org/10.1090/memo/1299

Memoirs of the American Mathematical Society

This journal is devoted entirely to research in pure and applied mathematics.

Subscription information. Beginning with the January 2010 issue, *Memoirs* is accessible from www.ams.org/journals. The 2020 subscription begins with volume 263 and consists of six mailings, each containing one or more numbers. Subscription prices for 2020 are as follows: for paper delivery, US$1085 list, US$868 institutional member; for electronic delivery, US$955 list, US$764 institutional member. Upon request, subscribers to paper delivery of this journal are also entitled to receive electronic delivery. If ordering the paper version, add US$20 for delivery within the United States; US$80 for outside the United States. Subscription renewals are subject to late fees. See www.ams.org/help-faq for more journal subscription information. Each number may be ordered separately; *please specify number* when ordering an individual number.

Back number information. For back issues see www.ams.org/backvols.

Subscriptions and orders should be addressed to the American Mathematical Society, P. O. Box 845904, Boston, MA 02284-5904 USA. *All orders must be accompanied by payment.* Other correspondence should be addressed to 201 Charles Street, Providence, RI 02904-2213 USA.

Copying and reprinting. Individual readers of this publication, and nonprofit libraries acting for them, are permitted to make fair use of the material, such as to copy select pages for use in teaching or research. Permission is granted to quote brief passages from this publication in reviews, provided the customary acknowledgment of the source is given.

Republication, systematic copying, or multiple reproduction of any material in this publication is permitted only under license from the American Mathematical Society. Requests for permission to reuse portions of AMS publication content are handled by the Copyright Clearance Center. For more information, please visit www.ams.org/publications/pubpermissions.

Send requests for translation rights and licensed reprints to reprint-permission@ams.org.

Excluded from these provisions is material for which the author holds copyright. In such cases, requests for permission to reuse or reprint material should be addressed directly to the author(s). Copyright ownership is indicated on the copyright page, or on the lower right-hand corner of the first page of each article within proceedings volumes.

Memoirs of the American Mathematical Society (ISSN 0065-9266 (print); 1947-6221 (online)) is published bimonthly (each volume consisting usually of more than one number) by the American Mathematical Society at 201 Charles Street, Providence, RI 02904-2213 USA. Periodicals postage paid at Providence, RI. Postmaster: Send address changes to Memoirs, American Mathematical Society, 201 Charles Street, Providence, RI 02904-2213 USA.

© 2020 by the American Mathematical Society. All rights reserved.
This publication is indexed in *Mathematical Reviews*®, *Zentralblatt MATH*, *Science Citation Index*®, *Science Citation Index*[TM]*-Expanded*, *ISI Alerting Services*[SM], *SciSearch*®, *Research Alert*®, *CompuMath Citation Index*®, *Current Contents*®/*Physical, Chemical & Earth Sciences*.
This publication is archived in *Portico* and *CLOCKSS*.
Printed in the United States of America.

∞ The paper used in this book is acid-free and falls within the guidelines
established to ensure permanence and durability.
Visit the AMS home page at https://www.ams.org/

10 9 8 7 6 5 4 3 2 1   25 24 23 22 21 20

Contents

Introduction		1
Chapter 1.	Almost complex manifolds	7
Chapter 2.	Elements of c-projective geometry	13
Chapter 3.	Tractor bundles and BGG sequences	33
Chapter 4.	Metrisability of almost c-projective structures	51
Chapter 5.	Metrisability, conserved quantities and integrability	73
Chapter 6.	Metric c-projective structures and nullity	91
Chapter 7.	Global results	115
Chapter 8.	Outlook	129
Bibliography		133

Abstract

We develop in detail the theory of (almost) c-projective geometry, a natural analogue of projective differential geometry adapted to (almost) complex manifolds. We realise it as a type of parabolic geometry and describe the associated Cartan or tractor connection. A Kähler manifold gives rise to a c-projective structure and this is one of the primary motivations for its study. The existence of two or more Kähler metrics underlying a given c-projective structure has many ramifications, which we explore in depth. As a consequence of this analysis, we prove the Yano–Obata Conjecture for complete Kähler manifolds: if such a manifold admits a one parameter group of c-projective transformations that are not affine, then it is complex projective space, equipped with a multiple of the Fubini–Study metric.

Received by the editor June 7, 2016, and, in revised form, September 13, 2017.

Article electronically published on January 4, 2021.

DOI: https://doi.org/10.1090/memo/1299

2010 *Mathematics Subject Classification*. Primary 53B10, 53B35, 32Q60, 53C24, 58J60; Secondary 53A20, 53C25, 53C55, 53D25, 58J70.

The first author is affiliated with the Department of Mathematical Sciences, University of Bath, Bath BA2 7AY, United Kingdom. Email: D.M.J.Calderbank@bath.ac.uk.

The second author is affiliated with the School of Mathematical Sciences, University of Adelaide, SA 5005 Australia. Email: michael.eastwood@adelaide.edu.au.

The third author is affiliated with the Institute of Mathematics, Friedrich Schiller University Jena, 07737 Jena, Germany. Email: vladimir.matveev@uni-jena.de.

The fourth author is affiliated with the Department of Mathematics and Statistics, Masaryk University, 611 37 Brno, Czech Republic. Email: neusser@math.muni.cz.

This article was initiated when its authors participated in a workshop at the Kioloa campus of the Australian National University in March 2013. We would like to thank the The Edith and Joy London Foundation for providing the excellent facilities at Kioloa. We would also like to thank the Group of Eight, Deutscher Akademischer Austausch Dienst, Australia-Germany Joint Research Cooperation Scheme for financially supporting the workshop in 2013 and a subsequent Kioloa workshop in 2014; for the latter, we thank, in addition, FSU Jena and the Deutsche Forschungsgemeinschaft (GK 1523/2) for their financial support. The fourth author was also supported during part of this project by Grant P201/12/G028 from the Czech Grant Agency. We would also like to thank the latter for supporting a meeting of the first, second, and fourth authors at the Mathematical Institute at Charles University in July 2014.

©2020 American Mathematical Society

Introduction

C-projective geometry is a natural analogue of real projective differential geometry for complex manifolds. Like projective geometry, it has many facets, which have been discovered and explored independently and repeatedly over the past sixty years. Our aim in this work is to develop in detail a unified theory of c-projective geometry, which highlights its relation with real projective geometry as well as its differences.

Projective geometry is a classical subject concerned with the behaviour of straight lines, or more generally, (unparametrised) geodesic curves of a metric or affine connection. It has been known for some time [66, 100] that two non-proportional metrics can have the same geodesic curves: central projection takes great circles on the n-sphere, namely the geodesics for the round metric, to straight lines in Euclidean n-space, namely geodesics for the flat metric. The quotient of the round n-sphere under the antipodal identification may be identified with the *flat model* for n-dimensional projective geometry: the real projective n-space $\mathbb{RP}^n$, viewed as a homogeneous space under the group $\mathrm{PGL}(n+1,\mathbb{R})$ of projective transformations, which preserve the family of (linearly embedded) projective lines $\mathbb{RP}^1 \hookrightarrow \mathbb{RP}^n$. More generally, a *projective structure* on a smooth n-manifold (for $n \geq 2$) is an equivalence class of torsion-free affine connections having the same geodesic curves. In this setting, it is a nontrivial and interesting question whether these curves are the geodesic curves of a (pseudo-)Riemannian metric, i.e. whether any connection in the projective equivalence class preserves a nondegenerate metric, possibly of indefinite signature. Such projective structures are called *metrisable* and the corresponding metrics *compatible*. Rather surprisingly, the partial differential equations controlling the metrisability of a given projective structure can be set up as a *linear* system [13, 40, 70, 95]. More precisely, there is a projectively invariant linear differential operator acting on symmetric contravariant 2-tensors such that the nondegenerate elements of its kernel correspond to compatible metrics.

In modern language, a projective structure determines a canonical *Cartan connection* [39] modelled on $\mathbb{RP}^n$, and hence projective geometry is a *parabolic geometry* [36, 42]. In these terms, the metrisability operator is a first *BGG (Bernstein–Gelfand–Gelfand) operator*, which is a differential operator of finite type [43]. Its solutions correspond to parallel sections of a bundle with connection, which is, up to curvature corrections, a linear representation of the Cartan connection. The kernel is thus finite-dimensional; it is zero for generic projective structures, with the maximal dimension attained on the flat model $\mathbb{RP}^n$. The parabolic viewpoint on projective geometry has proven to be very useful, for example in understanding projective compactifications of Einstein metrics [31, 34], the geometry of holonomy reductions of projective structures [6], and (solving problems posed by Sophus Lie in 1882) projective vector fields on surfaces [22, 75].

Projective geometry has been linked to the theory of finite dimensional integrable systems with great success: the equation for symmetric Killing tensors is projectively invariant [42], and (consequently) the existence of two projectively equivalent metrics on a manifold implies the existence of nontrivial integrals for the geodesic flows of both metrics. This method has been effectively employed when the manifold is closed or complete (see e.g. [72, 74]). Moreover, the integrability of many classically studied integrable geodesic flows (e.g., on ellipsoids) is closely related to the existence of a projectively compatible metric, and many geometric structures that lead to such integrable geodesic flows have been directly related to the existence of a projectively compatible metric, see e.g. [9, 13].

C-projective geometry arises when one retells this story, mutatis mutandis, for complex or, indeed, almost complex manifolds, i.e. smooth manifolds equipped with an almost complex structure J, which is a smooth endomorphism of the tangent bundle such that $J^2 = -\operatorname{Id}$. On such a manifold M, the relevant (pseudo-) Riemannian metrics are *Hermitian* with respect to J, i.e. J-invariant, and the relevant affine connections are those which preserve J, called *complex connections*. Such connections cannot be torsion-free unless the almost complex structure is *integrable*, i.e. its Nijenhuis tensor vanishes identically [90]. This holds in particular if the Levi-Civita connection of a Hermitian metric g preserves J, in which case g is called a (pseudo-)*Kähler* metric.

In 1947, Bochner [12, Theorem 2] observed that any two metrics that are Kähler with respect to the same complex structure cannot be projectively equivalent (i.e. have the same geodesic curves) unless they are affinely equivalent (i.e. have the same Levi-Civita connection). This led Otsuki and Tashiro [91] to introduce a broader class of curves, which they called "holomorphically flat", and which depend on both the connection and the (almost) complex structure. We refer to these curves as *J-planar*: whereas a geodesic curve for an affine connection ∇ is a curve c whose acceleration $\nabla_{\dot c}\dot c$ is proportional to its velocity $\dot c$, a J-planar curve is one whose acceleration is in the linear span of $\dot c$ and $J\dot c$. On a Riemann surface, therefore, all curves are J-planar. The other common manifold where it is possible to see all J-planar curves without computation is complex projective space with its Fubini–Study connection. The point here is that the linearly embedded complex lines $\mathbb{CP}^1 \hookrightarrow \mathbb{CP}^n$ are totally geodesic. Therefore, the J-planar curves on $\mathbb{CP}^n$ are precisely the smooth curves lying within such complex lines. Viewed in a standard affine coordinate patch $\mathbb{C}^n \hookrightarrow \mathbb{CP}^n$, the J-planar curves are again the smooth curves lying inside an arbitrary complex line $\{az + b\} \subset \mathbb{C}^n$ but otherwise unconstrained. Evidently, these are the intrinsic J-planar curves for the flat connection on $\mathbb{C}^n$.

The J-planar curves provide a nontrivial notion of projective equivalence in complex differential geometry, due to Otsuki and Tashiro [91] in the Kähler setting, and Tashiro [99] for almost complex manifolds in general. Two complex connections on an almost complex manifold (M, J) are *c-projectively equivalent* if they have the same torsion and the same J-planar curves. An *almost c-projective manifold* is a complex manifold (M, J) equipped with a c-projective equivalence class of such connections. If J is integrable, we follow the usual convention and drop the word "almost" to arrive at the notion of a *c-projective manifold*. We caution the reader that Otsuki and Tashiro [91], and many later researchers, refer to "holomorphically projective correspondences", rather than c-projective equivalences, and many authors use the terminology "h-projective" or "holomorphic(ally) projective" instead of "c-projective". We avoid their terminology because the connections in a c-projective class are typically not holomorphic, even if the complex structure is integrable; similarly, we avoid the term "complex projective structure", which is often used for the holomorphic analogue of a real projective structure, or related concepts.

During the decades following Otsuki and Tashiro's 1954 paper, c-projective structures provided a prominent research direction in Japanese and Soviet differential geometry. Many of the researchers involved had some background in projective geometry, and the dominant line of investigation sought to generalise methods and results from projective geometry to the c-projective setting. This was a very productive direction, with more than 300 publications appearing in the relatively short period from 1960 to 1990. One can compare, for example, the surveys by Mikeš [83, 84], or the papers of Hiramatu [53, 54], to see how successfully c-projective analogues of results in projective geometry were found. In particular, the linear system for c-projectively equivalent Kähler metrics was obtained by Domashev and Mikeš [41], and its finite type prolongation to a connection was given by Mikeš [82].

Relatively recently, the linear system for c-projectively equivalent Kähler metrics has been rediscovered, under different names and with different motivations. On a fixed complex manifold, a compatible (pseudo-)Kähler metric is determined uniquely by its Kähler form (a compatible symplectic form), and under this correspondence, c-projectively equivalent Kähler metrics are essentially the same as *Hamiltonian 2-forms* defined and investigated in Apostolov et al. [2–5]: the defining equation [2, (12)] for a Hamiltonian 2-form is actually algebraically equivalent to the metrisability equation (125). In dimension ≥ 6, c-projectively equivalent metrics are also essentially the same as conformal Killing (or twistor) $(1,1)$-forms studied in [87, 93, 94], see [2, Appendix A] or [79, §1.3] for details.

The work of [2, 3] provides, *a postiori*, local and global classification results for c-projectively equivalent Kähler metrics, although the authors were unaware of this interpretation and of the pre-existing literature. Instead, as explained in [2, 3] and [26], the notion and study of Hamiltonian 2-forms was motivated by their natural appearance in many interesting problems in Kähler geometry, and the unifying role they play in the construction of explicit Kähler metrics on projective bundles. In subsequent papers, e.g. [4, 5], Hamiltonian 2-form methods were used to construct many new examples of Kähler manifolds and orbifolds with interesting properties.

Another independent line of research closely related to c-projectively equivalent metrics (and perhaps underpinning the utility of Hamiltonian 2-forms) appeared within the theory of finitely dimensional integrable systems. C-projectively equivalent metrics are closely related (see e.g. [61]) to the so-called Kähler–Liouville integrable systems of type A introduced and studied by Kiyohara in [59]. In fact, Topalov [101] (see also [60]) shows that generic c-projectively equivalent Kähler metrics have integrable geodesic flows, cf. [102] for the analogous result in the projective case. On the one hand, integrability provides, as in projective geometry, a number of new methods that can be used in c-projective geometry. On the other hand, examples from c-projective geometry turn out to be interesting for the theory of integrable systems, since there are only a few known examples of Kähler metrics with integrable geodesic flows.

Despite the many analogies between results in projective and c-projective geometry, there seem to be very few attempts in the literature to explain why these two subjects are so similar. In 1978, it was noted by Yoshimatsu [104] that c-projective manifolds admit canonical Cartan connections, and this was generalised to almost c-projective manifolds by Hrdina [55] in 2009. Thus c-projective geometry, like projective geometry, is a parabolic geometry; its flat model is $\mathbb{CP}^n$, viewed as a homogeneous space under the group $\mathrm{PSL}(n+1,\mathbb{C})$ of projective transformations, which preserve the J-planar curves described above. Despite this, c-projective structures have received very little attention in the parabolic geometry literature: apart from the work of Hrdina, and some work in dimension 4 [28, 81], they have only been studied in [6], where they appear as holonomy reductions of projective geometries. A possible explanation for this oversight is that $\mathrm{PSL}(n+1,\mathbb{C})$ appears in c-projective geometry as a real Lie group and, as such, its complexification is semisimple, but not simple. This is related to the subtle point that most interesting c-projective structures are not holomorphic.

The development of c-projective geometry, as described above, has been rather nonlinear until relatively recently, when a number of independent threads have converged on a coherent set of ideas. However, a thorough description of almost c-projective manifolds in the framework of parabolic geometries is lacking in the literature. We therefore believe it is timely to lay down the fundamentals of such a theory.

The article is organised as follows. In Chapter 1, we survey the background on almost complex manifolds and complex connections. As we review in Chapter 1.2, the torsion of any complex connection on an almost complex manifold, of real dimension $2n \geq 4$, naturally decomposes into five irreducible pieces, one of which is invariantly defined and can be identified as the Nijenhuis tensor. All other pieces can be eliminated by a suitable choice of complex connection, which we call *minimal*. In first four sections of the article we carry along the Nijenhuis tensor in almost all calculations and discussions.

Chapter 2 begins with the classical viewpoint on almost c-projective structures, based on J-planar curves and equivalence classes of minimal complex connections [91]. We then recall the notion of parabolic geometries and establish, in Theorem 2.8, an equivalence of categories between almost c-projective manifolds and parabolic geometries with a normal Cartan connection, modelled on $\mathbb{CP}^n$ viewed as a homogeneous space of $\mathrm{PSL}(n+1,\mathbb{C})$.

As a consequence of this parabolic viewpoint, we can associate a fundamental local invariant to an almost c-projective manifold, namely the curvature κ of its normal Cartan connection; furthermore, $\kappa \equiv 0$ if and only if the almost c-projective manifold is locally isomorphic to $\mathbb{CP}^n$ equipped with its standard c-projective structure. Since the Cartan connection is normal (for this we need the complex connections to be minimal), its curvature is a 2-cycle for Lie algebra homology, and is uniquely determined by its homology class, also known as the *harmonic curvature*. We construct and discuss this curvature in section 2.7. For almost c-projective structures there are three irreducible parts to the harmonic curvature. One of the pieces is the Nijenhuis tensor, which is precisely the obstruction to the underlying almost complex manifold actually being complex. One of the other two parts is precisely the obstruction to there being a holomorphic connection in the c-projective class. When it vanishes we end up with *holomorphic projective geometry*, i.e. ordinary projective differential geometry but in the holomorphic category. The remaining piece can then be identified with the classical projective Weyl curvature (for $n \geq 3$) or Liouville curvature (for $n = 2$).

Another consequence of the parabolic perspective is that representation theory is brought to the fore, both as the appropriate language for discussing natural bundles on almost c-projective manifolds, and also as the correct tool for understanding invariant differential operators on the flat model, and their curved analogues. The various *BGG complexes* on $\mathbb{CP}^n$ and their curved analogues are systematically introduced and discussed in Chapter 3.

In particular, there is a BGG operator that controls the *metrisability* of a c-projective structure just as happens in the projective setting. A large part of this article is devoted to the metrisability equation, which we introduce in Chapter 4, where we also obtain its prolongation to a connection, not only for compatible (pseudo-)Kähler metric, but also in the non-integrable case of quasi-Kähler or (2,1)-symplectic structures. For the remainder of the article, we suppose that the Nijenhuis tensor vanishes, in other words that we are starting with a complex manifold. In this case, a compatible metric is exactly a (pseudo-)Kähler metric (and a *normal solution* of the metrisability equation corresponds to a (pseudo-)Kähler–Einstein metric). We also restrict our attention to *metric c-projective structures*, i.e. to the metrisable case where the c-projective structure arises from a (pseudo-)Kähler metric. Borrowing terminology from the projective case, we refer to the dimension of the solution space of the metrisability equation as the (*degree of*) *mobility* of the metric c-projective structure (or of any compatible (pseudo-)Kähler metric). We are mainly interested in understanding when the metric c-projective structure has mobility at least two, and the consequences this has for the geometry and topology of the manifold.

In Chapter 5, we develop the consequences of mobility for integrability, by showing that a pencil (two dimensional family) of solutions to the metrisability equation generates a family of holomorphic Killing vector fields and Hermitian symmetric Killing tensors, which together provide commuting linear and quadratic integrals for the geodesic flow of any metric in the pencil. In Chapter 6, we study an important, but somewhat mysterious, phenomenon in which tractor bundles for metric c-projective geometries are naturally equipped with congenial connections, which are neither induced by the normal Cartan connection nor equal to the prolongation connection, but which have the property that their covariant constant sections nevertheless correspond to solutions of the corresponding first BGG operator.

We bring these tools together in Chapter 7, where we establish the Yano–Obata Conjecture for complete Kähler manifolds, namely that the identity component of the group of c-projective transformations of the manifold consists of affine transformations unless the manifold is complex projective space equipped with a multiple of the Fubini–Study metric. This result is an analogue of the *the Projective Lichnerowicz Conjecture* obtained in [**73, 74**], but the proof given there does not generalise directly to the c-projective situation. Our proof also differs from the proof for closed manifolds given in [**78**], and makes use of many preliminary results obtained by the methods of parabolic geometry, which also apply in the projective case.

Here, and throughout the article, we see that not only results from projective geometry, but also methods and proofs, can be generalised to the c-projective case, and we explain why and how. We hope that this article will set the scene for what promises to be an interesting series of further developments in c-projective geometry. In fact, several such developments already appeared during our work on this article, which we discuss in Chapter 8.

Acknowledgement We would like to thank the referee for a careful reading of our manuscript resulting in many helpful suggestions for its improvement.

CHAPTER 1

Almost complex manifolds

Recall that an *almost complex structure* on a smooth manifold M is a smooth endomorphism J of the tangent bundle TM of M that satisfies $J^2 = -\operatorname{Id}$. Equivalently, an almost complex structure makes TM into a complex vector bundle in which multiplication by i is decreed to be the real endomorphism J. In particular, the dimension of M is necessarily even, say $2n$, and an almost complex structure is yet equivalently a reduction of structure group to $\operatorname{GL}(n, \mathbb{C}) \subset \operatorname{GL}(2n, \mathbb{R})$.

1.1. Real and complex viewpoints

If M is a complex manifold in the usual sense of being equipped with holomorphic transition functions, then TM is a complex vector bundle and multiplication by i defines a real endomorphism $TM \to TM$, which we write as J. This is enough to define the holomorphic structure on M: holomorphic functions may be characterised amongst all smooth complex-valued functions $f = u + iv$ as satisfying $Xu = (JX)v$ for all vector fields X (the *Cauchy–Riemann equations*).

Thus, complex manifolds may be regarded as a subclass of almost complex manifolds and the celebrated Newlander–Nirenberg Theorem tells us how to recognise them:

THEOREM 1.1 (Newlander–Nirenberg, [**90**]). *An almost complex manifold* (M, J) *is a complex manifold if and only if the tensor*

(1) $$N^J(X, Y) := [X, Y] - [JX, JY] + J([JX, Y] + [X, JY])$$

vanishes for all vector fields X and Y on M, where $[\ ,\ ]$ denotes the Lie bracket of vector fields.

Note that $N^J \colon TM \times TM \to TM$ is a 2-form with values in TM, which satisfies $N^J(JX, Y) = -JN^J(X, Y)$. It is called the *Nijenhuis tensor* of J. When N^J vanishes we say that the almost complex structure J is *integrable*. This viewpoint on complex manifolds, as even-dimensional smooth manifolds equipped with integrable almost complex structures, turns out to be very useful especially from the differential geometric viewpoint.

It is useful to complexify the tangent bundle of M and decompose the result into eigenbundles under the action of J. Specifically,

(2) $$\mathbb{C}TM = T^{1,0}M \oplus T^{0,1}M = \{X \text{ s.t. } JX = iX\} \oplus \{X \text{ s.t. } JX = -iX\}.$$

Notice that $T^{0,1}M = \overline{T^{1,0}M}$. There is a corresponding decomposition of the complexified cotangent bundle, which we write as $\wedge^1 M$ or simply $\wedge^1$ if M is understood. Specifically,

(3) $$\wedge^1 = \wedge^{0,1} \oplus \wedge^{1,0} = \{\omega \text{ s.t. } J\omega = -i\omega\} \oplus \{\omega \text{ s.t. } J\omega = i\omega\},$$

where sections of $\wedge^{1,0}$ respectively of $\wedge^{0,1}$ are known as 1-forms of *type* $(1,0)$ respectively $(0,1)$, see e.g. [**62**]. Notice that the canonical complex linear pairing between $\mathbb{C}TM$ and $\wedge^1 M$ induces natural isomorphisms $\wedge^{0,1} = (T^{0,1})^*$ and $\wedge^{1,0} = (T^{1,0})^*$ of complex vector bundles.

It is convenient to introduce abstract indices [**92**] for real or complex tensors on M and also for sections of the bundles $T^{1,0}M$, $\wedge^{0,1}$, and so on. Let us write X^α for real or complex fields and ω_α for real or complex 1-forms on M. In local coordinates α would range over $1, 2, \ldots, 2n$ where $2n$ is the real dimension of M. Let us denote by X^a a section of $T^{1,0}M$. In any frame, the index a would then range over $1, 2, \ldots, n$. Similarly, let us write $X^{\bar{a}}$ for a section of $T^{0,1}M$ and the conjugate linear isomorphism $T^{0,1}M = \overline{T^{1,0}M}$ as $X^a \mapsto \overline{X^a} = \overline{X}^{\bar{a}}$. Accordingly, sections of $\wedge^{1,0}$ and $\wedge^{0,1}$ will be denoted by ω_a and $\omega_{\bar{a}}$ respectively, and the canonical pairings between $T^{1,0}M$ and $\wedge^{1,0}$, respectively $T^{0,1}M$ and $\wedge^{0,1}$, written as $X^a \omega_a$, respectively $X^{\bar{a}} \omega_{\bar{a}}$, an abstract index counterpart to the *Einstein summation convention*.

We shall need the complex linear homomorphism $\mathbb{C}TM \to T^{1,0}M$ defined as projection onto the first summand in the decomposition (2) and given explicitly as $X \mapsto \frac{1}{2}(X - iJX)$. It is useful to write it in abstract indices as

$$X^\alpha \mapsto \Pi^a_\alpha X^\alpha.$$

It follows that the dual homomorphism $\wedge^{1,0} \hookrightarrow \wedge^1$ is given in abstract indices by

$$\omega_a \mapsto \Pi^a_\alpha \omega_a$$

and also that the homomorphisms $\mathbb{C}TM \to T^{0,1}M$ and $\wedge^{0,1} \hookrightarrow \wedge^1$ are given by

$$X^\alpha \mapsto \overline{\Pi}^{\bar{a}}_\alpha X^\alpha \quad \text{and} \quad \omega_{\bar{a}} \mapsto \overline{\Pi}^{\bar{a}}_\alpha \omega_{\bar{a}},$$

respectively.

Let us denote by $X^a \mapsto \Pi^\alpha_a X^a$, the inclusion $T^{1,0}M \hookrightarrow \mathbb{C}TM$, paying attention to the distinction in their indices between Π^a_α and Π^α_a. Various identities follow, such as $\Pi^\alpha_a \Pi^b_\alpha = \delta_a{}^b$, where the Kronecker delta $\delta_a{}^b$ denotes the identity transformation on $T^{1,0}M$. The symbol Π^α_a also gives us access to the dual and conjugate homomorphisms. Thus,

$$\omega_\alpha \mapsto \overline{\Pi}^\alpha_{\bar{a}} \omega_\alpha$$

extracts the $(0,1)$-part of a complex-valued 1-form ω_α on M. The following identities are immediate from the definitions

(4)
$$\begin{aligned} \Pi^a_\alpha \Pi^\beta_a &= \tfrac{1}{2}(\delta_\alpha{}^\beta - iJ_\alpha{}^\beta) & \overline{\Pi}^{\bar{a}}_\alpha \overline{\Pi}^\beta_{\bar{a}} &= \tfrac{1}{2}(\delta_\alpha{}^\beta + iJ_\alpha{}^\beta) \\ \Pi^\alpha_a J_\alpha{}^\beta &= i\Pi^\beta_a & \overline{\Pi}^\alpha_{\bar{a}} J_\alpha{}^\beta &= -i\overline{\Pi}^\beta_{\bar{a}} \\ J_\alpha{}^\beta \Pi^a_\beta &= i\Pi^a_\alpha & J_\alpha{}^\beta \overline{\Pi}^{\bar{a}}_\beta &= -i\overline{\Pi}^{\bar{a}}_\alpha. \end{aligned}$$

They are indispensable for the calculations in the following sections. Further useful abstract index conventions are as follows. Quantities endowed with several indices denote sections of the tensor product of the corresponding vector bundles. Thus, a section of $TM \otimes TM$ would be denoted $X^{\alpha\beta}$ whilst $\Phi_\alpha{}^\beta$ is necessarily a section of $T^*M \otimes TM$ or, equivalently, an endomorphism of TM, namely $X^\beta \mapsto \Phi_\alpha{}^\beta X^\alpha$, yet equivalently an endomorphism of T^*M, namely $\omega_\alpha \mapsto \Phi_\alpha{}^\beta \omega_\beta$. We have already seen this notation for an almost complex structure $J_\alpha{}^\beta$. But it is unnecessary notationally to distinguish between real- and complex-valued tensors. Thus, by ω_α we can mean a section of T^*M or of $\wedge^1 M := \mathbb{C}T^*M$ and if a distinction is warranted, then it can be made in words or by context. For example, an almost complex structure $J_\alpha{}^\beta$ is a real endomorphism whereas Π_α^a is necessarily complex.

Symmetry operations can also be written in abstract index notation. For example, the skew part of a covariant 2-tensor $\phi_{\alpha\beta}$ is $\frac{1}{2}(\phi_{\alpha\beta} - \phi_{\beta\alpha})$, which we write as $\phi_{[\alpha\beta]}$. Similarly, we write $\phi_{(\alpha\beta)} = \frac{1}{2}(\phi_{\alpha\beta} + \phi_{\beta\alpha})$ for the symmetric part and then $\phi_{\alpha\beta} = \phi_{(\alpha\beta)} + \phi_{[\alpha\beta]}$ realises the decomposition of vector bundles $\wedge^1 \otimes \wedge^1 = S^2 \wedge^1 \oplus \wedge^2$. In general, round brackets symmetrise over the indices they enclose whilst square brackets take the skew part, e.g.

$$R_{\alpha\beta}{}^\gamma{}_\delta \mapsto R_{[\alpha\beta}{}^\gamma{}_{\delta]} = \tfrac{1}{6}(R_{\alpha\beta}{}^\gamma{}_\delta + R_{\beta\delta}{}^\gamma{}_\alpha + R_{\delta\alpha}{}^\gamma{}_\beta - R_{\beta\alpha}{}^\gamma{}_\delta - R_{\alpha\delta}{}^\gamma{}_\beta - R_{\delta\beta}{}^\gamma{}_\alpha).$$

By (3) differential forms on almost complex manifolds can be naturally decomposed according to *type* (see e.g. [**62**]). We pause to examine the decomposition of 2-forms, especially from the abstract index point of view. From (3) it follows that the bundle $\wedge^2$ of complex-valued 2-forms decomposes into types according to

(5) $$\wedge^2 = \wedge^2(\wedge^{0,1} \oplus \wedge^{1,0}) = \wedge^{0,2} \oplus \wedge^{1,1} \oplus \wedge^{2,0}$$

and, as we shall make precise in Chapter 3.3, there is no finer decomposition available (it is a decomposition into *irreducibles*). Using the projectors Π_α^a and Π_a^α, we can explicitly execute this decomposition:

$$\omega_{\alpha\beta} \mapsto \left(\overline{\Pi}_{\bar a}^\alpha \overline{\Pi}_{\bar b}^\beta \omega_{\alpha\beta}, \Pi_a^\alpha \overline{\Pi}_{\bar b}^\beta \omega_{\alpha\beta}, \Pi_a^\alpha \Pi_b^\beta \omega_{\alpha\beta}\right)$$

$$\overline{\Pi}_\alpha^{\bar a} \overline{\Pi}_\beta^{\bar b} \omega_{\bar a\bar b} + 2\Pi_{[\alpha}^a \overline{\Pi}_{\beta]}^{\bar b} \omega_{a\bar b} + \Pi_\alpha^a \Pi_\beta^b \omega_{ab} \hookleftarrow (\omega_{\bar a \bar b}, \omega_{a\bar b}, \omega_{ab})$$

in accordance with (4). Notice that we made a choice here, namely to identify $\wedge^{1,1}$ as $\wedge^{1,0} \otimes \wedge^{0,1}$ *in this order* and, consequently, write forms of type $(1,1)$ as $\omega_{a\bar b}$. We could equally well choose the opposite convention or, indeed, use both conventions simultaneously representing a $(1,1)$ form as $\omega_{a\bar b}$ and/or $\omega_{\bar a b}$ but now subject to $\omega_{a\bar b} = -\omega_{\bar b a}$. Strictly speaking, this goes against the conventions of the abstract index notation [**92**] but we allow ourselves this extra leeway when it is useful. For example, the reconstructed form $\omega_{\alpha\beta}$ may then be written as

$$\omega_{\alpha\beta} = \Pi_\alpha^a \overline{\Pi}_\beta^{\bar b} \omega_{a\bar b} + \overline{\Pi}_\alpha^{\bar a} \Pi_\beta^b \omega_{\bar a b}.$$

Two-forms of various types may be characterised as

(6) $$\begin{aligned}\omega_{\alpha\beta} \text{ is type } (0,2) &\iff J_\alpha{}^\gamma \omega_{\beta\gamma} = i\omega_{\alpha\beta} \\ \omega_{\alpha\gamma} \text{ is type } (1,1) &\iff J_{[\alpha}{}^\gamma \omega_{\beta]\gamma} = 0 \\ \omega_{\alpha\beta} \text{ is type } (2,0) &\iff J_\alpha{}^\gamma \omega_{\beta\gamma} = -i\omega_{\alpha\beta}\end{aligned}$$

but already this is a little awkward and becomes more so for higher forms and extremely awkward when attempting to decompose more general tensors as we shall have cause to do when considering torsion and curvature. Notice that forms of type $(1,1)$ in (6) are characterised by a real condition. Indeed, the complex bundle $\wedge^{1,1}$ is the complexification of a real irreducible bundle whose sections are the real 2-forms satisfying $J_{[\alpha}{}^\gamma \omega_{\beta]\gamma} = 0$. As for forms of types $(0,2)$ and $(2,0)$, there is a real bundle whose sections satisfy

$$J_\alpha{}^\gamma J_\beta{}^\delta \omega_{\gamma\delta} = -\omega_{\alpha\beta}$$

(as opposed to $J_\alpha{}^\gamma J_\beta{}^\delta \omega_{\gamma\delta} = \omega_{\alpha\beta}$ for sections of $\wedge^{1,1}$) and whose complexification is $\wedge^{0,2} \oplus \wedge^{2,0}$. Thus, the real 2-forms split irreducibly into just two kinds but the complex 2-forms split into three types (5).

Notice that if E is a complex vector bundle on M, then we can decompose 2-forms with values in E into types by using the same formulae (6). In particular, we can do this on an almost complex manifold when $E = TM$, regarded as a complex bundle via the action of J. Writing this out explicitly, a real tensor $T_{\alpha\beta}{}^\gamma = T_{[\alpha\beta]}{}^\gamma$ is said to be

(7)
$$\begin{aligned}\text{of type } (0,2) &\iff J_\alpha{}^\gamma T_{\beta\gamma}{}^\delta = T_{\alpha\beta}{}^\gamma J_\gamma{}^\delta \\ \text{of type } (1,1) &\iff J_{[\alpha}{}^\gamma T_{\beta]\gamma}{}^\delta = 0 \\ \text{of type } (2,0) &\iff J_\alpha{}^\gamma T_{\beta\gamma}{}^\delta = -T_{\alpha\beta}{}^\gamma J_\gamma{}^\delta.\end{aligned}$$

For example, as the Nijenhuis tensor (1) satisfies $N^J(Y, JX) = JN^J(X, Y)$, it is of type $(0, 2)$. Further to investigate this decomposition (7), it is useful to apply the projectors Π_α^a and $\overline{\Pi}_a^\alpha$ to obtain

$$T_{\bar{a}\bar{b}}{}^{\bar{c}} \equiv \overline{\Pi}_{\bar{a}}^\alpha \overline{\Pi}_{\bar{b}}^\beta \overline{\Pi}_\gamma^{\bar{c}} T_{\alpha\beta}{}^\gamma \qquad T_{a\bar{b}}{}^{\bar{c}} \equiv \Pi_a^\alpha \overline{\Pi}_{\bar{b}}^\beta \overline{\Pi}_\gamma^{\bar{c}} T_{\alpha\beta}{}^\gamma \qquad T_{ab}{}^{\bar{c}} \equiv \Pi_a^\alpha \Pi_b^\beta \overline{\Pi}_\gamma^{\bar{c}} T_{\alpha\beta}{}^\gamma$$
$$T_{\bar{a}\bar{b}}{}^c \equiv \overline{\Pi}_{\bar{a}}^\alpha \overline{\Pi}_{\bar{b}}^\beta \Pi_\gamma^c T_{\alpha\beta}{}^\gamma \qquad T_{a\bar{b}}{}^c \equiv \Pi_a^\alpha \overline{\Pi}_{\bar{b}}^\beta \Pi_\gamma^c T_{\alpha\beta}{}^\gamma \qquad T_{ab}{}^c \equiv \Pi_a^\alpha \Pi_b^\beta \Pi_\gamma^c T_{\alpha\beta}{}^\gamma$$

satisfying

$$T_{ab}{}^c = T_{[ab]}{}^c \qquad T_{\bar{a}\bar{b}}{}^c = T_{[\bar{a}\bar{b}]}{}^c \qquad T_{ab}{}^{\bar{c}} = T_{[ab]}{}^{\bar{c}} \qquad T_{\bar{a}\bar{b}}{}^{\bar{c}} = T_{[\bar{a}\bar{b}]}{}^{\bar{c}}$$
$$\overline{T_{ab}{}^c} = T_{\bar{a}\bar{b}}{}^{\bar{c}} \qquad \overline{T_{a\bar{b}}{}^c} = -T_{b\bar{a}}{}^{\bar{c}} \qquad \overline{T_{ab}{}^{\bar{c}}} = T_{\bar{a}\bar{b}}{}^c$$

and from which we can recover $T_{\alpha\beta}{}^\gamma$ according to

$$T_{\alpha\beta}{}^\gamma = \overline{\Pi}_\alpha^{\bar{a}} \overline{\Pi}_\beta^{\bar{b}} \overline{\Pi}_{\bar{c}}^\gamma T_{\bar{a}\bar{b}}{}^{\bar{c}} + 2\Pi_{[\alpha}^a \overline{\Pi}_{\beta]}^{\bar{b}} \overline{\Pi}_{\bar{c}}^\gamma T_{a\bar{b}}{}^{\bar{c}} + \Pi_\alpha^a \Pi_\beta^b \overline{\Pi}_{\bar{c}}^\gamma T_{ab}{}^{\bar{c}}$$
$$+ \overline{\Pi}_\alpha^{\bar{a}} \overline{\Pi}_\beta^{\bar{b}} \Pi_c^\gamma T_{\bar{a}\bar{b}}{}^c + 2\Pi_{[\alpha}^a \overline{\Pi}_{\beta]}^{\bar{b}} \Pi_c^\gamma T_{a\bar{b}}{}^c + \Pi_\alpha^a \Pi_\beta^b \Pi_c^\gamma T_{ab}{}^c.$$

From (4) and (7), the splitting of $T_{\alpha\beta}{}^\gamma$ into types corresponds exactly to components

(8)
$$\begin{aligned}\text{type } (0,2) &\leftrightarrow (T_{ab}{}^{\bar{c}}, T_{\bar{a}\bar{b}}{}^c) \\ \text{type } (1,1) &\leftrightarrow (T_{a\bar{b}}{}^c, T_{a\bar{b}}{}^{\bar{c}}) \\ \text{type } (2,0) &\leftrightarrow (T_{ab}{}^c, T_{\bar{a}\bar{b}}{}^{\bar{c}}).\end{aligned}$$

Notice that, for each of types $(1,1)$ and $(2,0)$, a complex-valued 1-form can be invariantly extracted:

$$\text{type } (1,1): \phi_\alpha \equiv \Pi_\alpha^a T_{a\bar{b}}{}^{\bar{b}} = \tfrac{1}{2}\big(T_{\alpha\beta}{}^\beta + iT_{\alpha\beta}{}^\gamma J_\gamma{}^\beta\big)$$
$$\text{type } (2,0): \psi_\alpha \equiv \Pi_\alpha^a T_{ab}{}^b = \tfrac{1}{2}\big(T_{\alpha\beta}{}^\beta - iT_{\alpha\beta}{}^\gamma J_\gamma{}^\beta\big).$$

On the other hand, just from the index structure, tensors $T_{\alpha\beta}{}^\gamma = T_{[\alpha\beta]}{}^\gamma$ of type $(0, 2)$ seemingly cannot be further decomposed (and this is confirmed in Chapter 3.3). In any case, it follows easily from $J_\alpha{}^\gamma T_{\beta\gamma}{}^\delta = T_{\alpha\beta}{}^\gamma J_\gamma{}^\delta$ that $T_{\alpha\beta}{}^\gamma$ of type $(0,2)$ satisfy $T_{\alpha\beta}{}^\beta = 0 = T_{\alpha\beta}{}^\gamma J_\gamma{}^\beta$.

1.2. Complex connections

The geometrically useful affine connections ∇ on an almost complex manifold (M, J) are those that preserve $J_\alpha{}^\beta$, i.e. $\nabla_\alpha J_\beta{}^\gamma = 0$. We call them *complex connections*. The space of complex connections is an affine space over the vector space that consists of 1-forms with values in the complex endomorphisms $\mathfrak{gl}(TM, J)$ of TM. A complex connection ∇ naturally extends to a linear connection on $\mathbb{C}TM$ that preserves the decomposition into types (2). Indeed, preservation of type is also a sufficient condition for an affine connection to be complex.

1.2. COMPLEX CONNECTIONS

Given a complex connection ∇, we denote by $T_{\alpha\beta}{}^\gamma$ its torsion, which is a 2-form with values in TM. As such $T_{\alpha\beta}{}^\gamma$ naturally splits according to type into a direct sum of three components as in (7). A straightforward computation shows that the $(0,2)$-component of the torsion of any complex connection equals $-\frac{1}{4}N^J$. In particular, this component is an invariant of the almost complex structure and cannot be eliminated by a suitable choice of complex connection. However, all other components can be removed. To see this, suppose $\hat{\nabla}$ is another complex connection. Then there is an element $v \in T^*M \otimes \mathfrak{gl}(TM, J)$ such that $\hat{\nabla} = \nabla + v$. It follows that their torsions are related by the formula $\hat{T} = T + \partial v$, where ∂ is the composition

$$T^*M \otimes \mathfrak{gl}(TM, J) \hookrightarrow T^*M \otimes T^*M \otimes TM \to \wedge^2 T^*M \otimes TM$$
$$v_{\alpha\beta}{}^\gamma \mapsto 2v_{[\alpha\beta]}{}^\gamma.$$

Notice that the image of ∂ is spanned by 2-forms of type $(2,0)$ and $(1,1)$. Consequently, its cokernel can be identified with forms of type $(0,2)$. Hence, any complex connection can be deformed in such a way that its torsion is of type $(0,2)$. We have shown the following classical result:

PROPOSITION 1.2 ([**62, 69**]). *On any almost complex manifold (M, J) there is a complex connection such that $T = -\frac{1}{4}N^J$.*

Since ∂ is not injective such a complex connection is not unique. Complex connections ∇ with $T = -\frac{1}{4}N^J$ form an affine space over

(9) $$\ker \partial = (S^2 T^*M \otimes TM) \cap (T^*M \otimes \mathfrak{gl}(TM, J))$$

and are called *minimal connections*.

From Proposition 1.2 and the above discussion one also deduces immediately:

COROLLARY 1.3. *There exists a complex torsion-free connection on an almost complex manifold (M, J) if and only if $N^J \equiv 0$.*

REMARK 1.1. We have already noted that the cokernel of ∂ can be identified with tensors $T_{\alpha\beta}{}^\gamma$ such that

$$T_{(\alpha\beta)}{}^\gamma = 0 \qquad J_\alpha{}^\gamma T_{\beta\gamma}{}^\delta = T_{\alpha\beta}{}^\gamma J_\gamma{}^\delta.$$

Consequently, $T_{\alpha\beta}{}^\beta = T_{\alpha\beta}{}^\gamma J_\gamma{}^\beta = 0$. As we shall see in Chapter 3.3, such tensors are irreducible. More precisely, the natural vector bundles on an almost complex manifold (M, J) correspond to representations of $\mathrm{GL}(n, \mathbb{C})$ and we shall see that $\operatorname{coker} \partial$ corresponds to an irreducible representation of $\mathrm{GL}(n, \mathbb{C})$. On the other hand, its kernel (9) decomposes into two irreducible components, namely a trace-free part and a trace part. We shall see in the next section that deforming a complex connection by an element from the latter space exactly corresponds to changing a connection c-projectively.

CHAPTER 2

Elements of c-projective geometry

We now introduce almost c-projective structures, first from the classical perspective of J-planar curves and equivalence classes of complex affine connections [91], then from the modern viewpoint of parabolic geometries [36, 55, 104]. The (categorical) equivalence between these approaches is established in Theorem 2.8. This leads us to study the intrinsic curvature of an almost c-projective manifold, namely the harmonic curvature of its canonical normal Cartan connection.

2.1. Almost c-projective structures

Recall that affine connections ∇ and $\hat{\nabla}$ on a manifold M are projectively equivalent if there is a 1-form Υ_α on M such that

(10) $$\hat{\nabla}_\alpha X^\gamma = \nabla_\alpha X^\gamma + \Upsilon_\alpha X^\gamma + \delta_\alpha{}^\gamma \Upsilon_\beta X^\beta.$$

Suppose now that (M, J) is an almost complex manifold. Then ∇ and $\hat{\nabla}$ are called *c-projectively equivalent*, if there is a (real) 1-form Υ_α on M such that

(11) $$\hat{\nabla}_\alpha X^\gamma = \nabla_\alpha X^\gamma + \upsilon_{\alpha\beta}{}^\gamma X^\beta,$$

where $\upsilon_{\alpha\beta}{}^\gamma := \tfrac{1}{2}(\Upsilon_\alpha \delta_\beta{}^\gamma + \delta_\alpha{}^\gamma \Upsilon_\beta - J_\alpha{}^\delta \Upsilon_\delta J_\beta{}^\gamma - J_\alpha{}^\gamma \Upsilon_\delta J_\beta{}^\delta).$

Note that $\upsilon_{\alpha\beta}{}^\gamma J_\gamma{}^\delta = \upsilon_{\alpha\gamma}{}^\delta J_\beta{}^\gamma$. In other words $\upsilon_{\alpha\beta}{}^\gamma$ is a 1-form on M with values in $\mathfrak{gl}(TM, J)$, which implies that if ∇ is a complex connection, then so is $\hat{\nabla}$. Moreover $\upsilon_{\alpha\beta}{}^\gamma = \upsilon_{(\alpha\beta)}{}^\gamma$ and so c-projectively equivalent connections have the same torsion. In particular, if ∇ is minimal, then so is $\hat{\nabla}$.

A smooth curve $c\colon (a,b) \to M$ is called a *J-planar curve* with respect to a complex connection ∇, if $\nabla_{\dot c}\dot c$ lies in the span of $\dot c$ and $J\dot c$. The notion of J-planar curves gives rise to the following geometric interpretation of a c-projective equivalence class of complex connections.

PROPOSITION 2.1 ([55, 85, 91, 99]). *Suppose (M, J) is an almost complex manifold and let ∇ and $\hat{\nabla}$ be complex connections on M with the same torsion. Then ∇ and $\hat{\nabla}$ are c-projectively equivalent if and only if they have the same J-planar curves.*

PROOF. Suppose ∇ and $\hat{\nabla}$ are complex connections with the same torsion. If ∇ and $\hat{\nabla}$ are c-projectively equivalent, then they clearly have the same J-planar curves. Conversely, assume that ∇ and $\hat{\nabla}$ share the same J-planar curves and consider the difference tensor $A_{\alpha\beta}{}^\gamma Y^\beta = \hat{\nabla}_\alpha Y^\gamma - \nabla_\alpha Y^\gamma$. As both connections are complex and have the same torsion, the difference tensor satisfies $A_{\alpha\beta}{}^\gamma = A_{(\alpha\beta)}{}^\gamma$ and $A_{\alpha\beta}{}^\gamma J_\gamma{}^\delta = A_{\alpha\gamma}{}^\delta J_\beta{}^\gamma$. The fact that $\hat{\nabla}$ and ∇ have the same J-planar curves and that any tangent vector can be realised as the derivative of such a curve implies

that at any point $x \in M$ and for any nonzero vector $Y \in T_xM$ there exist uniquely defined real numbers $\gamma(Y)$ and $\mu(Y)$ such that
$$A(Y,Y) = \gamma(Y)Y + \mu(Y)JY. \tag{12}$$
Note that γ and μ give rise to well-defined smooth functions on $TM \setminus 0$. Extending γ and μ to functions on all of TM by setting $\gamma(0) = \mu(0) = 0$, formula (12) becomes valid for any tangent vector, and by construction γ and μ are then clearly homogeneous of degree one. From $A(Y,Y) = -A(JY,JY)$ we deduce that $\mu(X) = -\gamma(JX)$ whence
$$A(Y,Y) = \gamma(Y)Y - \gamma(JY)JY.$$
By polarisation we have for any tangent vectors X and Y
$$\begin{aligned}(13)\quad A(X,Y) &= \tfrac{1}{2}\big(A(X+Y,X+Y) - A(X,X) - A(Y,Y)\big) \\ &= \tfrac{1}{2}\big((\gamma(X+Y) - \gamma(X))X + (\gamma(X+Y) - \gamma(Y))Y\big) \\ &\quad - \tfrac{1}{2}\big((\gamma(JX+JY) - \gamma(JX))JX - (\gamma(JX+JY) - \gamma(JY))JY\big).\end{aligned}$$
Suppose that X and Y are linearly independent and expand the identity $A(X,tY) = tA(X,Y)$ for all $t \in \mathbb{R}$ using (13). Then a comparison of coefficients shows that
$$\gamma(X+tY) - t\gamma(Y) = \gamma(X+Y) - \gamma(Y).$$
Taking the limit $t \to 0$, shows that $\gamma(X+Y) = \gamma(X) + \gamma(Y)$. Hence, γ defines a (smooth) 1-form and
$$\begin{aligned}A(X,Y) &= \tfrac{1}{2}\big(A(X+Y,X+Y) - A(X,X) - A(Y,Y)\big) \\ &= \tfrac{1}{2}\big(\gamma(X)Y + \gamma(Y)X - \gamma(JX)(JY) - \gamma(JY)JX\big),\end{aligned}$$
for any tangent vectors X and Y as desired. $\square$

DEFINITION 2.1. Suppose that M is manifold of real dimension $2n \geq 4$.
(1) An *almost c-projective structure* on M consists of an almost complex structure J on M and a c-projective equivalence class $[\nabla]$ of minimal complex connections.
(2) The *torsion* of an almost c-projective structure $(M, J, [\nabla])$ is the torsion T of one, hence any, of the connections in $[\nabla]$, i.e. $T = -\tfrac{1}{4}N^J$.
(3) An almost c-projective structure $(M, J, [\nabla])$ is called a *c-projective structure*, if J is integrable. (This is the case if and only if some and hence all connections in the c-projective class are torsion-free.)

REMARK 2.1. If M is a 2-dimensional manifold, any almost complex structure J is integrable and any two torsion free complex connections are c-projectively equivalent. Therefore, in this case one needs to modify the definition of a c-projective structure in order to have something nontrivial (cf. [**23, 24**]). We shall not pursue this here.

REMARK 2.2. Recall that the geodesics of an affine connection can be also realised as the geodesics of a torsion-free connection; hence the definition of a projective structure as an equivalence class of torsion-free connections does not constrain the considered families of geodesics. The analogous statement for J-planar curves does not hold: the J-planar curves of a complex connection cannot in general be realised as the J-planar curves of a minimal connection. We discuss the motivation for the restriction to minimal connections in the definition of almost c-projective manifolds in Remark 2.9.

DEFINITION 2.2. Let $(M, J_M, [\nabla^M])$ and $(N, J_N, [\nabla^N])$ be almost c-projective manifolds of dimension $2n \geq 4$. A diffeomorphism $\Phi\colon M \to N$ is called *c-projective transformation or automorphism*, if Φ is complex (i.e. $T\Phi \circ J_M = J_N \circ T\Phi$) and for a (hence any) connection $\nabla^N \in [\nabla^N]$ the connection $\Phi^*\nabla^N$ is a connection in $[\nabla^M]$.

From Proposition 2.1 one deduces straightforwardly that also the following characterisation of c-projective transformations holds:

PROPOSITION 2.2. *Let $(M, J_M, [\nabla^M])$ and $(N, J_N, [\nabla^N])$ be almost c-projective manifolds of dimension $2n \geq 4$. Then a complex diffeomorphism $\Phi\colon M \to N$ is a c-projective transformation if and only if Φ maps J_M-planar curves to J_N-planar curves.*

Suppose that $(M, J, [\nabla])$ is an almost c-projective manifold. Let $\hat\nabla$ and ∇ be connections of the c-projective class $[\nabla]$ that differ by Υ_α as in (11). Then $\hat\nabla$ and ∇ give rise to linear connections on $\mathbb{C}TM = T^{1,0}M \oplus T^{0,1}M$ that preserve the decomposition into types. Hence, they induce connections on the complex vector bundles $T^{1,0}M$ and $T^{0,1}M$. To deduce the difference between the connections $\hat\nabla$ and ∇ on $T^{1,0}M$ (respectively $T^{0,1}M$), we just need to apply the splittings Π_α^a and Π_a^α (respectively their conjugates) from the previous section to (11). Using the identities (4), we obtain

$$\begin{aligned}
\Pi_a^\alpha \Pi_b^\beta \upsilon_{\alpha\beta}{}^\gamma \Pi_\gamma^c &= \tfrac{1}{2}\Pi_a^\alpha \Pi_b^\beta (\Upsilon_\alpha \delta_\beta{}^\gamma + \delta_\alpha{}^\gamma \Upsilon_\beta - J_\alpha{}^\delta \Upsilon_\delta J_\beta{}^\gamma - J_\alpha{}^\gamma \Upsilon_\delta J_\beta{}^\delta)\Pi_\gamma^c \\
&= \tfrac{1}{2}\Pi_a^\alpha \Pi_b^\beta \big((\Upsilon_\alpha - iJ_\alpha{}^\delta \Upsilon_\delta)\Pi_\beta^c + (\Upsilon_\beta - iJ_\beta{}^\delta \Upsilon_\delta)\Pi_\alpha^c\big) \\
&= \tfrac{1}{2}\Pi_a^\alpha (\Upsilon_\alpha - iJ_\alpha{}^\delta \Upsilon_\delta)\delta_b{}^c + \tfrac{1}{2}\Pi_b^\beta (\Upsilon_\beta - iJ_\beta{}^\delta \Upsilon_\delta)\delta_a{}^c \\
&= \Upsilon_a \delta_b{}^c + \Upsilon_b \delta_a{}^c, \qquad \text{where} \quad \Upsilon_a \equiv \Pi_a^\alpha \Upsilon_\alpha.
\end{aligned}$$

Similarly, we find that $\overline\Pi_{\bar a}^\alpha \Pi_b^\beta \upsilon_{\alpha\beta}{}^\gamma \Pi_\gamma^c = 0$. These identities are the key to the following:

PROPOSITION 2.3. *Suppose $(M, J, [\nabla])$ is an almost c-projective manifold of dimension $2n \geq 4$. Assume two connections $\hat\nabla$ and ∇ in $[\nabla]$ differ by Υ_α as in (11), and set $\Upsilon_a := \Pi_a^\alpha \Upsilon_\alpha$ and $\Upsilon_{\bar a} := \Pi_{\bar a}^\alpha \Upsilon_\alpha$. Then we have the following transformation rules for the induced connections on $T^{1,0}M$ and $T^{0,1}M$.*

(1) $\hat\nabla_a X^c = \nabla_a X^c + \Upsilon_a X^c + \delta_a{}^c \Upsilon_b X^b$ *and* $\hat\nabla_{\bar a} X^c = \nabla_{\bar a} X^c$,

(2) $\hat\nabla_{\bar a} X^{\bar c} = \nabla_{\bar a} X^{\bar c} + \Upsilon_{\bar a} X^{\bar c} + \delta_{\bar a}{}^{\bar c} \Upsilon_{\bar b} X^{\bar b}$ *and* $\hat\nabla_a X^{\bar c} = \nabla_a X^{\bar c}$.

PROOF. We compute

$$\begin{aligned}
\hat\nabla_a X^c &= \Pi_a^\alpha \hat\nabla_\alpha (\Pi_\gamma^c \Pi_b^\gamma X^b) = \Pi_a^\alpha \Pi_\gamma^c \hat\nabla_\alpha (\Pi_b^\gamma X^b) \\
&= \Pi_a^\alpha \Pi_\gamma^c \nabla_\alpha (\Pi_b^\gamma X^b) + \Pi_a^\alpha \Pi_\gamma^c \upsilon_{\alpha\beta}{}^\gamma \Pi_b^\beta X^b \\
&= \Pi_a^\alpha \nabla_\alpha (\Pi_\gamma^c \Pi_b^\gamma X^b) + (\Pi_a^\alpha \Pi_b^\beta \upsilon_{\alpha\beta}{}^\gamma \Pi_\gamma^c) X^b \\
&= \nabla_a X^c + (\Upsilon_a \delta_b{}^c + \Upsilon_b \delta_a{}^c) X^b = \nabla_a X^c + \Upsilon_a X^c + \delta_a{}^c \Upsilon_b X^b,
\end{aligned}$$

as required. The remaining calculations are similar. $\square$

REMARK 2.3. The differential operator $\nabla_{\bar a} \colon T^{1,0}M \to \wedge^{0,1}M \otimes T^{1,0}M$ is c-projectively invariant, as is its conjugate $\nabla_a \colon T^{0,1}M \to \wedge^{1,0}M \otimes T^{0,1}M$. (Here and throughout, the domain and codomain of a differential operator are declared as bundles, although the operator is a map between corresponding spaces of sections.)

This is unsurprising: it is the usual $\bar{\partial}$-operator on an almost complex manifold whose kernel (in the integrable case) comprises the holomorphic vector fields.

In contrast, the transformation rules for $\nabla_a \colon T^{1,0}M \to \wedge^{1,0}M \otimes T^{1,0}M$ and its conjugate are analogues of projective equivalence (10) in the $(1,0)$ and $(0,1)$ directions respectively. When (M,J) is real-analytic and the c-projective class contains a real-analytic connection ∇, this can be made precise by extending J and $[\nabla]$ to a complexification $M^{\mathbb{C}}$ of M, so that $T^{1,0}M$ and $T^{0,1}M$ extend to distributions on $M^{\mathbb{C}}$. If J is integrable, these distributions integrate to two foliations of $M^{\mathbb{C}}$, and $[\nabla]$ induces projective structures on the leaves of these foliations.

Taking the trace in equation (11) and in the formulae in Proposition 2.3, we deduce:

COROLLARY 2.4. *On an almost c-projective manifold $(M, J, [\nabla])$, the transformation rules for the induced linear connections on $\wedge^{2n}TM$, $\wedge^n T^{1,0}M$, and $\wedge^n T^{0,1}M$ are:*
  (1) $\hat{\nabla}_\alpha \Sigma = \nabla_\alpha \Sigma + (n+1)\Upsilon_\alpha \Sigma$, *for* $\Sigma \in \Gamma(\wedge^{2n}TM)$
  (2) $\hat{\nabla}_a \sigma = \nabla_a \sigma + (n+1)\Upsilon_a \sigma$ *and* $\hat{\nabla}_{\bar{a}}\sigma = \nabla_{\bar{a}}\sigma$, *for* $\sigma \in \Gamma(\wedge^n T^{1,0}M)$
  (3) $\hat{\nabla}_{\bar{a}}\bar{\sigma} = \nabla_{\bar{a}}\bar{\sigma} + (n+1)\Upsilon_{\bar{a}}\bar{\sigma}$ *and* $\hat{\nabla}_a \bar{\sigma} = \nabla_a \bar{\sigma}$, *for* $\bar{\sigma} \in \Gamma(\wedge^n T^{0,1}M)$.

For the convenience of the reader, let us also record the transformation rules for the induced connections on T^*M, respectively $\wedge^{1,0}$ and $\wedge^{0,1}$. If two complex connections $\hat{\nabla}$ and ∇ are related via Υ_α as in (11), then the induced connections on T^*M are related by

$$(14) \quad \hat{\nabla}_\alpha \phi_\gamma = \nabla_\alpha \phi_\gamma - \tfrac{1}{2}(\Upsilon_\alpha \phi_\gamma + \Upsilon_\gamma \phi_\alpha - J_\alpha{}^\beta \Upsilon_\beta J_\gamma{}^\delta \phi_\delta - J_\alpha{}^\beta \phi_\beta J_\gamma{}^\delta \Upsilon_\delta).$$

Therefore, we obtain:

PROPOSITION 2.5. *Suppose $(M, J, [\nabla])$ is an almost c-projective manifold of dimension $2n \geq 4$. Assume two connections $\hat{\nabla}$ and ∇ in $[\nabla]$ differ by Υ_α as in (11) and set $\Upsilon_a := \Pi_a^\alpha \Upsilon_\alpha$ and $\Upsilon_{\bar{a}} := \Pi_{\bar{a}}^\alpha \Upsilon_\alpha$. Then we have the following transformation rules for the induced connections on $\wedge^{1,0}$ and $\wedge^{0,1}$:*
  (1) $\hat{\nabla}_a \phi_c = \nabla_a \phi_c - \Upsilon_a \phi_c - \phi_a \Upsilon_c$ *and* $\hat{\nabla}_{\bar{a}}\phi_c = \nabla_{\bar{a}}\phi_c$,
  (2) $\hat{\nabla}_{\bar{a}}\phi_{\bar{c}} = \nabla_{\bar{a}}\phi_{\bar{c}} - \Upsilon_{\bar{a}}\phi_{\bar{c}} - \phi_{\bar{a}}\Upsilon_{\bar{c}}$ *and* $\hat{\nabla}_a \phi_{\bar{c}} = \nabla_a \phi_{\bar{c}}$.

Note that the real line bundle $\wedge^{2n}TM$ is oriented and hence admits oriented roots. We denote $(\wedge^{2n}TM)^{\frac{1}{n+1}}$ by $\mathcal{E}_{\mathbb{R}}(1,1)$ and for any $k \in \mathbb{N}$ we set $\mathcal{E}_{\mathbb{R}}(k,k) := \mathcal{E}(1,1)^{\otimes k}$ and $\mathcal{E}_{\mathbb{R}}(-k,-k) := \mathcal{E}_{\mathbb{R}}(k,k)^*$. It follows from Corollary 2.4 that for any $k \in \mathbb{Z}$ and for any section Σ of $\mathcal{E}_{\mathbb{R}}(k,k)$, we have

$$(15) \quad \hat{\nabla}_\alpha \Sigma = \nabla_\alpha \Sigma + k\Upsilon_\alpha \Sigma.$$

In particular, we immediately deduce the following result.

PROPOSITION 2.6. *Suppose $(M, J, [\nabla])$ is an almost c-projective manifold of dimension $2n \geq 4$. The map sending an affine connection to its induced connection on $\mathcal{E}_{\mathbb{R}}(1,1)$ induces a bijection from connections in $[\nabla]$ to linear connections on $\mathcal{E}_{\mathbb{R}}(1,1)$.*

Since $\wedge^{2n}TM$ and $\mathcal{E}_{\mathbb{R}}(1,1)$ are oriented, they can be trivialised by choosing a positive section. Such a positive section τ of $\mathcal{E}_{\mathbb{R}}(1,1)$ gives rise to a linear connection on $\mathcal{E}_{\mathbb{R}}(1,1)$ by decreeing that τ is parallel and therefore, by Proposition 2.6, to a connection in the c-projective class. We call a connection $\nabla \in [\nabla]$ that arises in this way a *special connection*. Suppose $\hat\tau$ and τ are two nowhere vanishing sections of $\mathcal{E}_{\mathbb{R}}(1,1)$ and denote by $\hat\nabla$ and ∇ the corresponding connections. Then $\hat\tau = e^{-f}\tau$ for some smooth function f on M and any $\sigma \in \Gamma(\mathcal{E}_{\mathbb{R}}(1,1))$ can be written as $\sigma = h\tau = he^f \hat\tau$ for a smooth function h on M. Since $\nabla \sigma = dh \otimes \tau$, we have

$$\hat\nabla \sigma = d(he^f) \otimes \hat\tau = dh \otimes \tau + df \otimes \sigma = \nabla \sigma + (\nabla f)\sigma.$$

Therefore, $\hat\nabla$ and ∇ differ by an exact 1-form, namely $\Upsilon_\alpha \equiv \nabla_\alpha f$.

In some of the following sections, such as Chapter 3.1, we shall assume also that the complex line bundle $\wedge^n T^{1,0} M$ admits a $(n+1)^{\text{st}}$ root and that we have chosen one, which we denote by $\mathcal{E}(1,0)$ (following a standard notation on $\mathbb{CP}^n$). In that case we define, for any $k \in \mathbb{N}$, $\mathcal{E}(k,0) := \mathcal{E}(1,0)^{\otimes k}$, with dual bundle $\mathcal{E}(-k,0) := \mathcal{E}(k,0)^*$ and, for any $k \in \mathbb{Z}$, conjugate bundles $\mathcal{E}(0,k) := \overline{\mathcal{E}(k,0)}$. In general, for any $k,\ell \in \mathbb{Z}$, we write $\mathcal{E}(k,\ell) := \mathcal{E}(k,0) \otimes \mathcal{E}(0,\ell)$ and refer to its sections as *c-projective densities of weight* (k,ℓ). By Corollary 2.4 we see that, for a c-projective density σ of weight (k,ℓ), we have

(16) $$\hat\nabla_a \sigma = \nabla_a \sigma + k\Upsilon_a \sigma \qquad \hat\nabla_{\bar a} \sigma = \nabla_{\bar a} \sigma + \ell \Upsilon_{\bar a} \sigma.$$

Our notion of c-projective density means, in particular, that we may identify $\wedge^{n,0}$ with $\mathcal{E}(-n-1,0)$ and it is useful to have a notation for this change of viewpoint. Precisely, we may regard our identification $\mathcal{E}(-n-1,0) \xrightarrow{\simeq} \wedge^{n,0}$ as a tautological section $\varepsilon_{ab\cdots c}$ of $\wedge^{n,0}(n+1,0)$, such that a c-projective density ρ of weight $(-n-1,0)$ corresponds to $\rho \varepsilon_{ab\cdots c}$, a form of type $(n,0)$. Note that $\mathcal{E}(k,k) \cong \mathcal{E}_{\mathbb{R}}(k,k) \otimes \mathbb{C}$.

2.2. Parabolic geometries

For the convenience of the reader we recall here some basics of parabolic geometries; for a comprehensive introduction see [**36**].

A *parabolic geometry* on a manifold M is a Cartan geometry of type (G,P), where G is a semisimple Lie group and $P \subset G$ a so-called *parabolic subgroup*. Hence, it is given by the following data:

- a principal P-bundle $p\colon \mathcal{G} \to M$
- a Cartan connection $\omega \in \Omega^1(\mathcal{G}, \mathfrak{g})$—that is, a P-equivariant 1-form on $\mathcal{G}$ with values in $\mathfrak{g}$ defining a trivialisation $T\mathcal{G} \cong \mathcal{G} \times \mathfrak{g}$ and reproducing the generators of the fundamental vector fields,

where $\mathfrak{g}$ denotes the Lie algebra of G. Note that the projection $G \to G/P$, equipped with the (left) Maurer–Cartan form $\omega_G \in \Omega^1(G,\mathfrak{g})$ of G, defines a parabolic geometry on G/P, which is called the *homogeneous* or *flat model* for parabolic geometries of type (G,P).

The *curvature* of a parabolic geometry $(\mathcal{G} \xrightarrow{p} M, \omega)$ is a 2-form K on $\mathcal{G}$ with values in $\mathfrak{g}$, defined by

$$K(\chi,\xi) = d\omega(\chi,\xi) + [\omega(\chi), \omega(\xi)] \text{ for vector fields } \chi \text{ and } \xi \text{ on } \mathcal{G},$$

where d denotes the exterior derivative and $[\ ,\ ]$ the Lie bracket of $\mathfrak{g}$.

The curvature of the homogeneous model $(G \to G/P, \omega_G)$ vanishes identically. Furthermore, the curvature K of a parabolic geometry of type (G,P) vanishes identically if and only if it is locally isomorphic to $(G \to G/P, \omega_G)$. Thus, the curvature K measures the extent to which the geometry differs from its homogeneous model.

Given a parabolic geometry $(\mathcal{G} \xrightarrow{p} M, \omega)$ of type (G,P), any representation $\mathbb{E}$ of P gives rise to an associated vector bundle $E := \mathcal{G} \times_P \mathbb{E}$ over M. These are the natural vector bundles on a parabolic geometry. Notice that the Cartan connection ω induces an isomorphism

$$\mathcal{G} \times_P \mathfrak{g}/\mathfrak{p} \cong TM$$
$$[u, X + \mathfrak{p}] \mapsto T_u p(\omega^{-1}(X)),$$

where $\mathfrak{p}$ denotes the Lie algebra of P and the action of P on $\mathfrak{g}/\mathfrak{p}$ is induced by the adjoint action of G. Similarly, ω allows us to identify all tensor bundles on M with associated vector bundles. The vector bundles corresponding to P-modules obtained by restricting a representation of G to P are called *tractor bundles*. These bundles play an important role in the theory of parabolic geometries, since the Cartan connection induces linear connections, called *tractor connections*, on these bundles. An important example of a tractor bundle is the *adjoint tractor* bundle $\mathcal{A}M = \mathcal{G} \times_P \mathfrak{g}$, which has a canonical projection to TM corresponding to the P-equivariant projection $\mathfrak{g} \to \mathfrak{g}/\mathfrak{p}$.

REMARK 2.4. The abstract theory of tractor bundles and connections even provides an alternative description of parabolic geometries (see [30]).

By normalising the curvature of a parabolic geometry, the prolongation procedures of [35, 86, 96] leads to an equivalence of categories between so-called *regular normal* parabolic geometries and certain underlying structures, which may be described in more conventional geometric terms. Among the most prominent of these are conformal structures, projective structures, and CR-structures of hypersurface type. In the next section we shall see that almost c-projective manifolds form another class of examples.

From the defining properties of a Cartan connection it follows immediately that the curvature K of a parabolic geometry of type (G, P) is P-equivariant and horizontal. Hence, K can be identified with a section of the vector bundle $\wedge^2 T^*M \otimes \mathcal{A}M$ and therefore corresponds via ω to a section κ of the vector bundle

$$\mathcal{G} \times_P \wedge^2(\mathfrak{g}/\mathfrak{p})^* \otimes \mathfrak{g} \cong \mathcal{G} \times_P \wedge^2 \mathfrak{p}_+ \otimes \mathfrak{g},$$

where $\mathfrak{p}_+$ is the nilpotent radical of $\mathfrak{p}$ and the latter isomorphism is induced by the Killing form of $\mathfrak{g}$. Now consider the complex for computing the Lie algebra homology $H_*(\mathfrak{p}_+, \mathfrak{g})$ of $\mathfrak{p}_+$ with values in $\mathfrak{g}$:

$$0 \leftarrow \mathfrak{g} \xleftarrow{\partial^*} \mathfrak{p}_+ \otimes \mathfrak{g} \xleftarrow{\partial^*} \wedge^2 \mathfrak{p}_+ \otimes \mathfrak{g} \leftarrow \ldots$$

Since the linear maps ∂^* are P-equivariant, they induce vector bundle maps between the corresponding associated vector bundles. Moreover, the homology spaces $H_i(\mathfrak{p}_+, \mathfrak{g})$ are naturally P-modules and therefore give rise to natural vector bundles. A parabolic geometry is called *normal*, if $\partial^* \kappa = 0$. In this case, we can project κ to a section κ_h of $\mathcal{G} \times_P H_2(\mathfrak{p}_+, \mathfrak{g})$, called the *harmonic curvature*. The spaces $H_i(\mathfrak{p}_+, \mathfrak{g})$ are completely reducible P-modules and hence arise as completely reducible representations of the reductive Levi factor G_0 of P via the projection $P \to P/\exp(\mathfrak{p}_+) = G_0$. In particular, the harmonic curvature is a section of a completely reducible vector bundle, which makes it a much simpler object than the full curvature. Moreover, using the Bianchi identities of κ, it can be shown that the harmonic curvature is still a complete obstruction to local flatness:

PROPOSITION 2.7 (see e.g. [36]). *Suppose that* $(\mathcal{G} \to M, \omega)$ *is a regular normal parabolic geometry. Then* $\kappa \equiv 0$ *if and only if* $\kappa_h \equiv 0$.

REMARK 2.5. The machinery of BGG sequences shows that the curvature of a regular normal parabolic geometry can be reconstructed from the harmonic curvature by applying a BGG splitting operator (see [**25**]).

2.3. Almost c-projective manifolds as parabolic geometries

It is convenient for our purposes to realise the Lie algebra $\mathfrak{g} := \mathfrak{sl}(n+1, \mathbb{C})$ of complex trace-free linear endomorphisms of $\mathbb{C}^{n+1}$ as block matrices of the form

$$(17) \qquad \mathfrak{g} = \left\{ \begin{pmatrix} -\operatorname{tr} A & Z \\ X & A \end{pmatrix} : A \in \mathfrak{gl}(n, \mathbb{C}), X \in \mathbb{C}^n, Z \in (\mathbb{C}^n)^* \right\},$$

where $\operatorname{tr}: \mathfrak{gl}(n, \mathbb{C}) \to \mathbb{C}$ denotes the trace. The block form equips $\mathfrak{g}$ with the structure of a graded Lie algebra:

$$\mathfrak{g} = \mathfrak{g}_{-1} \oplus \mathfrak{g}_0 \oplus \mathfrak{g}_1,$$

where $\mathfrak{g}_0$ is the block diagonal subalgebra isomorphic to $\mathfrak{gl}(n, \mathbb{C})$ and $\mathfrak{g}_{-1} \cong \mathbb{C}^n$, respectively $\mathfrak{g}_1 \cong (\mathbb{C}^n)^*$, as $\mathfrak{g}_0$-modules. Note that the subspace $\mathfrak{p} := \mathfrak{g}_0 \oplus \mathfrak{g}_1$ is a subalgebra of $\mathfrak{g}$ (with $\mathfrak{p} \cong \mathfrak{g}_0 \ltimes \mathfrak{g}_1$ as Lie algebra). Furthermore, $\mathfrak{p}$ is a parabolic subalgebra with Abelian nilpotent radical $\mathfrak{p}_+ := \mathfrak{g}_1$ and Levi factor isomorphic to $\mathfrak{g}_0$. For later purposes let us remark here that we may conveniently decompose an element $A \in \mathfrak{g}_0$ into its trace-free part and into its trace part as follows

$$(18) \qquad \begin{pmatrix} 0 & 0 \\ 0 & A - \frac{\operatorname{tr} A}{n} \operatorname{Id}_n \end{pmatrix} + \frac{n+1}{n} \operatorname{tr} A \begin{pmatrix} -\frac{n}{n+1} & 0 \\ 0 & \frac{1}{n+1} \operatorname{Id}_n \end{pmatrix}.$$

Now set $G := \operatorname{PSL}(n+1, \mathbb{C})$ and let P be the stabiliser in G of the complex line generated by the first standard basis vector of $\mathbb{C}^{n+1}$. Let G_0 be the subgroup of P that consists of all elements $g \in P$ whose adjoint action $\operatorname{Ad}(g) \colon \mathfrak{g} \to \mathfrak{g}$ preserve the grading. Hence, it consists of equivalence classes of matrices of the form

$$\begin{pmatrix} (\det_\mathbb{C} C)^{-1} & 0 \\ 0 & C \end{pmatrix} \quad \text{where } C \in \operatorname{GL}(n, \mathbb{C}),$$

and the adjoint action of G_0 on $\mathfrak{g}$ induces an isomorphism

$$G_0 \cong \operatorname{GL}(\mathfrak{g}_{-1}, \mathbb{C}) \cong \operatorname{GL}(n, \mathbb{C}).$$

Obviously, the subgroups G_0 and P of G have corresponding Lie algebras $\mathfrak{g}_0$ and $\mathfrak{p}$, respectively.

Henceforth we view $G_0 \subset P \subset G$ as real Lie groups in accordance with the identification of $\operatorname{GL}(n+1, \mathbb{C})$ with the real subgroup of $\operatorname{GL}(2n+2, \mathbb{R})$ that is given by

$$\operatorname{GL}(2n+2, \mathbb{J}_{2(n+1)}) = \left\{ A \in \operatorname{GL}(2(n+1), \mathbb{R}) : A\mathbb{J}_{2(n+1)} = \mathbb{J}_{2(n+1)} A \right\},$$

where $\mathbb{J}_{2(n+1)}$ is the following complex structure on $\mathbb{R}^{2n+2}$:

$$\mathbb{J}_{2(n+1)} = \begin{pmatrix} \mathbb{J}_2 & & \\ & \ddots & \\ & & \mathbb{J}_2 \end{pmatrix} \quad \text{with } \mathbb{J}_2 = \begin{pmatrix} 0 & -1 \\ 1 & 0 \end{pmatrix}.$$

Suppose now that $(M, J, [\nabla])$ is an almost c-projective manifold of real dimension $2n \geq 4$. Then J reduces the frame bundle $\mathcal{F}M$ of M to a principal bundle $p_0 \colon \mathcal{G}_0 \to M$ with structure group G_0 corresponding to the group homomorphism

$$G_0 \cong \operatorname{GL}(n, \mathbb{C}) \cong \operatorname{GL}(2n, \mathbb{J}_{2n}) \hookrightarrow \operatorname{GL}(2n, \mathbb{R}).$$

The general prolongation procedures of [**35**, **86**, **96**] further show that $\mathcal{G}_0 \to M$ can be canonically extended to a principal P-bundle $p\colon \mathcal{G} \to M$, equipped with a normal Cartan connection $\omega \in \Omega^1(\mathcal{G}, \mathfrak{g})$ of type (G, P). Moreover, $(\mathcal{G} \xrightarrow{p} M, \omega)$ is uniquely defined up to isomorphism and these constructions imply:

THEOREM 2.8 (see also [**55**, **104**]). *There is an equivalence of categories between almost c-projective manifolds of real dimension $2n \geq 4$ and normal parabolic geometries of type (G, P), where G and P are viewed as real Lie groups. The homogeneous model $(G \to G/P, \omega_G)$ corresponds to the c-projective manifold*
$$(\mathbb{CP}^n, J_{\mathrm{can}}, [\nabla^{g_{FS}}]),$$
where J_{can} denotes the canonical complex structure on $\mathbb{CP}^n$ and $\nabla^{g_{FS}}$ the Levi-Civita connection of the Fubini–Study metric g_{FS}.

Let us explain briefly how the Cartan bundle $\mathcal{G}$ and the normal Cartan connection ω of an almost c-projective manifold $(M, J, [\nabla])$ of dimension $2n \geq 4$ are constructed. The reduction $\mathcal{G}_0 \to \mathcal{F}M$ is determined by the pullback of the soldering form on $\mathcal{F}M$ and hence can be encoded by a strictly horizontal G_0-equivariant 1-form $\theta \in \Omega^1(\mathcal{G}_0, \mathfrak{g}_{-1})$. Recall also that any connection $\nabla \in [\nabla]$ can be equivalently viewed as a principal connection $\gamma^\nabla \in \Omega^1(\mathcal{G}_0, \mathfrak{g}_0)$ on $\mathcal{G}_0$. Then $\mathcal{G}$ is defined to be the disjoint union $\sqcup_{u \in \mathcal{G}_0} \mathcal{G}_u$, where
$$\mathcal{G}_u := \{\theta(u) + \gamma^\nabla(u) : \nabla \in [\nabla]\} \quad \text{for any } u \in \mathcal{G}_0.$$

The projection $p := p_0 \circ q\colon \mathcal{G} \to M$, where $\mathcal{G} \xrightarrow{q} \mathcal{G}_0 \xrightarrow{p_0} M$, naturally acquires the structure of a P-principal bundle. Any element $p \in P$ can be uniquely written as $p = g_0 \exp(Z)$, where $g_0 \in G_0$ and $Z \in \mathfrak{g}_1$. The right action of an element $g_0 \exp(Z) \in P$ on an element $\theta(u) + \gamma^\nabla(u) \in \mathcal{G}_u$ is given by

$$(19) \quad (\theta(u) + \gamma^\nabla(u)) \cdot g_0 \exp(Z) := \theta(u \cdot g_0)(\cdot) + \gamma^\nabla(u \cdot g_0)(\cdot) + [Z, \theta(u \cdot g_0)(\cdot)],$$

where $[\ ,\ ]$ denotes the Lie bracket $\mathfrak{g}_1 \times \mathfrak{g}_{-1} \to \mathfrak{g}_0$.

REMARK 2.6. The soldering form $\theta \in \Omega^1(\mathcal{G}_0, \mathfrak{g}_{-1})$ gives rise to isomorphisms $TM \cong \mathcal{G}_0 \times_{G_0} \mathfrak{g}_{-1}$ and $T^*M \cong \mathcal{G}_0 \times_{G_0} \mathfrak{g}_1$. For elements $X \in \mathfrak{g}_{-1}$ and $Z \in \mathfrak{g}_1$, the Lie bracket $[Z, X] \in \mathfrak{g}_0 \cong \mathfrak{gl}(\mathfrak{g}_{-1}, \mathbb{J}_{2n})$ evaluated on an element $Y \in \mathfrak{g}_{-1}$ is given by

$$(20) \quad [[Z, X], Y] = -(ZXY + ZYX - Z\mathbb{J}_{2n}X\mathbb{J}_{2n}Y - Z\mathbb{J}_{2n}Y\mathbb{J}_{2n}X).$$

This shows that changing a connection form $\theta + \gamma^\nabla$ by a G_0-equivariant function $Z\colon \mathcal{G}_0 \to \mathfrak{g}_1$ according to (19) corresponds precisely to changing it c-projectively (cf. formula (11)).

The definition of $\mathcal{G}$ easily implies that the following holds.

COROLLARY 2.9. *The projection $q\colon \mathcal{G} \to \mathcal{G}_0$ is a trivial principal bundle with structure group $P_+ := \exp(\mathfrak{p}_+)$ and its global G_0-equivariant sections, called Weyl structures, are in bijection with principal connections in the c-projective class. Moreover, any Weyl structure $\sigma\colon \mathcal{G}_0 \to \mathcal{G}$ induces an vector bundle isomorphism*
$$\mathcal{G}_0 \times_{G_0} \mathbb{E} \cong \mathcal{G} \times_P \mathbb{E}$$
$$[u, X] \mapsto [\sigma(u), X],$$
for any P-module $\mathbb{E}$.

Note that there is a tautological 1-form $\nu \in \Omega^1(\mathcal{G}, \mathfrak{g}_{-1} \oplus \mathfrak{g}_0)$ on $\mathcal{G}$ given by

(21) $$\nu(\theta(u) + \gamma^\nabla(u))(\xi) := (\theta(u) + \gamma^\nabla(u))((Tq)\xi).$$

Extending this form to a normal Cartan connection $\omega \in \Omega^1(\mathcal{G}, \mathfrak{g})$ establishes the equivalence of categories in Theorem 2.8.

REMARK 2.7. In Chapter 2.1 we observed that there are always so-called special connections in the c-projective class. A Weyl structure corresponding to a special connection is precisely what in the literature on parabolic geometries is called an *exact Weyl structure* (see [36, 37]). The name is due to the fact that they form an affine space over the space of exact 1-forms on M.

Note also that the almost complex structure J on M induces an almost complex structure $J^{\mathcal{G}_0}$ on the complex frame bundle $\mathcal{G}_0$ of M. If J is integrable, so is $J^{\mathcal{G}_0}$ and $\mathcal{G}_0$ is a holomorphic vector bundle over M. Moreover, the complex structure on $\mathfrak{g}$ induces, by means of the isomorphism $\omega \colon T\mathcal{G} \cong \mathcal{G} \times \mathfrak{g}$, an almost complex structure $J^{\mathcal{G}}$ on $\mathcal{G}$, satisfying $Tp \circ J^{\mathcal{G}} = J \circ Tp$ and $Tq \circ J^{\mathcal{G}} = J^{\mathcal{G}_0} \circ Tq$. Note that the definition of the almost complex structure on $J^{\mathcal{G}_0}$ and $J^{\mathcal{G}}$ implies that θ and ω are of type $(1,0)$.

Let us also remark that an immediate consequence of Theorem 2.8 and the Liouville Theorem for Cartan geometries (see e.g. [36, Proposition 1.5.3]) is the following classical result.

PROPOSITION 2.10. *For $n \geq 2$ the c-projective transformations of $(\mathbb{CP}^n, J_{\mathrm{can}}, [\nabla^{g_{FS}}])$ (which by Proposition 2.2 are the complex diffeomorphisms of $\mathbb{CP}^n$ that map complex lines to complex lines) are precisely given by the left multiplications of elements in $\mathrm{PSL}(n+1, \mathbb{C})$. Moreover, any local c-projective transformation of $(\mathbb{CP}^n, J_{\mathrm{can}}, [\nabla^{g_{FS}}])$ uniquely extends to a global one.*

We finish this section by introducing some notation. The P-module $\mathfrak{g}$ admits an invariant filtration $\mathfrak{g} \supset \mathfrak{p} \supset \mathfrak{g}_1$ and hence the adjoint tractor bundle $\mathcal{A}M := \mathcal{G} \times_P \mathfrak{g}$ is naturally filtered

$$\mathcal{A}M = \mathcal{A}^{-1}M \supset \mathcal{A}^0 M \supset \mathcal{A}^1 M,$$

with $\mathcal{A}^1 M \cong T^*M$ and $\mathcal{A}M/\mathcal{A}^0 M \cong TM$. Hence, the associated graded vector bundle to $\mathcal{A}M$ is given by

(22) $\mathrm{gr}(\mathcal{A}M) = \mathrm{gr}_{-1}(\mathcal{A}M) \oplus \mathrm{gr}_0(\mathcal{A}M) \oplus \mathrm{gr}_1(\mathcal{A}M) = TM \oplus \mathfrak{gl}(TM, J) \oplus T^*M$,

which can be identified with $\mathcal{G}_0 \times_{G_0} \mathfrak{g}$.

2.4. The curvature of the canonical Cartan connection

Suppose $\sigma \colon \mathcal{G}_0 \to \mathcal{G}$ is a Weyl structure and let γ^∇ be the corresponding principal connection in the c-projective class. Since the normal Cartan connection ω is P-equivariant and σ is G_0-equivariant, the pullback $\sigma^*\omega \in \Omega^1(\mathcal{G}_0, \mathfrak{g})$ is G_0-equivariant and hence decomposes according to the grading on $\mathfrak{g}$ into three components. Since ω extends the tautological form ν on $\mathcal{G}$, defined by (21), we deduce that

(23) $$\sigma^*\omega = \theta + \gamma^\nabla - \mathsf{p}^\nabla,$$

where $\mathsf{p}^\nabla \in \Omega^1(\mathcal{G}_0, \mathfrak{g}_1)$ is horizontal and G_0-equivariant and hence can be viewed as a section P^∇ of $T^*M \otimes T^*M$, called the *Rho tensor* of ∇. Via σ, the curvature

$\kappa \in \Omega^2(M, \mathcal{A}M)$ of ω can be identified with a section κ^σ of

$$\wedge^2 T^*M \otimes \mathrm{gr}(\mathcal{A}M)$$
$$= (\wedge^2 T^*M \otimes TM) \oplus (\wedge^2 T^*M \otimes \mathfrak{gl}(TM, J)) \oplus (\wedge^2 T^*M \otimes T^*M),$$

which decomposes according to this splitting into three components

$$\kappa^\sigma = T + W^\nabla - C^\nabla.$$

One computes straightforwardly that $T \in \Gamma(\wedge^2 T^*M \otimes TM)$ is the torsion of the almost c-projective structure and that $C^\nabla = d^\nabla \mathsf{P}^\nabla \in \Gamma(\wedge^2 T^*M \otimes T^*M)$, where d^∇ denotes the covariant exterior derivative on differential forms with values in T^*M induced by ∇. The tensor C^∇ is called the *Cotton–York tensor* of ∇. To describe the component $W^\nabla \in \Gamma(\wedge^2 T^*M \otimes \mathfrak{gl}(TM, J))$, called the (c-projective) *Weyl curvature* of ∇, let us denote by $R^\nabla \in \Omega^2(M, \mathfrak{gl}(TM, J))$ the curvature of ∇. Then one has

$$W^\nabla = R^\nabla - \partial \mathsf{P}^\nabla,$$

where

(24) $\quad (\partial \mathsf{P}^\nabla)_{\alpha\beta}{}^\gamma{}_\epsilon := \delta_{[\alpha}{}^\gamma \mathsf{P}^\nabla_{\beta]\epsilon} - J_{[\alpha}{}^\gamma \mathsf{P}^\nabla_{\beta]\zeta} J_\epsilon{}^\zeta - \mathsf{P}^\nabla_{[\alpha\beta]} \delta_\epsilon{}^\gamma - J_{[\alpha}{}^\zeta \mathsf{P}^\nabla_{\beta]\zeta} J_\epsilon{}^\gamma.$

REMARK 2.8. The map $\partial \colon T^*M \otimes T^*M \to \wedge^2 T^*M \otimes \mathfrak{gl}(TM, J)$ given by (24) is related to Lie algebra cohomology. It is easy to see that the Lie algebra differentials in the complex computing the Lie algebra cohomology of the Abelian real Lie algebra $\mathfrak{g}_{-1}$ with values in the representation $\mathfrak{g}$ are G_0-equivariant and that ∂ is induced by the restriction to $\mathfrak{g}_{-1}^* \otimes \mathfrak{g}_1 \cong \mathfrak{g}_1 \otimes \mathfrak{g}_1$ of half of the second differential in this complex.

The normal Cartan connection ω is characterised as the unique extension of ν to a Cartan connection such that $\partial^* \kappa^\sigma = 0$ for all Weyl structures $\sigma \colon \mathcal{G}_0 \to \mathcal{G}$. Analysing $\ker \partial^*$ shows that $T_{\alpha\beta}{}^\gamma$ is in there, since forms of type $(0, 2)$ are, and C^∇ is, since $\wedge^2 T^*M \otimes T^*M \subset \ker \partial^*$. Hence, P^∇ is uniquely determined by requiring that W^∇ be in the kernel of ∂^*.

REMARK 2.9. Recall that in Definition 2.1 we restricted our definition of almost c-projective structures to c-projective equivalence classes of minimal connections. Since the kernel of

$$\partial^* \colon \wedge^2 T^*M \otimes TM \to T^*M \otimes \mathfrak{gl}(TM, J)$$

consists precisely of all the 2-forms with values in TM of type $(0, 2)$, the discussion of the construction of the Cartan connection above shows that the minimality condition is forced by the normalisation condition of the Cartan connection. The requirement for the almost c-projective structure to be minimal is however not necessary in order to construct a canonical Cartan connection. In fact, starting with any complex connection, one can show that there is a complex connection with the same J-planar curves whose torsion has only two components, namely the $(0, 2)$-component $-\frac{1}{4} N_J$ and a component in the subspace of $(1, 1)$-tensors in $\wedge^2 T^*M \otimes TM$ that are trace and J-trace free. Imposing this normalisation condition on an almost c-projective structure allows then analogously as above to associate a canonical Cartan connection (see [65]).

2.4. THE CURVATURE OF THE CANONICAL CARTAN CONNECTION

PROPOSITION 2.11. *Suppose $(M, J, [\nabla])$ is an almost c-projective manifold of dimension $2n \geq 4$. Let $\nabla \in [\nabla]$ be a connection in the c-projective class. Then the Rho tensor corresponding to ∇ is given by*

$$\mathsf{P}^\nabla_{\alpha\beta} = \tfrac{1}{n+1}(\mathrm{Ric}^\nabla_{\alpha\beta} + \tfrac{1}{n-1}(\mathrm{Ric}^\nabla_{(\alpha\beta)} - J_{(\alpha}{}^\gamma J_{\beta)}{}^\delta \mathrm{Ric}^\nabla_{\gamma\delta})), \tag{25}$$

where $\mathrm{Ric}^\nabla_{\alpha\beta} := R^\nabla_{\gamma\alpha}{}^\gamma{}_\beta$ is the Ricci tensor of ∇. Moreover, if $\hat\nabla \in [\nabla]$ is another connection in the class, related to ∇ according to (11), then

$$\mathsf{P}^{\hat\nabla}_{\alpha\beta} = \mathsf{P}^\nabla_{\alpha\beta} - \nabla_\alpha \Upsilon_\beta + \tfrac{1}{2}(\Upsilon_\alpha \Upsilon_\beta - J_\alpha{}^\gamma J_\beta{}^\delta \Upsilon_\gamma \Upsilon_\delta). \tag{26}$$

PROOF. The map $\partial^* \colon \wedge^2 T^*M \otimes \mathfrak{gl}(TM, J) \to T^*M \otimes T^*M$ is a multiple of a Ricci-type contraction. Hence, the normality of ω implies

$$R^\nabla_{\alpha\beta}{}^\alpha{}_\epsilon = (\partial \mathsf{P}^\nabla)_{\alpha\beta}{}^\alpha{}_\epsilon = (n + \tfrac{1}{2})\mathsf{P}^\nabla_{\beta\epsilon} - \tfrac{1}{2}\mathsf{P}^\nabla_{\epsilon\beta} + J_{(\beta}{}^\gamma J_{\epsilon)}{}^\zeta \mathsf{P}^\nabla_{\gamma\zeta}. \tag{27}$$

Therefore, $\mathrm{Ric}^\nabla_{[\beta\epsilon]} = (n+1)\mathsf{P}^\nabla_{[\beta\epsilon]}$ and $\mathrm{Ric}^\nabla_{(\beta\epsilon)} = n\mathsf{P}^\nabla_{(\beta\epsilon)} + J_{(\beta}{}^\gamma J_{\epsilon)}{}^\zeta \mathsf{P}^\nabla_{\gamma\zeta}$, which implies that

$$\mathrm{Ric}^\nabla_{(\beta\epsilon)} - J_{(\beta}{}^\gamma J_{\epsilon)}{}^\zeta \mathrm{Ric}^\nabla_{\gamma\zeta} = (n-1)(\mathsf{P}^\nabla_{(\beta\epsilon)} - J_{(\beta}{}^\gamma J_{\epsilon)}{}^\zeta \mathsf{P}^\nabla_{\gamma\zeta}).$$

Using these identities one verifies immediately that formula (25) holds. The formula (26) for the change of the Rho tensor can easily be verified directly or follows from the general theory of Weyl structures for parabolic geometries established in [37] taking into account that the Rho tensor in [37] is $-\tfrac{1}{2}$ times the Rho tensor given by (25) and our conventions for the definition of Υ as in (11). □

As an immediate consequence (writing out (26) in terms of its components using the various projectors $\Pi^a_\alpha, \ldots$ and the formulae (4)) we have:

COROLLARY 2.12.
- $\overline{\mathsf{P}^\nabla_{ab}} = \mathsf{P}^\nabla_{\bar a \bar b}$ and $\overline{\mathsf{P}^\nabla_{a \bar b}} = \mathsf{P}^\nabla_{\bar a b}$
- $\mathsf{P}^{\hat\nabla}_{ab} = \mathsf{P}^\nabla_{ab} - \nabla_a \Upsilon_b + \Upsilon_a \Upsilon_b$
- $\mathsf{P}^{\hat\nabla}_{\bar a b} = \mathsf{P}^\nabla_{\bar a b} - \nabla_{\bar a} \Upsilon_b$

For any connection $\nabla \in [\nabla]$, its Weyl curvature W^∇ is, by construction, a section of $\wedge^2 T^*M \otimes \mathfrak{gl}(TM, J)$ that satisfies $W^\nabla_{\alpha\beta}{}^\alpha{}_\gamma \equiv 0$. This implies that also $J_\zeta{}^\alpha W^\nabla_{\alpha\beta}{}^\zeta{}_\epsilon - W^\nabla_{\alpha\beta}{}^\alpha{}_\zeta J_\epsilon{}^\zeta \equiv 0$. In the sequel we shall often simply write W instead of W^∇, and similarly for other tensors such as the Rho tensor, the dependence of ∇ being understood. Viewing W as a 2-form with values in the complex bundle vector bundle $\mathfrak{gl}(TM, J) \cong \mathfrak{gl}(T^{1,0}M, \mathbb{C})$, it decomposes according to (p,q)-types into three components:

$$W_{ab}{}^c{}_d \qquad W_{a\bar b}{}^c{}_d \qquad W_{\bar a \bar b}{}^c{}_d.$$

The vanishing of the trace and J-trace above, then imply that

$$W_{ab}{}^a{}_d = W_{a\bar b}{}^a{}_d \equiv 0.$$

In these conclusions and in Corollary 2.12 we begin to see the utility of writing our expressions in using the barred and unbarred indices introduced in Chapter 1. In the following discussion we pursue this systematically, firstly by describing exactly how the curvature of a complex connection decomposes. We analyse these decompositions from the perspective of c-projective geometry: some pieces are invariant whilst others transform simply. For the convenience of the reader, we reiterate some of our previous conclusions in the following theorem (but prove them more easily using barred and unbarred indices, as just advocated).

PROPOSITION 2.13. *Suppose $(M, J, [\nabla])$ is an almost c-projective manifold of dimension $2n \geq 4$. Let $T_{\bar{a}b}{}^c$ denote its torsion (already observed to be a constant multiple of the Nijenhuis tensor of (M, J)). Then the curvature R of a connection ∇ in the c-projective class decomposes as follows:*

$$\begin{aligned} R_{ab}{}^c{}_d &= W_{ab}{}^c{}_d + 2\delta_{[a}{}^c \mathsf{P}_{b]d} + \beta_{ab}\delta_d{}^c \\ R_{a\bar{b}}{}^c{}_d &= W_{a\bar{b}}{}^c{}_d + \delta_a{}^c \mathsf{P}_{\bar{b}d} + \delta_d{}^c \mathsf{P}_{\bar{b}a} \\ W_{a\bar{b}}{}^c{}_d &= H_{a\bar{b}}{}^c{}_d - \tfrac{1}{2(n+1)}\bigl(\delta_a{}^c T_{df}{}^{\bar{e}} T_{\bar{e}\bar{b}}{}^f + \delta_d{}^c T_{af}{}^{\bar{e}} T_{\bar{e}\bar{b}}{}^f\bigr) - \tfrac{1}{2} T_{ad}{}^{\bar{e}} T_{\bar{e}\bar{b}}{}^c \\ R_{\bar{a}\bar{b}}{}^c{}_d &= W_{\bar{a}\bar{b}}{}^c{}_d \;=\; \nabla_d T_{\bar{a}\bar{b}}{}^c \end{aligned} \tag{28}$$

where

$$W_{ab}{}^c{}_d = W_{[ab]}{}^c{}_d \qquad W_{[ab}{}^c{}_{d]} = 0 \qquad W_{ab}{}^a{}_d = 0 \qquad \beta_{ab} = -2\mathsf{P}_{[ab]}$$
$$H_{a\bar{b}}{}^c{}_d = H_{d\bar{b}}{}^c{}_a \qquad\qquad H_{a\bar{b}}{}^a{}_d = 0.$$

Let $\hat{\nabla}$ be another connection in the c-projective class, related to ∇ by (11), and denote its curvature components by $\hat{W}$, $\hat{H}$, and $\hat{\mathsf{P}}$. Then we have:

(1) $\hat{W}_{ab}{}^c{}_d = W_{ab}{}^c{}_d$ *and* $\hat{W}_{a\bar{b}}{}^c{}_d = W_{a\bar{b}}{}^c{}_d$ *and* $\hat{H}_{a\bar{b}}{}^c{}_d = H_{a\bar{b}}{}^c{}_d$,
(2) $\hat{W}_{\bar{a}\bar{b}}{}^c{}_d = W_{\bar{a}\bar{b}}{}^c{}_d + T_{\bar{a}\bar{b}}{}^e \Upsilon_{ed}{}^c$ *and if J is integrable, then* $W_{\bar{a}\bar{b}}{}^c{}_d \equiv 0$,
(3) $W_{ab}{}^c{}_c \equiv 0$,
(4) $W_{a\bar{b}}{}^c{}_c = T_{fa}{}^{\bar{e}} T_{\bar{e}\bar{b}}{}^f$,

whilst we recall that $\hat{\mathsf{P}}_{ab} = \mathsf{P}_{ab} - \nabla_a \Upsilon_b + \Upsilon_a \Upsilon_b$, $\hat{\mathsf{P}}_{\bar{b}d} = \mathsf{P}_{\bar{b}d} - \nabla_{\bar{b}} \Upsilon_d$.

The tensor $\beta_{ab} = -2\mathsf{P}_{[ab]}$ satisfies

$$\nabla_{[b}\beta_{ce]} = \mathsf{P}_{\bar{f}[b} T_{ce]}{}^{\bar{f}} - \tfrac{1}{n+1} T_{[bc}{}^{\bar{f}} T_{e]a}{}^{\bar{d}} T_{\bar{d}\bar{f}}{}^a. \tag{29}$$

Finally, the Cotton–York tensors C_{abc} and $C_{a\bar{b}c}$ are defined as

$$C_{abc} := \nabla_a \mathsf{P}_{bc} - \nabla_b \mathsf{P}_{ac} + T_{ab}{}^{\bar{d}} \mathsf{P}_{\bar{d}c} \quad \text{and} \quad C_{a\bar{b}c} := \nabla_a \mathsf{P}_{\bar{b}c} - \nabla_{\bar{b}} \mathsf{P}_{ac}. \tag{30}$$

The first of these satisfies a Bianchi identity

$$\nabla_a W_{bc}{}^a{}_e - (n-2) C_{bce} \tag{31}$$
$$= 2 T_{a[b}{}^{\bar{f}} H_{c]\bar{f}}{}^a{}_e + \tfrac{2}{n+1} T_{bc}{}^{\bar{f}} T_{ea}{}^{\bar{d}} T_{\bar{d}\bar{f}}{}^a - \tfrac{n}{n+1} T_{e[b}{}^{\bar{f}} T_{c]a}{}^{\bar{d}} T_{\bar{d}\bar{f}}{}^a$$

and transforms as

$$\hat{C}_{bce} = C_{bce} + \Upsilon_a W_{bc}{}^a{}_e \tag{32}$$

under c-projective change (11). Another part of the Bianchi identity reads

$$C_{a\bar{b}c} - C_{c\bar{b}a} = \tfrac{1}{n+1}\bigl(T_{\bar{b}\bar{f}}{}^d \nabla_d T_{ac}{}^{\bar{f}} + R_{\bar{b}\bar{f}}{}^d{}_d T_{ac}{}^{\bar{f}} - 2 R_{\bar{b}\bar{f}}{}^d{}_{[a} T_{c]d}{}^{\bar{f}}\bigr). \tag{33}$$

PROOF. In this proof we also take the opportunity to develop various useful formulae for torsion and curvature and for how these quantities transform under c-projective change (11). As in the statement of Proposition 2.13, we express all these formulae in terms of the abstract indices on almost complex manifolds developed in Chapter 1. Firstly, recall that since we are working with minimal connections (cf. Definition 2.1), their torsions are restricted to being of type $(0,2)$ and this means precisely that

$$\begin{array}{ll} (\nabla_a \nabla_b - \nabla_b \nabla_a) f + T_{ab}{}^{\bar{c}} \nabla_{\bar{c}} f = 0 & (\nabla_{\bar{a}} \nabla_b - \nabla_b \nabla_{\bar{a}}) f = 0 \\ (\nabla_{\bar{a}} \nabla_{\bar{b}} - \nabla_{\bar{b}} \nabla_{\bar{a}}) f + T_{\bar{a}\bar{b}}{}^c \nabla_c f = 0 & (\nabla_a \nabla_{\bar{b}} - \nabla_{\bar{b}} \nabla_a) f = 0, \end{array} \tag{34}$$

2.4. THE CURVATURE OF THE CANONICAL CARTAN CONNECTION

where $T_{ab}{}^{\bar{c}} \equiv \Pi_a^\alpha \Pi_b^\beta \overline{\Pi}_\gamma{}^c T_{\alpha\beta}{}^\gamma$, equivalently its complex conjugate $T_{\bar{a}\bar{b}}{}^c = \overline{T_{ab}{}^{\bar{c}}}$, represents the Nijenhuis tensor as in (8). Notice that the second line of (34) is the complex conjugate of the first. In this proof, we take advantage of this general feature by listing only one of such conjugate pairs, its partner being implicitly valid. For example, here are characterisations of sufficiently many components of the general curvature tensor $R_{\alpha\beta}{}^\gamma{}_\delta$.

$$(35) \quad \begin{aligned} (\nabla_a \nabla_b - \nabla_b \nabla_a) X^c + T_{ab}{}^{\bar{d}} \nabla_{\bar{d}} X^c &= R_{ab}{}^c{}_d X^d \\ (\nabla_a \nabla_{\bar{b}} - \nabla_{\bar{b}} \nabla_a) X^c &= R_{a\bar{b}}{}^c{}_d X^d \\ [\text{or} \quad (\nabla_{\bar{a}} \nabla_b - \nabla_b \nabla_{\bar{a}}) X^c &= R_{\bar{a}b}{}^c{}_d X^d, \quad \text{if preferred}] \\ (\nabla_a \nabla_b - \nabla_b \nabla_a) X^{\bar{c}} + T_{ab}{}^{\bar{d}} \nabla_{\bar{d}} X^{\bar{c}} &= R_{ab}{}^{\bar{c}}{}_{\bar{d}} X^{\bar{d}}. \end{aligned}$$

For convenience, the dual formulae are sometimes preferred: for example,

$$(36) \quad (\nabla_a \nabla_b - \nabla_b \nabla_a)\phi_d + T_{ab}{}^{\bar{e}} \nabla_{\bar{c}} \phi_d = -R_{ab}{}^c{}_d \phi_c.$$

The tensor $\upsilon_{\alpha\beta}{}^\gamma$ employed in a c-projective change of connection (11) was already broken into irreducible pieces in deriving Proposition 2.3, e.g.

(37)

$$\upsilon_{ab}{}^c = \Pi_a^\alpha \Pi_b^\beta \Pi_\gamma^c \upsilon_{\alpha\beta}{}^\gamma = \Upsilon_a \delta_b{}^c + \Upsilon_b \delta_a{}^c \quad \Rightarrow \hat{\nabla}_a X^c = \nabla_a X^c + \Upsilon_a X^c + \Upsilon_b X^b \delta_a{}^c$$

(38)
and $\upsilon_{\bar{a}b}{}^c = \overline{\Pi}_{\bar{a}}^\alpha \Pi_b^\beta \Pi_\gamma^c \upsilon_{\alpha\beta}{}^\gamma = 0 \quad \Rightarrow \hat{\nabla}_{\bar{a}} X^c = \nabla_{\bar{a}} X^c.$

It is an elementary matter, perhaps more conveniently executed in the dual formulation

$$(39) \quad \hat{\nabla}_a \phi_b = \nabla_a \phi_b - \Upsilon_a \phi_b - \Upsilon_b \phi_a \qquad \hat{\nabla}_{\bar{a}} \phi_b = \nabla_{\bar{a}} \phi_b,$$

to compute the effect of these changes on curvature, namely

$$(40) \quad \begin{aligned} \hat{R}_{ab}{}^c{}_d &= R_{ab}{}^c{}_d - 2\delta_{[a}{}^d(\nabla_{b]}\Upsilon_c) + 2\delta_{[a}{}^d \Upsilon_{b]}\Upsilon_c + 2(\nabla_{[a}\Upsilon_{b]})\delta_d{}^c \\ \hat{R}_{a\bar{b}}{}^c{}_d &= R_{a\bar{b}}{}^c{}_d - \delta_a{}^c \nabla_{\bar{b}} \Upsilon_d - \delta_d{}^c \nabla_{\bar{b}} \Upsilon_a \\ \hat{R}_{ab}{}^{\bar{c}}{}_{\bar{d}} &= R_{ab}{}^{\bar{c}}{}_{\bar{d}} + T_{ab}{}^{\bar{e}}\Upsilon_{\bar{d}} + \Upsilon_{\bar{e}} T_{ab}{}^{\bar{e}} \delta_{\bar{d}}{}^{\bar{c}} = R_{ab}{}^{\bar{c}}{}_{\bar{d}} + T_{ab}{}^{\bar{e}} \upsilon_{\bar{e}\bar{d}}{}^{\bar{c}}. \end{aligned}$$

We shall also need the Bianchi symmetries derived from (35) or, more conveniently in the dual formulation, as follows. Evidently,

$$\nabla_a(\nabla_b\nabla_c - \nabla_c\nabla_b)f + \nabla_b(\nabla_c\nabla_a - \nabla_a\nabla_c)f + \nabla_c(\nabla_a\nabla_b - \nabla_b\nabla_a)f$$
$$= (\nabla_a\nabla_b - \nabla_b\nabla_a)\nabla_c f + (\nabla_b\nabla_c - \nabla_c\nabla_b)\nabla_a f + (\nabla_c\nabla_a - \nabla_a\nabla_c)\nabla_b f,$$

which we may expand using (34) and (35) to obtain

$$(\nabla_a T_{bc}{}^{\bar{d}} + \nabla_b T_{ca}{}^{\bar{d}} + \nabla_c T_{ab}{}^{\bar{d}})\nabla_{\bar{d}} f = (R_{ab}{}^d{}_c + R_{bc}{}^d{}_a + R_{bc}{}^d{}_a)\nabla_d f$$

and hence that

$$(41) \quad \nabla_{[a} T_{bc]}{}^{\bar{d}} = 0 \qquad R_{[ab}{}^c{}_{d]} = 0.$$

Similarly, by looking at different orderings for the indices of $\nabla_a \nabla_{\bar{b}} \nabla_c f$, we find that

$$(42) \quad R_{a\bar{b}}{}^c{}_d - R_{d\bar{b}}{}^c{}_a + T_{ad}{}^{\bar{e}} T_{\bar{e}\bar{b}}{}^c = 0 \qquad R_{ab}{}^{\bar{c}}{}_{\bar{d}} = \nabla_{\bar{d}} T_{ab}{}^{\bar{c}}.$$

Already, the final statement of (28) is evident and if $T_{ab}{}^{\bar{c}} = 0$ then both $R_{ab}{}^{\bar{c}}{}_{\bar{d}}$ and its complex conjugate $R_{\bar{a}\bar{b}}{}^c{}_d$ vanish. Notice that $\partial\mathsf{P}$ does not contribute to this piece of curvature. Specifically, from (24)

$$(\partial\mathsf{P})_{\bar{a}\bar{b}}{}^c{}_d = \overline{\Pi}_{\bar{a}}^\alpha \overline{\Pi}_{\bar{b}}^\beta \Pi_\gamma^c \Pi_d^\epsilon (\partial\mathsf{P})_{\alpha\beta}{}^\gamma{}_\epsilon = -\mathsf{P}_{[\bar{a}\bar{b}]}\delta_d{}^c - \mathsf{P}_{[\bar{b}\bar{a}]}\delta_d{}^c = 0.$$

It follows that $W_{\bar{a}\bar{b}}{}^c{}_d = R_{\bar{a}\bar{b}}{}^c{}_d$ in general and that $W_{\bar{a}\bar{b}}{}^c{}_d = R_{\bar{a}\bar{b}}{}^c{}_d = 0$ in the integrable case. The rest of statement (2) also follows, either from the last line of (40) or, more easily, from the c-projective invariance of $T_{ab}{}^{\bar{c}}$ (depending only on the underlying almost complex structure), the second identity of (42), and the transformation rules (39).

Now let us consider the curvature $R_{ab}{}^c{}_d$. From (24), we compute that

$$(\partial\mathsf{P})_{ab}{}^c{}_d = \Pi_a^\alpha \Pi_b^\beta \Pi_\gamma^c \Pi_d^\epsilon (\partial\mathsf{P})_{\alpha\beta}{}^\gamma{}_\epsilon = 2\delta_{[a}{}^c \mathsf{P}_{b]d} - 2\mathsf{P}_{[ab]}\delta_d{}^c$$

and from (25) that

$$\mathsf{P}_{ab} = \Pi_a^\alpha \Pi_b^\beta \mathsf{P}_{\alpha\beta} = \tfrac{1}{n+1}\left(\mathrm{Ric}_{ab} + \tfrac{2}{n-1}\mathrm{Ric}_{(ab)}\right) = \tfrac{1}{n+1}\left(\mathrm{Ric}_{ab} + \tfrac{2}{n-1}\mathrm{Ric}_{(ab)}\right),$$

equivalently that $\mathrm{Ric}_{ab} = (n-1)\mathsf{P}_{ab} + 2\mathsf{P}_{[ab]}$. Bearing in mind the Bianchi symmetry (41) for $R_{ab}{}^c{}_d$, this means that we may write

$$(43) \qquad R_{ab}{}^c{}_d = W_{ab}{}^c{}_d + 2\delta_{[a}{}^c \mathsf{P}_{b]d} + \beta_{ab}\delta_d{}^c,$$

where

$$W_{ab}{}^c{}_d = W_{[ab]}{}^c{}_d \qquad W_{[ab}{}^c{}_{d]} = 0 \qquad W_{ab}{}^a{}_d = 0 \qquad \beta_{ab} = -2\mathsf{P}_{[ab]}.$$

Comparing this decomposition with the first line of (40) implies that $W_{ab}{}^c{}_d$ is invariant and confirms that P_{ab} transforms according to Corollary 2.12. In summary,

$$\hat{W}_{ab}{}^c{}_d = W_{ab}{}^c{}_d \qquad \hat{\mathsf{P}}_{ab} = \mathsf{P}_{ab} - \nabla_a \Upsilon_c + \Upsilon_b \Upsilon_c \qquad \hat{\beta}_{ab} = \beta_{ab} + 2\nabla_{[a}\Upsilon_{b]}.$$

We have shown (3) and the first statement of (1).

The remaining statements concern the curvature $R_{a\bar{b}}{}^c{}_d$. From (24), we compute that

$$(\partial\mathsf{P})_{a\bar{b}}{}^c{}_d = \Pi_a^\alpha \overline{\Pi}_{\bar{b}}^\beta \Pi_\gamma^c \Pi_d^\epsilon (\partial\mathsf{P})_{\alpha\beta}{}^\gamma{}_\epsilon = \delta_a{}^c \mathsf{P}_{\bar{b}d} + \delta_d{}^c \mathsf{P}_{\bar{b}a}$$

and from (25) that

$$\mathsf{P}_{\bar{b}d} = \overline{\Pi}_{\bar{b}}^\beta \Pi_d^\epsilon \mathsf{P}_{\beta\epsilon} = \tfrac{1}{n+1}\mathrm{Ric}_{\bar{b}d} = \tfrac{1}{n+1}R_{a\bar{b}}{}^a{}_d.$$

From (42) it now follows that

$$(44) \quad R_{a\bar{b}}{}^c{}_d + \tfrac{1}{2} T_{ad}{}^{\bar{e}} T_{\bar{e}\bar{b}}{}^c = H_{a\bar{b}}{}^c{}_d - \tfrac{1}{2(n+1)}\left(\delta_a{}^c T_{df}{}^{\bar{e}} T_{\bar{e}\bar{b}}{}^f + \delta_d{}^c T_{af}{}^{\bar{e}} T_{\bar{e}\bar{b}}{}^f\right) + (\partial\mathsf{P})_{a\bar{b}}{}^c{}_d,$$

where

$$H_{a\bar{b}}{}^c{}_d = H_{d\bar{b}}{}^c{}_a \qquad H_{a\bar{b}}{}^a{}_d = 0.$$

Recall that by definition

$$W_{a\bar{b}}{}^c{}_d = R_{a\bar{b}}{}^c{}_d - (\partial\mathsf{P})_{a\bar{b}}{}^c{}_d.$$

Therefore

$$W_{a\bar{b}}{}^c{}_d = H_{a\bar{b}}{}^c{}_d - \tfrac{1}{2(n+1)}\left(\delta_a{}^c T_{df}{}^{\bar{e}} T_{\bar{e}\bar{b}}{}^f + \delta_d{}^c T_{af}{}^{\bar{e}} T_{\bar{e}\bar{b}}{}^f\right) - \tfrac{1}{2} T_{ad}{}^{\bar{e}} T_{\bar{e}\bar{b}}{}^c.$$

Comparison with the formula for $\hat{R}_{a\bar{b}}{}^c{}_d$ in (40) immediately shows that $W_{a\bar{b}}{}^c{}_d$ and $H_{a\bar{b}}{}^c{}_d$ are c-projectively invariant and also that

$$W_{a\bar{b}}{}^c{}_c = -T_{af}{}^{\bar{e}} T_{\bar{e}\bar{b}}{}^f,$$

2.4. THE CURVATURE OF THE CANONICAL CARTAN CONNECTION

as required to complete (1) and (4). Next we demonstrate the behaviour of the Cotton–York tensor. For this, we need a Bianchi identity with torsion, which may be established as follows. Evidently,

$$\nabla_a(\nabla_b\nabla_c - \nabla_c\nabla_b)\phi_e + \nabla_b(\nabla_c\nabla_a - \nabla_a\nabla_c)\phi_e + \nabla_c(\nabla_a\nabla_b - \nabla_b\nabla_a)\phi_e$$
$$= (\nabla_a\nabla_b - \nabla_b\nabla_a)\nabla_c\phi_e + (\nabla_b\nabla_c - \nabla_c\nabla_b)\nabla_a\phi_e + (\nabla_c\nabla_a - \nabla_a\nabla_c)\nabla_b\phi_e,$$

the left hand side of which may be expanded by (36) as

$$\nabla_a(-R_{bc}{}^d{}_e\phi_d - T_{bc}{}^{\bar{d}}\nabla_{\bar{d}}\phi_e) + \cdots + \cdots,$$

where $\cdots$ represent similar terms where the indices abc are cycled around. On the other hand, the right hand side may be expanded as

$$\cdots - R_{bc}{}^d{}_a\nabla_d\phi_e - R_{bc}{}^d{}_e\nabla_a\phi_d - T_{bc}{}^{\bar{d}}\nabla_{\bar{d}}\nabla_a\phi_e - \cdots.$$

Comparison yields

$$(\nabla_{[a}R_{bc]}{}^d{}_e)\phi_d + (\nabla_{[a}T_{bc]}{}^{\bar{d}})\nabla_{\bar{d}}\phi_e - T_{[bc}{}^{\bar{d}}R_{a]\bar{d}}{}^f{}_e\phi_f = R_{[bc}{}^d{}_{a]}\nabla_d\phi_e$$

and, from the Bianchi symmetries (41), we conclude that

$$\nabla_{[a}R_{bc]}{}^d{}_e = T_{[ab}{}^{\bar{f}}R_{c]\bar{f}}{}^d{}_e.$$

Using (28) and tracing over a and d yields

$$\nabla_a W_{bc}{}^a{}_e - 2(n-2)\nabla_{[b}\mathsf{P}_{c]e} + 3\nabla_{[b}\beta_{ce]}$$
$$= 2T_{a[b}{}^{\bar{f}}H_{c]\bar{f}}{}^a{}_e + \frac{1}{n+1}T_{bc}{}^{\bar{f}}T_{ea}{}^{\bar{d}}T_{\bar{d}\bar{f}}{}^a - \frac{n+2}{n+1}T_{e[b}{}^{\bar{f}}T_{c]a}{}^{\bar{d}}T_{\bar{d}\bar{f}}{}^a$$
$$+ (n-2)T_{bc}{}^{\bar{f}}\mathsf{P}_{\bar{f}e} + 3\mathsf{P}_{\bar{f}[b}T_{ce]}{}^{\bar{f}}.$$

Skewing this identity over bce gives (29) and substituting back gives

$$\nabla_a W_{bc}{}^a{}_e - 2(n-2)\nabla_{[b}\mathsf{P}_{c]e}$$
$$= 2T_{a[b}{}^{\bar{f}}H_{c]\bar{f}}{}^a{}_e + \frac{2}{n+1}T_{bc}{}^{\bar{f}}T_{ea}{}^{\bar{d}}T_{\bar{d}\bar{f}}{}^a - \frac{n}{n+1}T_{e[b}{}^{\bar{f}}T_{c]a}{}^{\bar{d}}T_{\bar{d}\bar{f}}{}^a + (n-2)T_{bc}{}^{\bar{f}}\mathsf{P}_{\bar{f}e}.$$

The contracted Bianchi identity (31) follows from the definition (30) of the Cotton–York tensor. Notice that the right hand side of (31) is c-projectively invariant. Also, by computing that

$$\hat{\nabla}_a\hat{W}_{bc}{}^d{}_e = \hat{\nabla}_a W_{bc}{}^d{}_e$$
$$= \nabla_a W_{bc}{}^d{}_e - 2\Upsilon_a W_{bc}{}^d{}_e - \Upsilon_b W_{ac}{}^d{}_e - \Upsilon_c W_{ba}{}^d{}_e + \delta_a{}^d\Upsilon_f W_{bc}{}^f{}_e - \Upsilon_e W_{bc}{}^d{}_a$$

and tracing over a and d, we see that

$$\hat{\nabla}_a\hat{W}_{bc}{}^a{}_e = \nabla_a W_{bc}{}^a{}_e + (n-2)\Upsilon_a W_{bc}{}^a{}_e$$

and for $n > 2$ conclude that (32) is valid. The case $n = 2$ is somewhat degenerate. Although (32) is still valid, as we shall see below in Proposition 2.14, the Weyl curvature $W_{bc}{}^a{}_d$ vanishes by symmetry considerations and (32) reads $\hat{C}_{bce} = C_{bce}$, the straightforward verification of which is left to the reader. Similarly, by considering different orderings for the indices of $\nabla_a\nabla_{\bar{b}}\nabla_c\phi_e$, we are rapidly led to

$$\nabla_a R_{c\bar{b}}{}^d{}_e - \nabla_c R_{a\bar{b}}{}^d{}_e + \nabla_{\bar{b}}R_{ac}{}^d{}_e = T_{ac}{}^{\bar{f}}R_{\bar{b}\bar{f}}{}^d{}_e$$

as another piece of the Bianchi identity, which may then be further split into irreducible parts. In particular, tracing over d and e (equivalently, tracing over a and d and then skewing over c and e) gives (33). □

PROPOSITION 2.14. *Suppose that $W_{ab}{}^c{}_d \in \wedge^{1,0} \otimes \wedge^{1,0} \otimes T^{1,0}M \otimes \wedge^{1,0}$ has the following symmetries*:
$$W_{ab}{}^c{}_d = W_{[ab]}{}^c{}_d \qquad W_{[ab}{}^c{}_{d]} = 0 \qquad W_{ab}{}^a{}_d = 0.$$
If $2n = 4$, then $W_{ab}{}^c{}_d \equiv 0$.

PROOF. Fix a nonzero skew tensor V_{ab}. As $W_{ab}{}^c{}_d$ is skew in a and b, it follows that $W_{ab}{}^c{}_d = V_{ab} S^c{}_d$ for some unique tensor $S^c{}_d$. Now $W_{ab}{}^a{}_d = V_{ab} S^a{}_d$ but V_{ab} is also nondegenerate so $W_{ab}{}^a{}_d = 0$ implies $S^c{}_d = 0$. □

REMARK 2.10. When $n = 2$, the identity (31) is vacuous. Proposition 2.14 implies that the left hand side vanishes. For the right hand side, the vanishing of $T_{a[b}{}^{\bar f} H_{c]\bar f}{}^a{}_e$ follows by tracing the identity $T_{[ab}{}^{\bar f} H_{c]\bar f}{}^d{}_e = 0$ over a and d, bearing in mind that $H_{a\bar f}{}^d{}_e$ is trace-free in a and d. The remaining terms also evaporate because, when $n = 2$, the tensor $T_{bc}{}^{\bar f} T_{ea}{}^{\bar d}$ is symmetric in $\bar f \bar d$ whereas $T_{\bar d \bar f}{}^a$ is skew.

The torsion $T_{ab}{}^{\bar c}$ (equivalently, its complex conjugate $T_{\bar a \bar b}{}^c$) is c-projectively invariant. The same is true, not only of the Weyl curvature $W_{a\bar b}{}^c{}_d$, but also of its trace-free symmetric part $H_{a\bar b}{}^c{}_d$ (which will be identified as part of the *harmonic curvature* in Chapter 2.7). The Weyl curvature $W_{ab}{}^c{}_d$ is c-projectively invariant and forms the final piece of harmonic curvature except when $2n = 4$, in which case $W_{ab}{}^c{}_d$ necessarily vanishes, its role being taken by C_{abc}, the c-projectively invariant part of the Cotton–York tensor. In Chapter 2.7, we place this discussion in the context of general parabolic geometry but, before that, we collect in the following section some useful formulae for the various curvature operators on c-projective densities.

2.5. Curvature operators on c-projective densities

Suppose $X^{cd\cdots e} = X^{[cd\cdots e]}$ is a section of $\mathcal{E}(n+1, 0) = \wedge^n T^{1,0}M$ and $Y^{\bar c \bar d \cdots \bar e}$ a section of $\mathcal{E}(0, n+1) = \wedge^n T^{0,1}M$. Then it follows from (35) that
$$(\nabla_{\bar a}\nabla_b - \nabla_b\nabla_{\bar a})X^{cd\cdots e} = R_{\bar a b}{}^c{}_f X^{fd\cdots e} + R_{\bar a b}{}^d{}_f X^{cf\cdots e} + \cdots + R_{\bar a b}{}^e{}_f X^{cd\cdots f}$$
$$= R_{\bar a b}{}^f{}_f X^{cd\cdots e}.$$
$$(\nabla_{\bar a}\nabla_b - \nabla_b\nabla_{\bar a})Y^{\bar c \bar d \cdots \bar e} = R_{\bar a b}{}^{\bar c}{}_{\bar f} Y^{\bar f \bar d \cdots \bar e} + R_{\bar a b}{}^{\bar d}{}_{\bar f} Y^{\bar c \bar f \cdots \bar e} + \cdots + R_{\bar a b}{}^{\bar e}{}_{\bar f} Y^{\bar c \bar d \cdots \bar f}$$
$$= R_{\bar a b}{}^{\bar f}{}_{\bar f} Y^{\bar c \bar d \cdots \bar e}.$$

However, from Proposition 2.13 part (4), we find that
$$R_{\bar a b}{}^f{}_f = -R_{b \bar a}{}^f{}_f = -W_{b \bar a}{}^f{}_f - (\partial \mathsf{P})_{b \bar a}{}^f{}_f = -T_{fb}{}^{\bar e} T_{\bar e \bar a}{}^f - (n+1)\mathsf{P}_{\bar a b}$$
$$R_{\bar a b}{}^{\bar f}{}_{\bar f} = W_{\bar a b}{}^{\bar f}{}_{\bar f} + (\partial \mathsf{P})_{\bar a b}{}^{\bar f}{}_{\bar f} = T_{fb}{}^{\bar e} T_{\bar e \bar a}{}^f + (n+1)\mathsf{P}_{b \bar a}.$$

We conclude immediately that for a section σ of $\mathcal{E}(k, \ell)$ we have
(45) $\qquad (\nabla_{\bar a}\nabla_b - \nabla_b\nabla_{\bar a})\sigma = \tfrac{\ell - k}{n+1} T_{fb}{}^{\bar e} T_{\bar e \bar a}{}^f \sigma + \ell \mathsf{P}_{b \bar a}\sigma - k \mathsf{P}_{\bar a b}\sigma.$

Similarly, from (35) it also follows that
$$(\nabla_a \nabla_b - \nabla_b \nabla_a) X^{cd\cdots e} + T_{ab}{}^{\bar f} \nabla_{\bar f} X^{cd\cdots e} = R_{ab}{}^f{}_f X^{cd\cdots e}$$
$$(\nabla_a \nabla_b - \nabla_b \nabla_a) Y^{\bar c \bar d \cdots \bar e} + T_{ab}{}^{\bar f} \nabla_{\bar f} Y^{\bar c \bar d \cdots \bar e} = R_{ab}{}^{\bar f}{}_{\bar f} Y^{\bar c \bar d \cdots \bar e}.$$

From Proposition 2.13 we conclude that
$$R_{ab}{}^f{}_f = 2\mathsf{P}_{[ba]} + n\beta_{ab} = (n+1)\beta_{ab}$$
$$R_{ab}{}^{\bar f}{}_{\bar f} = \nabla_{\bar f} T_{ab}{}^{\bar f}.$$
Therefore, if σ is c-projective density of weight (k, ℓ), then
(46) $\qquad (\nabla_a \nabla_b - \nabla_b \nabla_a)\sigma + T_{ab}{}^{\bar f}\nabla_{\bar f}\sigma = k\beta_{ab}\sigma + \frac{\ell}{n+1}(\nabla_{\bar f}T_{ab}{}^{\bar f})\sigma$
and, accordingly,
(47) $\qquad (\nabla_{\bar a} \nabla_{\bar b} - \nabla_{\bar b} \nabla_{\bar a})\sigma + T_{\bar a \bar b}{}^f \nabla_f \sigma = \ell \beta_{\bar a \bar b} + \frac{k}{n+1}(\nabla_f T_{\bar a \bar b}{}^f)\sigma.$

Recall that for any connection $\nabla \in [\nabla]$ its Rho tensor, by definition, satisfies $\mathsf{P}_{\bar a b} = \frac{1}{n+1}\mathrm{Ric}_{\bar a b}$ and $\mathsf{P}_{[ab]} = \frac{1}{n+1}\mathrm{Ric}_{[ab]}$. Hence, the identities (45) and (46) imply that the Ricci tensor of a special connection $\nabla \in [\nabla]$ satisfies
- $\mathrm{Ric}_{\bar a b} = \mathrm{Ric}_{b \bar a}$
- $\mathrm{Ric}_{[ab]} = \frac{1}{2}\nabla_{\bar c} T_{ab}{}^{\bar c}$.

If $\nabla_{\bar c} T_{ab}{}^{\bar c}$ vanishes, the special connection has symmetric Ricci tensor. In particular, if J is integrable all special connections have symmetric Ricci tensor.

2.6. The curvature of complex projective space

In Chapter 2.4, and especially in Proposition 2.13, the curvature of a complex connection on a general almost complex manifold was decomposed into various irreducible pieces (irreducibility to be further discussed in Chapter 3.3). Here, we pause to examine this decomposition on complex projective space $\mathbb{CP}^n$ with its standard Fubini–Study metric.

LEMMA 2.15. *The Riemannian curvature tensor for the Fubini–Study metric $g_{\alpha\beta}$ on $\mathbb{CP}^n$ is given by*
(48) $\qquad R_{\alpha\beta\gamma\delta} = g_{\alpha\gamma}g_{\beta\delta} - g_{\beta\gamma}g_{\alpha\delta} + \Omega_{\alpha\gamma}\Omega_{\beta\delta} - \Omega_{\beta\gamma}\Omega_{\alpha\delta} + 2\Omega_{\alpha\beta}\Omega_{\gamma\delta}$
where $J_\alpha{}^\beta$ is the complex structure and $\Omega_{\alpha\gamma} \equiv J_\alpha{}^\beta g_{\beta\gamma}$ (the Kähler form).

PROOF. A direct calculation from the definition of the Fubini–Study metric (e.g. [**27**]). Alternatively, one can argue by invariant theory noting that (up to scale) the right hand side of (48) is the only covariant expression in $g_{\alpha\beta}$ and $\Omega_{\alpha\beta}$ such that
$$R_{\alpha\beta\gamma\delta} = R_{[\alpha\beta][\gamma\delta]} \qquad R_{[\alpha\beta\gamma]\delta} = 0 \qquad R_{\alpha\beta\gamma[\delta}J_{\epsilon]}{}^\gamma = 0,$$
the last condition following from the Kähler condition $d\Omega = 0$. More precisely, the last condition is a direct consequence of $\nabla_\alpha \Omega_{\beta\gamma} = 0$ and one can check, in case the almost complex structure $J_\alpha{}^\beta$ is orthogonal (i.e. $J_\alpha{}^\beta g_{\beta\gamma}$ is skew), that
$$2\nabla_\alpha \Omega_{\beta\gamma} = 3\nabla_{[\alpha}\Omega_{\beta\gamma]} - 3J_\alpha{}^\delta J_\beta{}^\epsilon \nabla_{[\alpha}\Omega_{\delta\epsilon]} - \Omega_{\alpha\delta}N_{\alpha\beta}{}^\delta.$$
Recall that $N_{\alpha\beta}{}^\gamma$ is the Nijenhuis tensor (1), which vanishes when the complex structure is integrable, as it is on $\mathbb{CP}^n$. $\square$

To apply the decompositions of Proposition 2.13 to (48) we should raise an index
$$R_{\alpha\beta}{}^\gamma{}_\epsilon = \delta_\alpha{}^\gamma g_{\beta\epsilon} - \delta_\beta{}^\gamma g_{\alpha\epsilon} + J_\alpha{}^\gamma \Omega_{\beta\epsilon} - J_\beta{}^\gamma \Omega_{\alpha\epsilon} - 2\Omega_{\alpha\beta}J_\epsilon{}^\gamma$$
and then apply the various projectors such as $\Pi_a^\alpha \Pi_b^\beta \Pi_\gamma^c \Pi_d^\epsilon$. However, firstly note that applying $\Pi_a^\alpha \Pi_c^\gamma$ to $J_\alpha{}^\beta g_{\beta\gamma} + J_\gamma{}^\beta g_{\beta\alpha} = 0$ implies that $g_{ac} = 0$ (consequently

$\Omega_{ac} = 0$) whilst applying $\Pi_a^\alpha \overline{\Pi}_{\bar c}^{\bar\gamma}$ to $\Omega_{\alpha\gamma} = J_\alpha{}^\beta g_{\beta\gamma}$ shows that $\Omega_{a\bar c} = i g_{a\bar c}$. We conclude that $R_{ab}{}^c{}_d = 0$ and

$$R_{a\bar b}{}^c{}_d \equiv \Pi_a^\alpha \overline{\Pi}_{\bar b}^{\bar\beta} \Pi_\gamma^c \Pi_d^\epsilon R_{\alpha\bar\beta}{}^\gamma{}_\epsilon = \delta_a{}^c g_{d\bar b} - i\delta_a{}^c \Omega_{d\bar b} - 2i\Omega_{a\bar b}\delta_d{}^c = 2\delta_a{}^c g_{d\bar b} + 2\delta_d{}^c g_{a\bar b}.$$

Thus, with reference to Proposition 2.13, we see that all irreducible pieces of curvature vanish save for $\mathsf{P}_{\bar b d} = 2g_{d\bar b}$. In particular, all invariant pieces

$$T_{\bar a \bar b}{}^c \qquad H_{a\bar b}{}^c{}_d \qquad W_{ab}{}^c{}_d$$

of *harmonic curvature* (as identified the following section) vanish. This is, of course, consistent with $\mathbb{CP}^n$, equipped with its standard complex structure and Fubini–Study connection, being the flat model of c-projective geometry, as discussed in Chapter 2.3 and especially Theorem 2.8.

Finally, observe that if we regard $\mathbb{CP}^n$ as

$$\mathrm{SL}(n+1, \mathbb{C}) \Big/ \left\{ \begin{pmatrix} \lambda & * & \cdots & * \\ 0 & * & \cdots & * \\ \vdots & \vdots & \ddots & \vdots \\ 0 & * & \cdots & * \end{pmatrix} \right\},$$

rather than as a homogeneous $\mathrm{PSL}(n+1, \mathbb{C})$-space as in Chapter 2.3, then the character $\lambda \mapsto \lambda^{-k}\bar\lambda^{-\ell}$ induces a homogeneous line bundle $\mathcal{E}(k, \ell)$ on $\mathbb{CP}^n$ as we were supposing earlier and as we shall soon suppose in Chapter 3.1. This observation also explains our copacetic choice of notation: on $\mathbb{CP}^n$ it is standard to write $\mathcal{O}(k)$ for the holomorphic bundle that is $\mathcal{E}(k, 0)$ just as a complex bundle (and then $\overline{\mathcal{E}(k, 0)} = \mathcal{E}(0, k)$).

2.7. The harmonic curvature

A normal Cartan connection gives rise to a simpler local invariant than the Cartan curvature κ, called the *harmonic curvature* κ_h, which still provides a full obstruction to local flatness, as discussed in Chapter 2.2 (cf. especially Proposition 2.7). The harmonic curvature κ_h of an almost c-projective manifold is the projection of $\kappa \in \ker \partial^*$ to its homology class in

$$\mathcal{G} \times_P H_2(\mathfrak{g}_1, \mathfrak{g}) \cong \mathcal{G}_0 \times_{G_0} H_2(\mathfrak{g}_1, \mathfrak{g}).$$

By Kostant's version of the Bott–Borel–Weil Theorem [64] the G_0-module $H_2(\mathfrak{g}_1, \mathfrak{g})$ can be naturally identified with a G_0-submodule in $\wedge^2 \mathfrak{g}_1 \otimes_\mathbb{R} \mathfrak{g} \cong \wedge^2 \mathfrak{g}_{-1}^* \otimes_\mathbb{R} \mathfrak{g}$ that decomposes into three irreducible components as follows:

- for $n = 2$
$$(\wedge^{0,2}\mathfrak{g}_{-1}^* \otimes_\mathbb{C} \mathfrak{g}_{-1}) \oplus (\wedge^{1,1}\mathfrak{g}_{-1}^* \odot_\mathbb{C} \mathfrak{sl}(\mathfrak{g}_{-1}, \mathbb{C})) \oplus (\wedge^{2,0}\mathfrak{g}_{-1}^* \otimes_\mathbb{C} \mathfrak{g}_1)$$

- for $n > 2$
$$(\wedge^{0,2}\mathfrak{g}_{-1}^* \otimes_\mathbb{C} \mathfrak{g}_{-1}) \oplus (\wedge^{1,1}\mathfrak{g}_{-1}^* \odot_\mathbb{C} \mathfrak{sl}(\mathfrak{g}_{-1}, \mathbb{C})) \oplus (\wedge^{2,0}\mathfrak{g}_{-1}^* \odot_\mathbb{C} \mathfrak{sl}(\mathfrak{g}_{-1}, \mathbb{C})),$$

where these are complex vector spaces but regarded as real, and where $\odot$ denotes the Cartan product. Correspondingly, we decompose the harmonic curvature as

$$\kappa_h = \tau + \psi + \chi$$

in case $n = 2$ and

$$\kappa_h = \tau + \psi_1 + \psi_2$$

in case $n > 3$.

Note that ∂^* preserves homogeneities, i.e. $\partial^*(\wedge^i \mathfrak{g}_1 \otimes \mathfrak{g}_j) \subset \wedge^{i-1}\mathfrak{g}_1 \otimes \mathfrak{g}_{j+1}$. In particular, the induced vector bundle map ∂^* maps $\wedge^3 T^*M \otimes \mathcal{A}M$ to $\wedge^2 T^*M \otimes \mathcal{A}^0 M$. Hence, we conclude that τ must equal the torsion $T_{\alpha\beta}{}^\gamma$. If $n = 2$, then ψ is the component $H_{a\bar{b}}{}^c{}_d$ in $(\wedge^{1,1} \odot_{\mathbb{C}} \mathfrak{sl}(T^{1,0}M))$ of the Weyl curvature of any connection in the c-projective class, and χ is the $(2,0)$-part of the Cotton–York tensor. If $n > 2$, then ψ_1, respectively ψ_2, is the totally trace-free $(1,1)$-part, respectively $(2,0)$-part, of the Weyl curvature of any connection in the class.

We now give a geometric interpretation of the three harmonic curvature components.

THEOREM 2.16. *Suppose $(M, J, [\nabla])$ is an almost c-projective manifold of dimension $2n \geq 4$ and denote by κ_h the harmonic curvature of its normal Cartan connection. Then the following statements hold.*
  (1) *$\kappa_h \equiv 0$ if and only if the almost c-projective manifold $(M, J, [\nabla])$ is locally isomorphic to $(\mathbb{C}P^n, J_{\mathrm{can}}, [\nabla^{g_{FS}}])$.*
  (2) *τ is the torsion of $(M, J, [\nabla])$. In particular, $\tau \equiv 0$ if and only if J is integrable, i.e. $(M, J, [\nabla])$ is a c-projective manifold. Moreover, in this case, $J^{\mathcal{G}}$ is integrable and the Cartan bundle $p \colon \mathcal{G} \to M$ is a holomorphic principal P-bundle.*
  (3) *Suppose $\tau \equiv 0$. Then $\psi_1 \equiv 0$ (resp. $\psi \equiv 0$) if and only if ω is a holomorphic Cartan connection on the holomorphic principal bundle $p \colon \mathcal{G} \to M$. This is the case if and only if $[\nabla]$ locally admits a holomorphic connection, i.e. for any connection $\nabla \in [\nabla]$ and any point $x \in M$ there is an open neighbourhood $U \ni x$ such that $\nabla|_U$ is c-projectively equivalent to a holomorphic connection on U.*

PROOF. We have already observed (1) and the first two assertions of (2). To prove the last statement of (2) and (3), assume that $\tau \equiv 0$, which says, in particular, that the Cartan geometry is torsion-free. Since P acts on the complex vector space $\wedge^2 \mathfrak{g}_1 \otimes_{\mathbb{R}} \mathfrak{g}$ by complex linear maps, P preserves the decomposition of this vector space into the three (p, q)-types. Therefore [28, Corollary 3.2] applies and hence κ_h has components of type (p, q) if and only if κ has components of type (p, q). Therefore, $\tau \equiv 0$ implies that κ has no $(0, 2)$-part, which by the proof of [28, Theorem 3.4] (cf. [36, Proposition 3.1.17]) implies that $J^{\mathcal{G}}$ is integrable and $p \colon \mathcal{G} \to M$ a holomorphic principal bundle. This finishes the proof of (2). We know that the component ψ_1 (respectively ψ) vanishes identically if and only if κ is of type $(2, 0)$, which by [28, Theorem 3.4] is the case if and only if $\omega \in \Omega^{1,0}(\mathcal{G}, \mathfrak{g})$ is holomorphic, i.e. $d\omega$ is of type $(2, 0)$. Hence, it just remains to prove the last assertion of (3). Assume firstly that ψ_1 (respectively ψ) vanishes identically and hence that $(p \colon \mathcal{G} \to M, \omega)$ is a holomorphic Cartan geometry. Then we can find around each point of M an open neighbourhood $U \subset M$ such that $\mathcal{G}$ and $\mathcal{G}_0$ trivialise as holomorphic principal bundles over U. Having chosen such trivialisations, the holomorphic inclusion $G_0 \hookrightarrow P$ induces a holomorphic G_0-equivariant section $\sigma \colon p_0^{-1}(U) \to p^{-1}(U)$. Since $d\omega$ is of type $(2, 0)$ and σ is holomorphic

$$\sigma^* d\omega = d\sigma^* \omega = d\theta + d\gamma^\nabla - d\mathsf{p}^\nabla$$

is also of type $(2, 0)$. In particular, $d\gamma^\nabla$ is of type $(2, 0)$ and it follows that $\gamma^\nabla \in \Omega^{1,0}(p_0^{-1}(U), \mathfrak{g}_0)$ is a holomorphic principal connection in the c-projective class. Conversely, assume that $U \subset M$ is an open set and that $\gamma^\nabla \in \Omega^{1,0}(p_0^{-1}(U), \mathfrak{g}_0)$ is a holomorphic principal connection that belongs to the c-projective class. Since the Lie bracket on $\mathfrak{g}$ is complex linear, the holomorphicity of γ^∇ implies that its curvature $d\gamma^\nabla + [\gamma^\nabla, \gamma^\nabla]$ is of type $(2,0)$. By definition of the Weyl curvature this implies that also its Weyl curvature is of type $(2,0)$ and hence so is $\kappa_h|_U$. By assumption there exists locally around any point a holomorphic connection and hence κ_h is of type $(2,0)$ on all of M. □

CHAPTER 3

Tractor bundles and BGG sequences

The normal Cartan connection of an almost c-projective manifold induces a canonical linear connection on all associated vector bundles corresponding to representations of $\mathrm{PSL}(n+1,\mathbb{C})$ (cf. Chapter 2.2). These, in the theory of parabolic geometries, so-called *tractor connections*, provide an efficient calculus, especially well suited for explicit constructions of local invariants and invariant differential operators. We develop in this section the basics of the theory of tractor connections for almost c-projective manifolds, and explain their relation to geometrically significant overdetermined systems of PDE and sequences of invariant differential operators.

3.1. Standard complex tractors

Suppose that $(M, J, [\nabla])$ is an almost c-projective manifold of dimension $2n \geq 4$. Further, assume that the complex line bundle $\wedge^n T^{1,0}M$ admits an $(n+1)^{\text{st}}$ root and choose one, denoted $\mathcal{E}(1,0)$, with conjugate $\mathcal{E}(0,1)$. More generally, we write $\mathcal{E}(k,\ell) = \mathcal{E}(1,0)^{\otimes k} \otimes \mathcal{E}(0,1)^{\otimes \ell}$ for any $(k,\ell) \in \mathbb{Z} \times \mathbb{Z}$ (cf. Chapter 2.1). Note that such a choice of a root $\mathcal{E}(1,0)$ is at least locally always possible and the assumption that such roots exists globally is a relatively minor constraint. The choice of $\mathcal{E}(1,0)$ canonically extends the Cartan bundle of $(M, J, [\nabla])$ to a $\tilde{P}$-principal bundle $\tilde{p}\colon \tilde{\mathcal{G}} \to M$, where $\tilde{P}$ is the stabiliser in $\mathrm{SL}(n+1,\mathbb{C})$ of the complex line generated by the first basis vector in $\mathbb{C}^{n+1}$, and the normal Cartan connection of $(M, J, [\nabla])$ naturally extends to a normal Cartan connection on $\tilde{\mathcal{G}}$ of type $(\mathrm{SL}(n+1,\mathbb{C}), \tilde{P})$, which we also denote by ω. The groups $\mathrm{SL}(n+1,\mathbb{C})$ and $\tilde{P}$ are here viewed as real Lie groups as in Chapter 2.3, and we obtain in this way, analogously to Theorem 2.8, an equivalence of categories between almost c-projective manifolds equipped with an $(n+1)$st root $\mathcal{E}(1,0)$ of $\wedge^n T^{1,0}M$ and normal Cartan geometries of type $(\mathrm{SL}(n+1,\mathbb{C}), \tilde{P})$. The homogeneous model of such structures is again $\mathbb{CP}^n$, but now viewed as a homogeneous space $\mathrm{SL}(n+1,\mathbb{C})/\tilde{P}$ with $\mathcal{E}(1,0)$ being dual to the tautological line bundle $\mathcal{O}(-1)$, cf. Chapter 2.6.

The extended normal Cartan geometry of type $(\mathrm{SL}(n+1,\mathbb{C}), \tilde{P})$ allows us to form the *standard complex tractor bundle*

$$\mathcal{T} := \tilde{\mathcal{G}} \times_{\tilde{P}} \mathbb{V}$$

of $(M, J, [\nabla], \mathcal{E}(1,0))$, where $\mathbb{V} = \mathbb{R}^{2n+2}$ is the defining representation of the real Lie group $\mathrm{SL}(2(n+1), \mathbb{J}_{2(n+1)}) \cong \mathrm{SL}(n+1,\mathbb{C})$. Note that the complex structure $\mathbb{J}_{2(n+1)}$ on $\mathbb{V}$ induces a complex structure $J^{\mathcal{T}}$ on $\mathcal{T}$. Analogously to the discussion of the tangent bundle of an almost complex manifold in Chapter 1, $(\mathcal{T}, J^{\mathcal{T}})$ can be identified with the $(1,0)$-part of its complexification $\mathcal{T}_{\mathbb{C}}$, on which $J^{\mathcal{T}}$ acts by multiplication by i. We shall implement this identification in the sequel without further comment, and similarly for all the other tractor bundles with complex structures in the following Chapters 3.1, 3.3 and 3.4. Since $\tilde P$ stabilises the complex line generated by the first basis vector in $\mathbb{C}^{n+1}$, this line defines a complex 1-dimensional submodule of $\mathbb{C}^{n+1}$. Correspondingly, the standard complex tractor bundle (identified with the $(1,0)$-part of its complexification $\mathcal{T}_\mathbb{C}$) is filtered as

$$(49) \qquad \mathcal{T} = \mathcal{T}^0 \supset \mathcal{T}^1$$

where $\mathcal{T}^1 \cong \mathcal{E}(-1,0)$ and $\mathcal{T}^0/\mathcal{T}^1 \cong T^{1,0}M(-1,0)$. Since $\mathcal{T}$ is induced by a representation of $\mathrm{SL}(n+1,\mathbb{C})$, the Cartan connection induces a linear connection $\nabla^\mathcal{T}$ on $\mathcal{T}$, called the *tractor connection* (see Chapter 2.2). Any choice of a linear connection $\nabla \in [\nabla]$, splits the filtration of the tractor bundle $\mathcal{T}$ and the splitting changes by

$$(50) \qquad \widehat{\begin{pmatrix} X^b \\ \rho \end{pmatrix}} = \begin{pmatrix} X^b \\ \rho - \Upsilon_c X^c \end{pmatrix}, \text{ where } \begin{cases} X^b \in T^{1,0}M(-1,0) \\ \rho \in \mathcal{E}(-1,0), \end{cases}$$

if one changes the connection according to (11). In terms of a connection $\nabla \in [\nabla]$, the tractor connection is given by

$$(51) \qquad \nabla^\mathcal{T}_\alpha \begin{pmatrix} X^b \\ \rho \end{pmatrix} = \begin{pmatrix} \nabla_\alpha X^b + \rho \Pi_\alpha^b \\ \nabla_\alpha \rho - \mathsf{P}_{\alpha b} X^b \end{pmatrix}.$$

Applying Π_a^α and $\overline{\Pi}_{\bar a}^\alpha$ to (51), we can write the tractor connection as

$$(52) \qquad \nabla^\mathcal{T}_a \begin{pmatrix} X^b \\ \rho \end{pmatrix} = \begin{pmatrix} \nabla_a X^b + \rho \delta_a{}^b \\ \nabla_a \rho - \mathsf{P}_{ab} X^b \end{pmatrix} \text{ and } \nabla^\mathcal{T}_{\bar a} \begin{pmatrix} X^b \\ \rho \end{pmatrix} = \begin{pmatrix} \nabla_{\bar a} X^b \\ \nabla_{\bar a} \rho - \mathsf{P}_{\bar a b} X^b \end{pmatrix}.$$

By (49) the dual (or "co-standard") complex tractor bundle $\mathcal{T}^* \cong \tilde{\mathcal{G}} \times_{\tilde P} \mathbb{V}^*$ admits a natural subbundle isomorphic to $\wedge^{1,0}(1,0)$ and the quotient $\mathcal{T}^*/\wedge^{1,0}(1,0)$ is isomorphic to $\mathcal{E}(1,0)$. One immediately deduces from (52) that in terms of a connection $\nabla \in [\nabla]$, the tractor connection on $\mathcal{T}^*$ is given by

$$(53) \qquad \nabla^{\mathcal{T}^*}_a \begin{pmatrix} \sigma \\ \mu_b \end{pmatrix} = \begin{pmatrix} \nabla_a \sigma - \mu_a \\ \nabla_a \mu_b + \mathsf{P}_{ab}\sigma \end{pmatrix} \text{ and } \nabla^{\mathcal{T}^*}_{\bar a} \begin{pmatrix} \sigma \\ \mu_b \end{pmatrix} = \begin{pmatrix} \nabla_{\bar a} \sigma \\ \nabla_{\bar a} \mu_b + \mathsf{P}_{\bar a b}\sigma \end{pmatrix}.$$

REMARK 3.1. If $(M, J, [\nabla])$ is a c-projective manifold the $(0,1)$-parts of (52) and (53) make $\mathcal{T}$ and $\mathcal{T}^*$ into holomorphic vector bundles. Parallel sections for $\nabla^\mathcal{T}$ or $\nabla^{\mathcal{T}^*}$ are then holomorphic sections of $\mathcal{T}$ or $\mathcal{T}^*$ such that the $(1,0)$-parts of (52) or (53) vanish (respectively). If, in addition, $(M, J, [\nabla])$ is a holomorphic c-projective manifold, these $(1,0)$-parts are complexifications of the formulae for standard and costandard tractor connections in real projective geometry [7]. A similar remark holds for tractor connections associated to any representation of $\mathrm{SL}(n+1,\mathbb{C})$ as a *complex* Lie group (of which $\mathbb{V}$ and $\mathbb{V}^*$ are examples).

For a choice of connection $\nabla \in [\nabla]$ consider now the following overdetermined system of PDE on sections σ of $\mathcal{E}(1,0)$:

$$(54) \qquad \text{(i) } \nabla_{\bar a} \sigma = 0 \qquad \text{(ii) } \nabla_{(a} \nabla_{b)} \sigma + \mathsf{P}_{(ab)} \sigma = 0.$$

Suppose $\hat\nabla \in [\nabla]$ is another connection in the c-projective class. Then the formulae (16) imply that $\hat\nabla_{\bar a} \sigma = \nabla_{\bar a} \sigma$. Moreover, we deduce from Proposition 2.5 and the formulae (16) that

$$\hat\nabla_a \hat\nabla_b \sigma = \nabla_a \nabla_b \sigma + (\nabla_a \Upsilon_b)\sigma - \Upsilon_a \Upsilon_b \sigma,$$

3.1. STANDARD COMPLEX TRACTORS

which together with Corollary 2.12 implies that

$$\hat\nabla_a\hat\nabla_b\sigma + \hat{\mathsf{P}}_{ab}\sigma = \nabla_a\nabla_b\sigma + \mathsf{P}_{ab}\sigma.$$

This shows that the overdetermined system (54) is c-projectively invariant. Note also that by equation (46) we have

(55) $$\nabla_{[a}\nabla_{b]}\sigma + P_{[ab]}\sigma = -\tfrac{1}{2}T_{ab}{}^{\bar c}\nabla_{\bar c}\sigma.$$

Therefore $\nabla_a\nabla_b\sigma + \mathsf{P}_{ab}\sigma$ is symmetric provided that J is integrable or that σ satisfies equation (i) of (54). The following proposition shows that, if J is integrable, the tractor connection on $\mathcal{T}^*$ can be viewed as the prolongation of (54):

PROPOSITION 3.1. *Suppose $(M, J, [\nabla], \mathcal{E}(1,0))$ is a c-projective manifold of dimension $2n \geq 4$. The projection $\pi\colon \mathcal{T}^* \to \mathcal{T}^*/\wedge^{1,0}(1,0) \cong \mathcal{E}(1,0)$ induces a bijection between sections of $\mathcal{T}^*$ parallel for $\nabla^{\mathcal{T}^*}$ and sections σ of $\mathcal{E}(1,0)$ that satisfy (54) for some (and hence any) connection $\nabla \in [\nabla]$. Moreover, suppose that $\sigma \in \mathcal{E}(1,0)$ is a nowhere vanishing section, then for any connection $\nabla \in [\nabla]$ the connection*

(56) $$\hat\nabla_a\sigma' = \nabla_a\sigma' - (\nabla_a\sigma)\sigma^{-1}\sigma'$$

is induced from a connection in the c-projective class, and σ with $\nabla_{\bar a}\sigma = 0$ is a solution of (54) if and only if (56) is Ricci-flat.

PROOF. Suppose s is a parallel section of $\mathcal{T}^*$ and set $\sigma := \pi(s) \in \Gamma(\mathcal{E}(1,0))$. It follows from (50) that changing from connection to another in $[\nabla]$ changes the splitting of $\mathcal{T}^*$ by $\widehat{(\sigma, \mu_b)} = (\sigma, \mu_b + \Upsilon_b\sigma)$. Hence, for any connection $\nabla \in [\nabla]$ we can identify s with a section of the form (σ, μ_b) for some $\mu_b \in \Gamma(\wedge^{1,0}(1,0))$ and by definition of the tractor connection $\sigma \in \Gamma(\mathcal{E}(1,0))$ satisfies (54) for any connection $\nabla \in [\nabla]$. So π induces a map from parallel sections of $\mathcal{T}^*$ to solutions of (54).

Conversely, let us contemplate the differential operator $L\colon \mathcal{E}(1,0) \to \mathcal{T}^*$, which, for a choice of connection in $[\nabla]$, is given by $L\sigma = (\sigma, \nabla_b\sigma)$. Suppose now σ is a solution of (54). Then obviously $\nabla_a^{\mathcal{T}^*}L\sigma = 0$, since (55) vanishes. By (45) we have

(57)
$$\begin{aligned}\nabla_{\bar a}\nabla_b\sigma + \mathsf{P}_{\bar a b}\sigma &= (\nabla_{\bar a}\nabla_b - \nabla_b\nabla_{\bar a})\sigma + \mathsf{P}_{\bar a b}\sigma \\ &= \tfrac{1}{n+1}T_{bf}{}^{\bar e}T_{\bar e \bar a}{}^f\sigma - \mathsf{P}_{\bar a b}\sigma + \mathsf{P}_{\bar a b}\sigma \\ &= \tfrac{1}{n+1}T_{bf}{}^{\bar e}T_{\bar e \bar a}{}^f\sigma\,,\end{aligned}$$

which vanishes, since J is integrable. Hence, we also have $\nabla_{\bar a}^{\mathcal{T}^*}L\sigma = 0$. Therefore L maps solutions $\sigma \in \Gamma(\mathcal{E}(1,0))$ of (54) to parallel sections of $\mathcal{T}^*$ and defines an inverse to the restriction of π to parallel section. For the second statement, assume now that σ is a section of $\mathcal{E}(1,0)$ that is nowhere vanishing and let $\nabla \in [\nabla]$ be a connection in the c-projective class. If we change ∇ according to (11) by $\Upsilon_a = -(\nabla_a\sigma)\sigma^{-1}$ to a connection $\hat\nabla \in [\nabla]$, then we deduce from Corollary 2.4 that the induced connection on $\mathcal{E}(1,0)$ is given by (56). Moreover, $\hat\nabla_a\sigma = 0$. Therefore, using that (57) vanishes, we deduce that σ with $\hat\nabla_{\bar a}\sigma = 0$ satisfies (54) if and only if $\hat{\mathsf{P}}_{ab}\sigma = 0$ and $\hat{\mathsf{P}}_{\bar a b}\sigma = 0$, and hence if and only if $\hat\nabla$ is Ricci-flat. □

More generally, we immediately conclude from equation (57) that, in the case of an almost c-projective manifold, i.e. J is not necessarily integrable, the corresponding proposition reads as follows:

PROPOSITION 3.2. *Let $(M, J, [\nabla], \mathcal{E}(1,0))$ be an almost c-projective manifold of dimension $2n \geq 4$. Then sections σ of $\mathcal{E}(1,0)$ that satisfy (54) are in bijection with sections of $\mathcal{T}^*$ that are parallel for the connection given by*

$$(58) \qquad \nabla_a^{\mathcal{T}^*} \quad \text{and} \quad \nabla_{\bar{a}}^{\mathcal{T}^*} \begin{pmatrix} \sigma \\ \mu_b \end{pmatrix} - \frac{1}{n+1} \begin{pmatrix} 0 \\ T_{bf}{}^{\bar{e}} T_{\bar{e}\bar{a}}{}^f \sigma \end{pmatrix}.$$

Moreover, suppose $\sigma \in \mathcal{E}(1,0)$ is a nowhere vanishing section with $\nabla_{\bar{a}}\sigma = 0$. Then σ is a solution of (54) if and only if $\hat{\nabla}$, defined as in (56), satisfies $\hat{\mathsf{P}}_{ab} = 0$ and $\hat{\mathsf{P}}_{\bar{a}b} = \frac{1}{n+1} T_{bf}{}^{\bar{e}} T_{\bar{e}\bar{a}}{}^f$.

REMARK 3.2. Recall that a parallel section of a linear connection of a vector bundle over a connected manifold, is already determined by its value at one point. The correspondences established in Propositions 3.1 and 3.2 between solutions of (54) and parallel sections of a linear connection on $\mathcal{T}^*$ therefore implies, that on a connected almost c-projective manifold

$$U := \{x \in M : \sigma(x) \neq 0\} \subset M$$

is a dense open subset for any nontrivial solution $\sigma \in \Gamma(\mathcal{E}(1,0))$ of (54). In particular, the second statement of Proposition 3.1, respectively of Proposition 3.2, holds always on the dense open subset U.

The equations (54) define an invariant differential operator of order two

$$D^{\mathcal{T}^*} : \mathcal{E}(1,0) \to \wedge^{0,1}(1,0) \oplus S^2 \wedge^{1,0}(1,0),$$

whose kernel are the solutions of (54). The differential operator $D^{\mathcal{T}^*}$ is the first operator in the BGG sequence corresponding to the co-standard complex tractor bundle; see [25, 38]. The proof of Proposition 3.2 implies that in order for a (nonzero) parallel section of the tractor connection on $\mathcal{T}^*$ to exist, it is necessary that Nijenhuis tensor of J satisfy $N_{bf}^{J}{}^{\bar{e}} N_{\bar{e}\bar{a}}^{J}{}^f \equiv 0$ and, in this case, parallel sections of the tractor connection are in bijection with sections in the kernel of $D^{\mathcal{T}^*}$.

Similarly, one may consider the first BGG operator in the sequence corresponding to the standard complex tractor bundle $\mathcal{T}$, which is a first order invariant differential operator, defined, for a choice of connection $\nabla \in [\nabla]$, by

$$D^{\mathcal{T}} : T^{1,0}M(-1,0) \to (\wedge^{0,1} \otimes T^{1,0}M(-1,0)) \oplus (\wedge^{1,0} \otimes_\circ T^{1,0}M(-1,0)),$$
$$(59) \qquad X^b \mapsto (\nabla_{\bar{a}} X^b, \nabla_a X^b - \tfrac{1}{n} \nabla_c X^c \delta_a{}^b),$$

where the subscript $\circ$ denotes the trace-free part.

PROPOSITION 3.3. *Suppose $(M, J, [\nabla], \mathcal{E}(1,0))$ is an almost c-projective manifold of dimension $2n \geq 4$. The projection $\pi : \mathcal{T} \to \mathcal{T}/\mathcal{E}(-1,0) \cong T^{1,0}M(-1,0)$ induces a bijection between sections of $\mathcal{T}$ that are parallel for the connection given by*

$$(60) \qquad \nabla_a^{\mathcal{T}} \quad \text{and} \quad \nabla_{\bar{a}}^{\mathcal{T}} \begin{pmatrix} X^b \\ \rho \end{pmatrix} - \frac{1}{n(n+1)} \begin{pmatrix} 0 \\ T_{bf}{}^{\bar{e}} T_{\bar{e}\bar{a}}{}^f X^b \end{pmatrix}$$

and sections $X^b \in \Gamma(T^{1,0}M(-1,0))$ that are in the kernel of $D^{\mathcal{T}}$. In particular, elements $X^b \in \ker D^{\mathcal{T}}$ with $N_{bf}^{J}{}^{\bar{e}} N_{\bar{e}\bar{a}}^{J}{}^f X^b = 0$ are in bijection with parallel sections of the tractor connection $\nabla^{\mathcal{T}}$.

PROOF. Suppose firstly that $s \in \Gamma(\mathcal{T})$ is parallel for the connection (60) and set $X^b := \pi(s)$. For a choice of connection $\nabla \in [\nabla]$ we can identify s with an element of the form (X^b, ρ), where $\rho \in \Gamma(\mathcal{E}(-1, 0))$. By assumption $\nabla_{\bar{a}} X^b = 0$ and $\nabla_a X^b = -\rho \delta_a{}^b$. Taking the trace of the latter equation shows that $\rho = -\frac{1}{n} \nabla_c X^c$. Hence, X^b is in the kernel of $D^{\mathcal{T}}$.

Conversely, suppose $X^b \in \ker D^{\mathcal{T}}$ and pick a connection $\nabla \in [\nabla]$. Then we deduce from Proposition 2.13 and equation (46) that

$$(61) \quad (\nabla_a \nabla_b - \nabla_b \nabla_a) X^c = (\nabla_a \nabla_b - \nabla_b \nabla_a) X^c + T_{ab}{}^{\bar{d}} \nabla_{\bar{d}} X^c$$
$$= R_{ab}{}^c{}_d X^d + 2\mathsf{P}_{[ab]} X^c = W_{ab}{}^c{}_d X^d + 2\delta_{[a}{}^c \mathsf{P}_{b]d} X^d.$$

By assumption $\nabla_a X^b = \frac{1}{n} \nabla_c X^c \delta_a{}^b$ and therefore (61) implies

$$(62) \quad \tfrac{1}{n}(\nabla_a \nabla_d X^d \delta_b{}^c - \nabla_b \nabla_d X^d \delta_a{}^c) = W_{ab}{}^c{}_d X^d + 2\delta_{[a}{}^c \mathsf{P}_{b]d} X^d.$$

Taking the trace in (62) over a and c shows that $-\frac{1}{n} \nabla_b \nabla_d X^d = \mathsf{P}_{bd} X^d$. Hence, $(X^b, -\frac{1}{n} \nabla_c X^c)$ defines a section s of $\mathcal{T}$ that satisfies $\nabla_a^{\mathcal{T}} s = 0$. Similarly, since $\nabla_{\bar{a}} X^b = 0$, Proposition 2.13 and equation (45) imply

$$(63) \quad \nabla_{\bar{a}} \nabla_b X^c = (\nabla_{\bar{a}} \nabla_b - \nabla_b \nabla_{\bar{a}}) X^c$$
$$= R_{\bar{a}b}{}^c{}_d X^d - \tfrac{1}{n+1} T_{bf}{}^{\bar{e}} T_{\bar{e}\bar{a}}{}^f X^c + \mathsf{P}_{\bar{a}b} X^c$$
$$= W_{\bar{a}b}{}^c{}_d X^d - 2\mathsf{P}_{\bar{a}(b} \delta_{d)}{}^c X^d + \mathsf{P}_{\bar{a}b} X^c - \tfrac{1}{n+1} T_{bf}{}^{\bar{e}} T_{\bar{e}\bar{a}}{}^f X^c.$$

Taking the trace in (63) over b and c implies that

$$-\tfrac{1}{n} \nabla_{\bar{a}} \nabla_c X^c - \mathsf{P}_{\bar{a}c} X^c = \tfrac{1}{n(n+1)} T_{cf}{}^{\bar{e}} T_{\bar{e}\bar{a}}{}^f X^c.$$

Hence, s is parallel for the connection (60) and it follows immediately that the differential operator $X^b \mapsto (X^b, -\frac{1}{n} \nabla_c X^c)$ defines an inverse to the restriction of π to parallel sections of (60). $\square$

3.2. Cone description of almost c-projective structures

For (real) projective structures there is an alternative description of the tractor connection as an affine connection on a cone manifold over the projective manifold [36, 46]. This point of view, which (at least in spirit) goes back to work of Tracy Thomas, is often convenient—it has for instance been used in [6] to classify holonomy reductions of projective structures. An analogue holds for almost c-projective manifolds, which we now sketch, following the presentation in [36] of the projective case. This canonical cone connection was used in [6] to realise c-projective structures as holonomy reductions of projective structures. It also underlies metric cone constructions [79, 84] which we discuss later.

Let $(M, J, [\nabla], \mathcal{E}(1,0))$ be an almost c-projective manifold and, as in Chapter 3.1, let $\tilde{P} \subset \tilde{G} = \mathrm{SL}(n+1, \mathbb{C})$ be the stabiliser of the complex line $\mathbb{V}^1$ through the first basis vector e_1 of $\mathbb{V} = \mathbb{C}^{n+1}$. Denote by $\tilde{Q} \subset \tilde{P}$ the stabiliser of e_1, which is the derived group of $\tilde{P}$, hence a normal complex Lie subgroup. Now set

$$\mathcal{C} := \tilde{\mathcal{G}}/\tilde{Q} = \tilde{\mathcal{G}} \times_{\tilde{P}} \tilde{P}/\tilde{Q}.$$

The natural projection $p_{\mathcal{C}} \colon \tilde{\mathcal{G}} \to \mathcal{C}$ defines a (real) principal bundle with structure group $\tilde{Q}$. Since the canonical Cartan connection ω of $(M, J, [\nabla], \mathcal{E}(1,0))$ can also be viewed as a Cartan connection of type $(\tilde{G}, \tilde{Q})$ for $p_{\mathcal{C}}$, it induces an isomorphism

$$T\mathcal{C} \cong \tilde{\mathcal{G}} \times_{\tilde{Q}} \mathfrak{g}/\tilde{\mathfrak{q}}.$$

Note that $\mathcal{C}$ inherits an almost complex structure $J^{\mathcal{C}}$ from the almost complex structure on $\tilde{\mathcal{G}}$ characterised by $Tp_{\mathcal{C}} \circ J^{\mathcal{G}} = J^{\mathcal{C}} \circ Tp_{\mathcal{C}}$. Furthermore, the projection $\pi_{\mathcal{C}} \colon \mathcal{C} \to M$ defines a principal bundle with structure group $\tilde{P}/\tilde{Q} \cong \mathbb{C}^{\times}$. Since $\tilde{P}/\tilde{Q}$ can be identified with the nonzero elements in the complex $\tilde{P}$-submodule $\mathbb{V}^1 \subset \mathbb{V}$, we see that $\mathcal{C}$ can be identified with the space of nonzero elements in $\mathcal{E}(-1, 0)$ or with the complex frame bundle of $\mathcal{E}(-1, 0)$. Note that, by construction (recall that $Tp \circ J^{\mathcal{G}} = J \circ Tp$), we have $T\pi_{\mathcal{C}} \circ J^{\mathcal{C}} = J \circ T\pi_{\mathcal{C}}$. By the compatibility of the almost complex structures $J^{\mathcal{G}}$, $J^{\mathcal{C}}$ and J with the various projections, it follows immediately that vanishing of the Nijenhuis tensor $N^{J^{\mathcal{G}}}$ of $J^{\mathcal{G}}$ implies vanishing of the Nijenhuis tensors $N^{J^{\mathcal{C}}}$, which in turn implies vanishing of N^J. Conversely, Theorem 2.16 shows that $N^J \equiv 0$ implies $N^{J^{\mathcal{G}}} \equiv 0$. This shows in particular that

$$(64) \qquad N^J \equiv 0 \iff N^{J^{\mathcal{C}}} \equiv 0,$$

in which case $\pi_{\mathcal{C}} \colon \mathcal{C} \to M$ is a holomorphic principal bundle with structure group $\mathbb{C}^{\times}$.

LEMMA 3.4. *There are canonical isomorphisms $T\mathcal{C} \cong \tilde{\mathcal{G}} \times_{\tilde{Q}} \mathbb{V} \cong \pi_{\mathcal{C}}^{*}\mathcal{T}$.*

PROOF. From the block decomposition (17) of $\mathfrak{g}$ it follows that $\mathfrak{g}/\tilde{\mathfrak{q}} = (\mathbb{V}^1)^* \otimes \mathbb{V}$ and hence $\mathfrak{g}/\tilde{\mathfrak{q}} \cong \mathbb{V}$ as $\tilde{Q}$-modules, i.e.

$$T\mathcal{C} \cong \tilde{\mathcal{G}} \times_{\tilde{Q}} \mathfrak{g}/\tilde{\mathfrak{q}} \cong \tilde{\mathcal{G}} \times_{\tilde{Q}} \mathbb{V} \cong \pi_{\mathcal{C}}^{*}(\tilde{\mathcal{G}} \times_{\tilde{P}} \mathbb{V}) = \pi_{\mathcal{C}}^{*}\mathcal{T}. \qquad \square$$

Hence the standard tractor connection induces an affine connection $\nabla^{\mathcal{C}}$ on $\mathcal{C}$ which preserves $J^{\mathcal{C}}$ and the complex volume form $\mathrm{vol}^{\mathcal{C}} \in \wedge^{n,0} T^*\mathcal{C}$ that is induced by the standard complex volume on $\mathbb{V} = \mathbb{C}^{n+1}$. Alternatively, note that ω can be extended to a principal connection on the principal G-bundle $\tilde{\mathcal{G}} \times_{\tilde{Q}} G$, and since $\mathbb{V} = \mathbb{C}^{n+1}$ is a G-module, we obtain an induced connection on $T\mathcal{C} \cong \tilde{\mathcal{G}} \times_{\tilde{Q}} G \times_G \mathbb{V}$.

If we identify a vector field $Y \in \mathfrak{X}(\mathcal{C})$ with a $\tilde{Q}$-equivariant function $f \colon \tilde{\mathcal{G}} \to \mathbb{V}$ via Lemma 3.4, then by [**36**, Theorem 1.5.8], the equivariant function corresponding to $\nabla^{\mathcal{C}}_X Y$ for a vector field $X \in \mathfrak{X}(\mathcal{C})$ is given by

$$(65) \qquad \tilde{X} \cdot f + \omega(\tilde{X})f$$

where $\tilde{X} \in \mathfrak{X}(\tilde{\mathcal{G}})$ is an arbitrary lift of X. Moreover [**36**, Corollary 1.5.7] shows that the curvature $R^{\mathcal{C}} \in \wedge^2 T^*\mathcal{C} \otimes \mathcal{AC}$ of $\nabla^{\mathcal{C}}$ is given by

$$(66) \qquad R^{\mathcal{C}}(X,Y)(Z) = \kappa(X,Y) \bullet Z,$$

where $\mathcal{AC} = \tilde{\mathcal{G}} \times_{\tilde{Q}} \mathfrak{g} \cong \mathfrak{sl}(T\mathcal{C}, J^{\mathcal{C}})$ and $\bullet \colon \mathcal{AC} \times T\mathcal{C} \to T\mathcal{C}$ denotes the vector bundle map induced by the action of $\mathfrak{g}$ on $\mathbb{V}$. Let us write $T^{\mathcal{C}} \in \wedge^2 T^*\mathcal{C} \otimes T\mathcal{C}$ for the torsion of $\nabla^{\mathcal{C}}$. It follows straightforwardly from (65) that $T^{\mathcal{C}}$ is the projection of $\kappa \in \Gamma(\wedge^2 T^*\mathcal{C} \otimes \mathcal{AC})$ to $\wedge^2 T^*\mathcal{C} \otimes T\mathcal{C}$, i.e. it is the torsion of ω viewed as a Cartan connection of type $(\tilde{Q}, \tilde{G})$. In particular, like κ, the 2-forms $T^{\mathcal{C}}$ and $R^{\mathcal{C}}$ vanish upon insertion of sections of the vertical bundle of $\pi_{\mathcal{C}}$, which is canonically trivialised by the fundamental vector fields E and $J^{\mathcal{C}} E$ generated by 1 and i respectively.

PROPOSITION 3.5. *Suppose $(M, J, [\nabla], \mathcal{E}(1,0))$ is an almost c-projective manifold. Then there is a unique affine connection $\nabla^{\mathcal{C}}$ on the total space of the principal bundle $\pi_{\mathcal{C}} \colon \mathcal{C} \to M$ with the following properties:*

(1) $\nabla^{\mathcal{C}} J^{\mathcal{C}} = 0$ *and* $\nabla^{\mathcal{C}} \mathrm{vol}^{\mathcal{C}} = 0$;
(2) $\nabla^{\mathcal{C}}_X E = X$ *for all* $X \in \mathfrak{X}(\mathcal{C})$;

(3) $\mathcal{L}_E \nabla^\mathcal{C} = 0$ and $\mathcal{L}_{J^\mathcal{C} E} \nabla^\mathcal{C} = 0$;
(4) $i_E T^\mathcal{C} = 0$ and $i_{J^\mathcal{C} E} T^\mathcal{C} = 0$;
(5) $T\pi \circ T^\mathcal{C}$ is of type $(0,2)$ and the $(2,0)$-part of $T^\mathcal{C}$ vanishes;
(6) $\nabla^\mathcal{C}$ is Ricci-flat;
(7) for any (local) section s of $\pi_\mathcal{C}$, the connection ∇^s on TM, defined by $\nabla^s_X Y := T\pi_\mathcal{C}(\nabla^\mathcal{C}_{Ts(X)}(Ts(Y))$ for any $X, Y \in \mathfrak{X}(M)$, belongs to $[\nabla]$.

Moreover, if J is integrable, $\pi_\mathcal{C} \colon \mathcal{C} \to M$ is a holomorphic principal bundle and $T^\mathcal{C} \equiv 0$.

PROOF. We already observed that (1) and (4) hold and (2) is an immediate consequence of (65). Since we have in addition $i_E R^\mathcal{C} = 0$ and $i_{J^\mathcal{C} E} R^\mathcal{C} = 0$ by (66), statement (3) follows from (78). The statements (5) and (6) are consequences of $\partial^* \kappa = 0$. More explicitly, note that (5) can be simply read off Proposition 2.13, which also shows that if J is integrable, $T^\mathcal{C} \equiv 0$. In this case $N^{J^\mathcal{C}}$, which is up to a constant multiple the $(0,2)$-part of $T^\mathcal{C}$, vanishes and $\pi_\mathcal{C}$ is a holomorphic principal bundle as calimed. Statement (6) follows because $T\pi \circ T^\mathcal{C}$, viewed as a section of $\wedge^{0,2} T^*M \otimes TM$, has vanishing trace and $\partial^* \colon \wedge^2 T^*M \otimes \mathfrak{gl}(TM, J) \to T^*M \otimes T^*M$ is a multiple of a Ricci-type contraction. The proof of statement (7) and the uniqueness of $\nabla^\mathcal{C}$ we leave to the reader, but note that (1)–(6) imply that $\nabla^\mathcal{C}$ descends to the normal Cartan connection on $T\mathcal{C}/\mathbb{C}^\times \cong \mathcal{T}$. □

3.3. BGG sequences

For a general parabolic geometry, it was shown in [**25, 38**] that there are natural sequences of invariant linear differential operators generalising the corresponding complexes on the flat model. These are the Bernstein–Gelfand–Gelfand (BGG) sequences, named after the constructors [**10**] of complexes of Verma modules, roughly dual to the current circumstances.

Here is not the place to say much about the general theory. Instead, we would like to like to present something of the theory as it applies in the c-projective case. The point is that the invariant operators that we have already encountered and are about to encounter, all can be seen as curved analogues of operators from the BGG complex on $\mathbb{CP}^n$ (as the flat model of c-projective geometry).

In fact, the main hurdle in presenting the BGG complex and sequences is in having a suitable notation for the vector bundles involved. Furthermore, this notation is already of independent utility since, as foretold in Remark 1.1, it neatly captures the natural irreducible bundles on an almost complex manifold.

Recall from Chapter 1 that the complexified tangent bundle on an almost complex manifold decomposes

$$\mathbb{C}TM = T^{1,0}M \oplus T^{0,1}M$$

as does its dual. An alternative viewpoint on these decompositions is that the tangent bundle TM on any $2n$-dimensional manifold is tautologically induced from its frame-bundle by the defining representation of $\mathrm{GL}(2n, \mathbb{R})$, that an almost complex structure is a reduction of structure group for the frame bundle to $\mathrm{GL}(n, \mathbb{C}) \subset \mathrm{GL}(2n, \mathbb{R})$, that the defining representation of $\mathrm{GL}(2n, \mathbb{R})$ on $\mathbb{R}^{2n}$ complexifies as $\mathrm{GL}(2n, \mathbb{R})$ acting on $\mathbb{C}^{2n}$ (as real matrices acting on complex vectors), and finally that this complex representation when restricted to $\mathrm{GL}(n, \mathbb{C})$ decomposes into two irreducibles inducing the bundles $T^{1,0}M$ and $T^{0,1}M$, respectively. Of course, the dual decomposition comes from the dual representation, namely $\mathrm{GL}(n, \mathbb{C})$ acting on $(\mathbb{C}^{2n})^*$. Our notation arises by systematically using the representation theory of $\mathrm{GL}(n,\mathbb{C})$ as a real Lie group but adapted to its embedding

$$\mathrm{GL}(n,\mathbb{C}) \cong G_0 \subset P \subset G = \mathrm{PSL}(n+1,\mathbb{C})$$

as described in Chapter 2.3.

For relatively simple bundles, there is no need for any more advanced notation. In several complex variables, for example, it is essential to break up the complex-valued differential forms into *types* but that's about it. Recall already with 2-forms

$$\wedge^2 = \wedge^{0,2} \oplus \wedge^{1,1} \oplus \wedge^{2,0}$$

that this complex decomposition is finer that the real decomposition

(67) $$\wedge^2 T^*M = \left[\wedge^{0,2} \oplus \wedge^{2,0}\right]_{\mathbb{R}} \oplus \wedge^{1,1}_{\mathbb{R}}$$

already discussed in Chapter 1 following (6). Of course, as soon as one speaks of holomorphic functions on a complex manifold one is obliged to work with complex-valued differential forms. However, even if one is concerned only with real-valued forms and tensors, it is convenient firstly to decompose the complex versions and then impose reality as, for example, in (67). In fact, this is already a feature of representation theory in general.

For more complicated bundles, we use Dynkin diagrams from [36] decorated in the style of [8]. The formal definitions will not be given here but the upshot is that the general complex irreducible bundle on an almost complex manifold will be denoted as

(68) $$\begin{array}{c} {}^{p}\;{}^{a}\;{}^{b}\;{}^{c}\;{}^{d} \\ \times\!\!-\!\!\bullet\!\!-\!\!\bullet\!\!-\!\!\bullet\!\!-\!\!\bullet \\ \updownarrow\;\updownarrow\;\updownarrow\;\updownarrow\;\updownarrow \\ \times\!\!-\!\!\bullet\!\!-\!\!\bullet\!\!-\!\!\bullet\!\!-\!\!\bullet \\ {}^{q}\;{}^{e}\;{}^{f}\;{}^{g}\;{}^{h} \end{array}$$ (in the 10-dimensional case ($2n$ nodes in general))

where a,b,c,d,e,f,g,h are nonnegative integers whilst, in the first instance, p,q are real numbers restricted by the requirement that

(69) $$p + 2a + 3b + 4c + 5d = q + 2e + 3f + 4g + 5h \bmod 6$$

(again, in the 10-dimensional case). For example,

$$T^{1,0}M = \begin{array}{c} {}^{1}\;{}^{0}\;{}^{0}\;{}^{0}\;{}^{1} \\ \times\!-\!\bullet\!-\!\bullet\!-\!\bullet\!-\!\bullet \\ \updownarrow\;\updownarrow\;\updownarrow\;\updownarrow\;\updownarrow \\ \times\!-\!\bullet\!-\!\bullet\!-\!\bullet\!-\!\bullet \\ {}^{0}\;{}^{0}\;{}^{0}\;{}^{0}\;{}^{0} \end{array} \qquad \wedge^{1,0} = \begin{array}{c} {}^{-2}\;{}^{1}\;{}^{0}\;{}^{0}\;{}^{0} \\ \times\!-\!\bullet\!-\!\bullet\!-\!\bullet\!-\!\bullet \\ \updownarrow\;\updownarrow\;\updownarrow\;\updownarrow\;\updownarrow \\ \times\!-\!\bullet\!-\!\bullet\!-\!\bullet\!-\!\bullet \\ {}^{0}\;{}^{0}\;{}^{0}\;{}^{0}\;{}^{0} \end{array} \qquad \wedge^{0,2} = \begin{array}{c} {}^{0}\;{}^{0}\;{}^{0}\;{}^{0}\;{}^{0} \\ \times\!-\!\bullet\!-\!\bullet\!-\!\bullet\!-\!\bullet \\ \updownarrow\;\updownarrow\;\updownarrow\;\updownarrow\;\updownarrow \\ \times\!-\!\bullet\!-\!\bullet\!-\!\bullet\!-\!\bullet \\ {}^{-3}\;{}^{0}\;{}^{1}\;{}^{0}\;{}^{0} \end{array}$$

but the point is that this notation captures all irreducible possiblities and, in particular, the various awkward bundles that have already arisen and will now arise in this article. In general, the integrality condition (69) is needed, as typified by

$$\det \wedge^{1,0} = \wedge^{5,0} = \begin{array}{c} {}^{-6}\;{}^{0}\;{}^{0}\;{}^{0}\;{}^{0} \\ \times\!-\!\bullet\!-\!\bullet\!-\!\bullet\!-\!\bullet \\ \updownarrow\;\updownarrow\;\updownarrow\;\updownarrow\;\updownarrow \\ \times\!-\!\bullet\!-\!\bullet\!-\!\bullet\!-\!\bullet \\ {}^{0}\;{}^{0}\;{}^{0}\;{}^{0}\;{}^{0} \end{array} \qquad \mathcal{E}(p,p) = \begin{array}{c} {}^{p}\;{}^{0}\;{}^{0}\;{}^{0}\;{}^{0} \\ \times\!-\!\bullet\!-\!\bullet\!-\!\bullet\!-\!\bullet \\ \updownarrow\;\updownarrow\;\updownarrow\;\updownarrow\;\updownarrow \\ \times\!-\!\bullet\!-\!\bullet\!-\!\bullet\!-\!\bullet \\ {}^{p}\;{}^{0}\;{}^{0}\;{}^{0}\;{}^{0} \end{array}$$

but, as already discussed at the end of Chapter 2.1, on an almost c-projective manifold we suppose that there is a bundle $\mathcal{E}(1,0)$ and an identification $\mathcal{E}(n+1,0) := \mathcal{E}(1,0)^{n+1} = \wedge^n T^{1,0} M$, in which case we add

$$\mathcal{E}(1,0) = \begin{array}{c} {}^{1}\;{}^{0}\;{}^{0}\;{}^{0}\;{}^{0} \\ \times\!-\!\bullet\!-\!\bullet\!-\!\bullet\!-\!\bullet \\ \updownarrow\;\updownarrow\;\updownarrow\;\updownarrow\;\updownarrow \\ \times\!-\!\bullet\!-\!\bullet\!-\!\bullet\!-\!\bullet \\ {}^{0}\;{}^{0}\;{}^{0}\;{}^{0}\;{}^{0} \end{array}$$

3.3. BGG SEQUENCES

to our notation and relax (69) to requiring merely that $p-q$ be integral. In fact, all of $p, q, a, b, c, d, e, f, g, h$ will be integral for the rest of this article.

Our Dynkin diagram notation is well suited to the barred and unbarred indices that we have already been using. Specifically, a section of

$$\begin{array}{ccccc} p & a & b & c & d \\ \times\!\!\!\!-\!\!\!\!\bullet\!\!\!\!-\!\!\!\!\bullet\!\!\!\!-\!\!\!\!\bullet\!\!\!\!-\!\!\!\!\bullet \\ \updownarrow & \updownarrow & \updownarrow & \updownarrow & \updownarrow \\ \times\!\!\!\!-\!\!\!\!\bullet\!\!\!\!-\!\!\!\!\bullet\!\!\!\!-\!\!\!\!\bullet\!\!\!\!-\!\!\!\!\bullet \\ 0 & 0 & 0 & 0 & 0 \end{array}$$

may be realised as tensors with $a + 2b + 3c + 4d$ unbarred covariant indices, having symmetries specified by the Young diagram

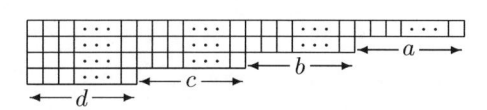

(Young diagram with arrows labeled a, b, c, d)

and of c-projective weight $(p + 2a + 3b + 4c + 5d, 0)$. Indeed, for those reluctant to trace through the conventions in [8], this suffices as a definition and then

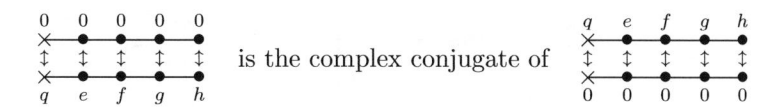

$$\begin{array}{ccccc} 0 & 0 & 0 & 0 & 0 \\ \times\!\!-\!\!\bullet\!\!-\!\!\bullet\!\!-\!\!\bullet\!\!-\!\!\bullet \\ \updownarrow & \updownarrow & \updownarrow & \updownarrow & \updownarrow \\ \times\!\!-\!\!\bullet\!\!-\!\!\bullet\!\!-\!\!\bullet\!\!-\!\!\bullet \\ q & e & f & g & h \end{array} \quad \text{is the complex conjugate of} \quad \begin{array}{ccccc} q & e & f & g & h \\ \times\!\!-\!\!\bullet\!\!-\!\!\bullet\!\!-\!\!\bullet\!\!-\!\!\bullet \\ \updownarrow & \updownarrow & \updownarrow & \updownarrow & \updownarrow \\ \times\!\!-\!\!\bullet\!\!-\!\!\bullet\!\!-\!\!\bullet\!\!-\!\!\bullet \\ 0 & 0 & 0 & 0 & 0 \end{array}$$

corresponding to tensors with barred indices in the obvious fashion and

$$\begin{array}{ccccc} p & a & b & c & d \\ \times\!\!-\!\!\bullet\!\!-\!\!\bullet\!\!-\!\!\bullet\!\!-\!\!\bullet \\ \updownarrow & \updownarrow & \updownarrow & \updownarrow & \updownarrow \\ \times\!\!-\!\!\bullet\!\!-\!\!\bullet\!\!-\!\!\bullet\!\!-\!\!\bullet \\ q & e & f & g & h \end{array} = \begin{array}{ccccc} p & a & b & c & d \\ \times\!\!-\!\!\bullet\!\!-\!\!\bullet\!\!-\!\!\bullet\!\!-\!\!\bullet \\ \updownarrow & \updownarrow & \updownarrow & \updownarrow & \updownarrow \\ \times\!\!-\!\!\bullet\!\!-\!\!\bullet\!\!-\!\!\bullet\!\!-\!\!\bullet \\ 0 & 0 & 0 & 0 & 0 \end{array} \otimes \begin{array}{ccccc} 0 & 0 & 0 & 0 & 0 \\ \times\!\!-\!\!\bullet\!\!-\!\!\bullet\!\!-\!\!\bullet\!\!-\!\!\bullet \\ \updownarrow & \updownarrow & \updownarrow & \updownarrow & \updownarrow \\ \times\!\!-\!\!\bullet\!\!-\!\!\bullet\!\!-\!\!\bullet\!\!-\!\!\bullet \\ q & e & f & g & h \end{array}.$$

Already, these bundles provide locations for the tensors we encountered earlier. For example,

$$T_{ab}{}^{\bar{c}} \in \Gamma\!\left(\begin{array}{ccccc} -3 & 0 & 1 & 0 & 0 \\ \times\!\!-\!\!\bullet\!\!-\!\!\bullet\!\!-\!\!\bullet\!\!-\!\!\bullet \\ \updownarrow & \updownarrow & \updownarrow & \updownarrow & \updownarrow \\ \times\!\!-\!\!\bullet\!\!-\!\!\bullet\!\!-\!\!\bullet\!\!-\!\!\bullet \\ 1 & 0 & 0 & 0 & 1 \end{array}\right) \quad \text{and} \quad H_{a\bar{b}}{}^c{}_d \in \Gamma\!\left(\begin{array}{ccccc} -3 & 2 & 0 & 0 & 1 \\ \times\!\!-\!\!\bullet\!\!-\!\!\bullet\!\!-\!\!\bullet\!\!-\!\!\bullet \\ \updownarrow & \updownarrow & \updownarrow & \updownarrow & \updownarrow \\ \times\!\!-\!\!\bullet\!\!-\!\!\bullet\!\!-\!\!\bullet\!\!-\!\!\bullet \\ -2 & 1 & 0 & 0 & 0 \end{array}\right).$$

Although the Dynkin diagram notation may at first seem arcane, it comes into its own when discussing invariant linear differential operators. The complex-valued de Rham complex

$$(70) \qquad \wedge^{0,0} \to \begin{array}{c} \wedge^{1,0} \\ \oplus \\ \wedge^{0,1} \end{array} \to \begin{array}{c} \wedge^{2,0} \\ \oplus \\ \wedge^{1,1} \\ \oplus \\ \wedge^{0,2} \end{array} \to \cdots$$

becomes

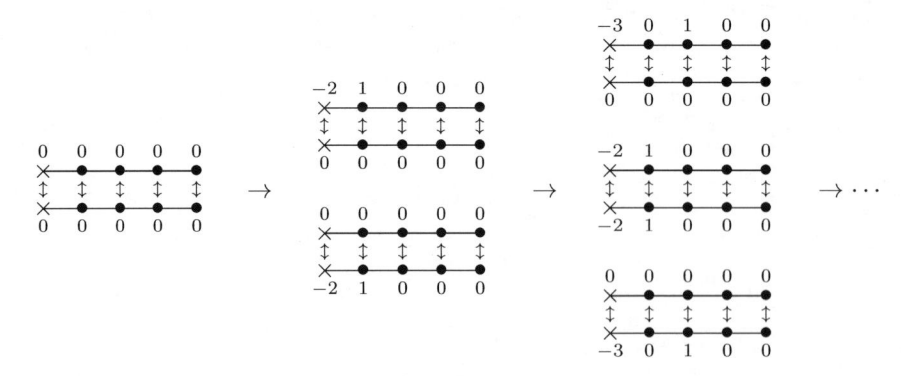

$$\begin{array}{c} \text{diagram} \end{array} \to \begin{array}{c} \text{diagram} \end{array} \to \begin{array}{c} \text{diagram} \end{array} \to \cdots$$

and in either of them one sees the torsion $T_{ab}{}^{\bar c}\colon \wedge^{0,1} \to \wedge^{2,0}$ and its complex conjugate $T_{\bar a\bar b}{}^c\colon \wedge^{1,0} \to \wedge^{0,2}$ as the restriction of the exterior derivative $d\colon \wedge^1 \to \wedge^2$ to the relevant bundles (note that

$$\operatorname{Hom}(\wedge^{0,1}, \wedge^{2,0}) = T^{0,1} \otimes \wedge^{2,0} = \begin{array}{c} \text{diagram} \end{array},$$

as expected). In the torsion-free case, the de Rham complex takes the form

(71) $$\wedge^{0,0} \;\nearrow\; \wedge^{1,0} \;\searrow\; \wedge^{2,0} \;\nearrow\; \wedge^{1,1} \;\searrow\; \wedge^{0,1} \;\nearrow\; \wedge^{0,2}$$

familiar from complex analysis and the remarkable fact about c-projectively invariant linear differential operators is firstly that this pattern is repeated on the flat model starting with any bundle (68) with $p, q \in \mathbb{N} = \{0, 1, 2, \ldots\}$, for example

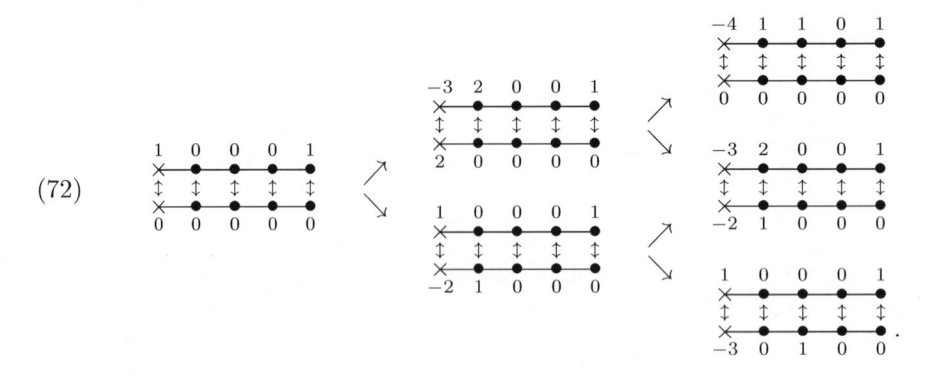

(72) $$\begin{array}{c} \text{diagram} \end{array}$$

The algorithm for determining the bundles in these patterns is detailed in [8] (it is the affine action $\lambda \mapsto w(\lambda + \rho) - \rho$ of the Weyl group for G along the Hasse diagram corresponding to the parabolic subgroup P). On G/P in general, these are complexes of differential operators referred to as *Bernstein–Gelfand–Gelfand*

3.3. BGG SEQUENCES

(*BGG*) complexes. In our case, i.e. on $\mathbb{CP}^n$, they provide resolutions of the finite-dimensional representations

$$\begin{array}{ccccc} p & a & b & c & d \\ \bullet\!\!\!&\!\!\!\bullet\!\!\!&\!\!\!\bullet\!\!\!&\!\!\!\bullet\!\!\!&\!\!\!\bullet \\ \updownarrow & \updownarrow & \updownarrow & \updownarrow & \updownarrow \\ \bullet\!\!\!&\!\!\!\bullet\!\!\!&\!\!\!\bullet\!\!\!&\!\!\!\bullet\!\!\!&\!\!\!\bullet \\ q & e & f & g & h \end{array} \quad \text{(in case } n = 5 \text{ (}2n\text{ nodes in general))}$$

of the group $G = \mathrm{PSL}(n+1,\mathbb{C})$ as a real Lie group. More precisely, any finite-dimensional representation $\mathbb{E}$ of G gives rise to a constant sheaf $G/P \times \mathbb{E}$, which may in turn be identified with the corresponding homogeneous bundle induced on G/P by means of

(73) $$G/P \times_P \mathbb{E} \ni [g, e] \mapsto ([g], ge) \in G/P \times \mathbb{E}.$$

Since the first bundle in the BGG complex is a quotient of this bundle, we obtain a mapping of $\mathbb{E}$ to the sections of this first bundle and to say that the complex is a resolution of $\mathbb{E}$ is to say that these sections are locally precisely the kernel of the first BGG operator (just as the locally constant functions are precisely the kernel of the first exterior derivative $d\colon \wedge^0 \to \wedge^1$). In our example (72), this means that

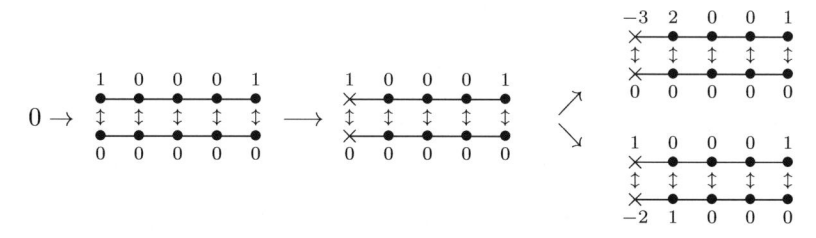

is exact, the G-module $\mathbb{E}$ in this case being the adjoint representation of $\mathrm{PSL}(n+1,\mathbb{C})$ as a complex Lie algebra. More generally, the BGG resolutions on $\mathbb{CP}^n$ as a homogeneous space for $\mathrm{PSL}(n+1,\mathbb{C})$ begin

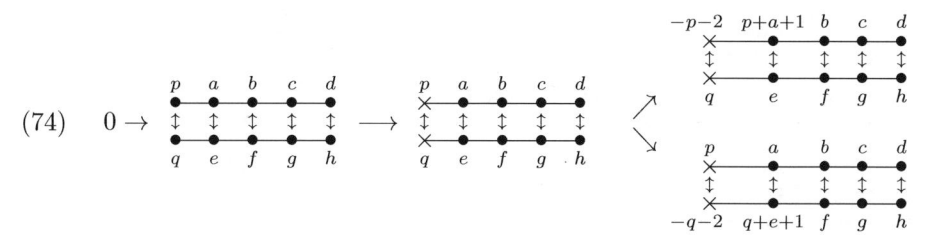

(74)

for nonnegative integers $p, a, b, c, d, q, e, f, g, h$ constrained by (69). We may drop the constraint (69) by considering $\mathbb{CP}^n$ instead as a homogeneous space for $\mathrm{SL}(n+1,\mathbb{C})$, as is perhaps more usual. Having done that, the standard representation of $\mathrm{SL}(n+1,\mathbb{C})$ on $\mathbb{C}^{n+1}$ gives rise to the BGG resolution

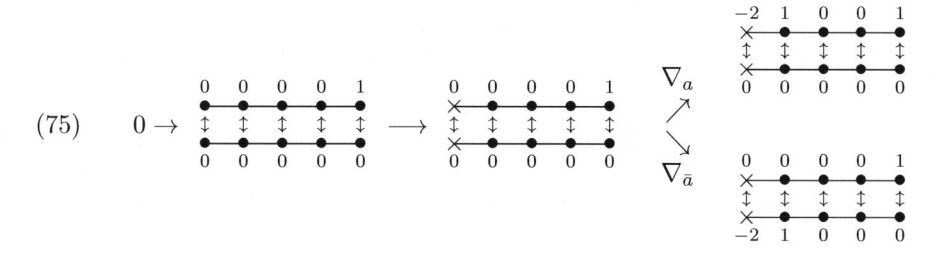

(75)

where the operators ∇_a and $\nabla_{\bar{a}}$ are, more explicitly and as noted in (59),

(76) $$X^b \mapsto (\nabla_a X^b)_\circ \quad \text{and} \quad X^b \mapsto \nabla_{\bar{a}} X^b$$

where X^b is a vector field of type $(1,0)$ and of c-projective weight $(-1,0)$ and the subscript $\circ$ means to take the trace-free part. Notice that these are exactly the operators implicitly encoded in the standard tractor connection (52). More precisely, the filtration (49) is equivalent to the short exact sequence of vector bundles

$$0 \to \underset{\mathcal{E}(-1,0)}{\underbrace{\text{[diagram]}}} \to \underset{\mathcal{T}}{\underbrace{\text{[diagram]}}} \to \underset{T^{1,0}M(-1,0)}{\underbrace{\text{[diagram]}}} \to 0$$

and on the flat model, namely $\mathbb{CP}^n$ as a homogeneous space for $\mathrm{SL}(n+1,\mathbb{C})$, the tractor connection is the exactly the flat connection induced by (73). In the c-projectively flat case, the remaining entries in (52), namely

$$\nabla_a \rho - \mathrm{P}_{ab} X^b \quad \text{and} \quad \nabla_{\bar{a}} \rho - \mathrm{P}_{\bar{a}b} X^b$$

may be regarded as quantities whose vanishing are differential consequences of setting

$$\nabla_a X^b + \rho \delta_a{}^b = 0 \quad \text{and} \quad \nabla_{\bar{a}} X^b = 0.$$

Hence, they add no further conditions to being in the kernel of the first BGG operator (76) and the exactness of (75) follows. The same reasoning pertains in the curved but torsion-free setting and leads to the standard tractor connection being obtained by prolongation of the first BGG operator. This is detailed in Proposition 3.3. For more complicated representations, the tractor connection may not be obtained by prolongation in the curved setting, even if torsion-free. This phenomenon will soon be seen in two key examples, specifically in the connection (85) and Proposition 3.9 concerned with infinitesimal automorphisms and in Proposition 4.5, Theorem 4.6, and Corollary 4.7 dealing with the metrisability of c-projective structures. With reference to the general first BGG operators (74), the following cases occur prominently in this article.

[diagram] $\xrightarrow{\pi}$ [diagram] This is the standard complex tractor bundle $\mathcal{T}$ and its canonical projection to $T^{1,0}M(-1,0)$.

[diagram] $\xrightarrow{\pi}$ [diagram] This is the adjoint tractor bundle $\mathcal{A}M$ to be considered in Chapter 3.4 and its canonical projection to $T^{1,0}M$. A first BGG operator acting on $T^{1,0}M$ is given in Remark 3.4.

[diagram] $\xrightarrow{\pi}$ [diagram] This is the tractor bundle arising in the metrisability of c-projective structures to be discussed in Chapter 4 and a first BGG operator is given in (121).

3.4. ADJOINT TRACTORS AND INFINITESIMAL AUTOMORPHISMS

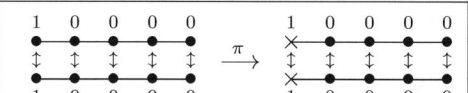

This is the dual of the previous case and arises in Chapter 4.6, which is concerned with the first BGG operator $\mathcal{D}^{\mathcal{W}}$ defined in (167) and acting on c-projective densities of weight $(1,1)$. It is a second order and c-projectively invariant operator.

In fact, there is quite a bit of flexibility in what one might allow as BGG operators, already for the first ones (74). For example, the operator $\mathcal{D}^{\mathcal{A}}$ in Remark 3.4 is rather different from the c-projectively invariant operators occurring as the left hand sides of (1) and (2) in Proposition 3.7. Even for the bundle $T^{1,0}M(-1,0)$ in (75) corresponding to standard complex tractors, there is the option of replacing the second operator in (76) by

$$X^c \mapsto \nabla_{\bar{a}} X^c + T_{\bar{a}\bar{b}}{}^c X^{\bar{a}}$$

in line with equation (1) in Proposition 3.7. Only in the torsion-free case do these operators agree (with each other and the usual $\bar{\partial}$-operator).

On the flat model, however, there is no choice. The operators occurring in the BGG complexes are unique up to scale. Moreover, there are no other c-projectively invariant linear differential operators: every such operator is determined by its symbol and the BGG operators comprise a classification. In the curved setting it is necessary to add curvature correction terms and there is almost always some choice. It is usual however, to restrict the term "BGG operator" to operators constructed in a systematic way from operators on tractor-valued differential forms, such as the constructions in [**25, 38**]. Even here, there is some ambiguity. In particular, in [**25**], two constructions were given, one equivalent to the construction in [**38**], the other differing in the presence of torsion, even for the BGG sequence associated to the trivial representation. Specifically, the construction in [**38**] follows the Hasse diagram beginning as in (71). In particular, there is no place for the torsion as an operator $\wedge^{0,1} \to \wedge^{2,0}$ whereas, in the second construction in [**25**], the first BGG sequence associated to the trivial representation for the case of $|1|$-graded geometry such as c-projective geometry, is just the de Rham complex (70).

In summary, the BGG operators on $\mathbb{CP}^n$ provide models for what one should expect in the curved setting. In the flat case, there is no choice. In the curved case, there is a certain degree of flexibility, more so when there is torsion. Finally, the general theory of parabolic geometry [**36**] provides a location for *harmonic curvature*, as already discussed in Chapters 2.2 and 2.7 and Kostant's Theorem [**64**] on Lie algebra cohomology provides the location for this curvature, namely the three bundles appearing in the second step of the BGG sequence (72) for the adjoint representation whilst the two bundles at the first step locate the infinitesimal deformations of an almost c-projective structure, in line with the general theory [**29**].

3.4. Adjoint tractors and infinitesimal automorphisms

For a vector field X on a manifold M we write $\mathcal{L}_X$ for the *Lie derivative* along X of tensor fields on M. Recall that there is also a notion of a *Lie derivative of an affine connection* ∇ along a vector field $X \in \Gamma(TM)$. It is given by the tensor field

$$\mathcal{L}_X \nabla \colon TM \to T^*M \otimes TM$$

characterised by
$$(\mathcal{L}_X \nabla)(Y) \equiv \mathcal{L}_X(\nabla Y) - \nabla \mathcal{L}_X Y$$
for any vector field $Y \in \Gamma(TM)$. In abstract index notation we adopt the convention that $(\mathcal{L}_X \nabla)_{\alpha\beta}{}^\gamma Y^\beta = \mathcal{L}_X(\nabla_\alpha Y^\gamma)$.

DEFINITION 3.1. A *c-projective vector field* or *infinitesimal automorphism* of an almost c-projective manifold $(M, J, [\nabla])$ of dimension $2n \geq 4$ is a vector field X on M that satisfies
- $\mathcal{L}_X J \equiv 0$ (i.e. $[X, JY] = J[X, Y]$ for all vector fields $Y \in \Gamma(TM)$)
- $(\mathcal{L}_X \nabla)_{\alpha\beta}{}^\gamma = \upsilon_{\alpha\beta}{}^\gamma$, where $\upsilon_{\alpha\beta}{}^\gamma \in \Gamma(S^2 T^*M \otimes TM)$ is a tensor of the form (11).

Note that $X \in \Gamma(TM)$ is an infinitesimal automorphism of an almost c-projective manifold precisely if its flow acts by local automorphisms thereof.

Let us rewrite the two conditions defining a c-projective vector field as a system of differential equations on a vector field X of M. Expressing the Lie bracket in terms of a connection $\nabla \in [\nabla]$ and its torsion shows that $\mathcal{L}_X J = 0$ is equivalent to

$$(77) \qquad T_{\alpha\beta}{}^\gamma X^\alpha = -\tfrac{1}{2}(\nabla_\beta X^\gamma + J_\epsilon{}^\gamma J_\beta{}^\zeta \nabla_\zeta X^\epsilon).$$

for one (and hence any) connection $\nabla \in [\nabla]$. Moreover, one deduces straightforwardly from the definition of the Lie derivative of a connection that for any connection $\nabla \in [\nabla]$ we have

$$(78) \qquad (\mathcal{L}_X \nabla)_{\beta\delta}{}^\gamma = R_{\alpha\beta}{}^\gamma{}_\delta X^\alpha + \nabla_\beta \nabla_\delta X^\gamma + \nabla_\beta(T_{\alpha\delta}{}^\gamma X^\alpha).$$

Via the isomorphism $TM \cong T^{1,0}M$ we may write the result as a differential equation on X^a: equation (77) then becomes

$$(79) \qquad \nabla_{\bar b} X^c + T_{\bar a \bar b}{}^c X^{\bar a} = 0.$$

Since a tensor $\upsilon_{\beta\delta}{}^\gamma$ of the form (11) satisfies $\upsilon_{\bar b d}{}^c = 0 = \upsilon_{bd}{}^{\bar c}$, the equation $(\mathcal{L}_X \nabla)_\beta{}^\gamma{}_\delta = \upsilon_{\beta\delta}{}^\gamma$ can be equivalently encoded by the three equations

$$(80) \qquad (\mathcal{L}_X \nabla)_{bd}{}^c = \upsilon_{bd}{}^c \qquad (\mathcal{L}_X \nabla)_{\bar b d}{}^c \equiv 0 \qquad (\mathcal{L}_X \nabla)_{\bar b \bar d}{}^c \equiv 0,$$

or, alternatively, their complex conjugates.

LEMMA 3.6. *If $X^a \in \Gamma(T^{1,0}M)$ satisfies the invariant differential equation (79), then*
$$(\mathcal{L}_X \nabla)_{\bar b d}{}^c \equiv 0 \qquad (\mathcal{L}_X \nabla)_{\bar b \bar d}{}^c \equiv 0,$$
for any connection $\nabla \in [\nabla]$.

PROOF. Equation (78) and the formulae of Proposition 2.13 imply
$$\begin{aligned}(\mathcal{L}_X \nabla)_{\bar b d}{}^c &= R_{a \bar b}{}^c{}_d X^a + R_{\bar a \bar b}{}^c{}_d X^{\bar a} + \nabla_{\bar b} \nabla_d X^c \\ &= -R_{\bar b a}{}^c{}_d X^a + R_{\bar b d}{}^c{}_a X^a + \nabla_d \nabla_{\bar b} X^c + R_{\bar a \bar b}{}^c{}_d X^{\bar a} \\ &= 2 R_{\bar b [d}{}^c{}_{a]} X^a + \nabla_d \nabla_{\bar b} X^c - (\nabla_d T_{\bar b \bar a}{}^c) X^{\bar a}.\end{aligned}$$
Hence, the Bianchi symmetry (42) shows that
$$(\mathcal{L}_X \nabla)_{\bar b d}{}^c = T_{da}{}^{\bar e} T_{\bar e \bar b}{}^c X^a + \nabla_d \nabla_{\bar b} X^c - (\nabla_d T_{\bar b \bar a}{}^c) X^{\bar a},$$
which evidently vanishes if $\nabla_{\bar b} X^c = T_{\bar b \bar a}{}^c X^{\bar a}$, and consequently also $\nabla_d X^{\bar e} = T_{da}{}^{\bar e} X^a$. As $(\mathcal{L}_X \nabla)_{\bar b \bar d}{}^c = \nabla_{\bar b} \nabla_{\bar d} X^c + \nabla_{\bar b}(T_{\bar a \bar d}{}^c X^{\bar a})$, the second assertion is obvious. □

3.4. ADJOINT TRACTORS AND INFINITESIMAL AUTOMORPHISMS

According to Lemma 3.6, it remains to rewrite $(\mathcal{L}_X \nabla)_b{}^c{}_d = v_{bd}{}^c$ as a differential equation on X^a. Note that we have

$$(81) \qquad (\mathcal{L}_X \nabla)_b{}^c{}_d = \nabla_b \nabla_d X^c + R_{ab}{}^c{}_d X^a + R_{\bar{a}b}{}^c{}_d X^{\bar{a}}.$$

The Bianchi symmetry (41) $R_{[bd}{}^c{}_{a]} \equiv 0$ implies $R_{bd}{}^c{}_a = -2R_{a[b}{}^c{}_{d]}$. Moreover,

$$\nabla_b \nabla_d X^c = \nabla_{(b} \nabla_{d)} X^c + \nabla_{[b} \nabla_{d]} X^c = \nabla_{(b} \nabla_{d)} X^c + \tfrac{1}{2}(R_{bd}{}^c{}_a X^a - T_{bd}{}^{\bar{e}} \nabla_{\bar{e}} X^c).$$

Therefore, we may rewrite (81) as

$$(82) \qquad \begin{aligned}(\mathcal{L}_X \nabla)_b{}^c{}_d &= \nabla_{(b} \nabla_{d)} X^c + R_{a(b}{}^c{}_{d)} X^a + R_{\bar{a}(b}{}^c{}_{d)} X^{\bar{a}} \\ &\quad - \tfrac{1}{2} T_{bd}{}^{\bar{e}} \nabla_{\bar{e}} X^c + \tfrac{1}{2} T_{bd}{}^{\bar{e}} T_{\bar{e}\bar{a}}{}^c X^{\bar{a}},\end{aligned}$$

where we used the Bianchi symmetry (42) given by $R_{\bar{a}[b}{}^c{}_{d]} = \tfrac{1}{2} T_{bd}{}^{\bar{e}} T_{\bar{e}\bar{a}}{}^c X^{\bar{a}}$. The torsion terms of (82) evidently cancel if X^a satisfies (79).

Suppose now that $2n \geq 6$. Then we deduce from Proposition 2.13 that

$$(83) \quad R_{a(b}{}^c{}_{d)} X^a = W_{a(b}{}^c{}_{d)} X^a + \mathsf{P}_{(bd)} X^c + \delta_{(b}{}^c \mathsf{P}_{d)a} X^a - \delta_b{}^c \mathsf{P}_{ad} X^a - \delta_d{}^c \mathsf{P}_{ab} X^a,$$

where the third term and the two last terms already define two tensors of the form (11). Moreover, we obtain by Proposition 2.13 that

$$(84) \qquad R_{\bar{a}(b}{}^c{}_{d)} X^{\bar{a}} = H_{\bar{a}b}{}^c{}_d X^{\bar{a}} + \tfrac{1}{n+1} \delta_{(b}{}^c T_{d)f}{}^{\bar{e}} T_{\bar{e}\bar{a}}{}^f X^{\bar{a}} - 2 \mathsf{P}_{\bar{a}(b} \delta_{d)}{}^c X^{\bar{a}},$$

where the last two terms are again already of the form (11). Therefore, we conclude:

PROPOSITION 3.7. *Suppose $(M, J, [\nabla])$ is an almost c-projective manifold of dimension $2n \geq 6$. A vector field $X^a \in \Gamma(T^{1,0}M)$ is c-projective if and only if it satisfies the following equations*

(1) $\nabla_{\bar{b}} X^c + T_{\bar{a}\bar{b}}{}^c X^{\bar{a}} = 0$
(2) $(\nabla_{(b} \nabla_{d)} X^c + \mathsf{P}_{(bd)} X^c + W_{a(b}{}^c{}_{d)} X^a + H_{\bar{a}b}{}^c{}_d X^{\bar{a}})_\circ = 0,$

where the subscript $\circ$ denotes the trace-free part.

Due to Proposition 2.14 for $2n = 4$ the equations take a simpler form:

PROPOSITION 3.8. *Suppose $(M, J, [\nabla])$ is an almost c-projective manifold of dimension $2n = 4$. A vector field $X^a \in \Gamma(T^{1,0}M)$ is c-projective if and only if it satisfies the following equations*

(1) $\nabla_{\bar{b}} X^c + T_{\bar{a}\bar{b}}{}^c X^{\bar{a}} = 0$
(2) $(\nabla_{(b} \nabla_{d)} X^c + \mathsf{P}_{(bd)} X^c + H_{\bar{a}b}{}^c{}_d X^{\bar{a}})_\circ = 0,$

where the subscript $\circ$ denotes the trace-free part.

The equations in Propositions 3.7 and 3.8 define an invariant differential operator

$$D^{\mathrm{aut}} : T^{1,0}M \to (\wedge^{1,0} \otimes T^{0,1}M) \oplus (S^2 \wedge^{1,0} \otimes T^{1,0}M)_\circ,$$

whose kernel comprises the infinitesimal automorphisms of $(M, J, [\nabla])$.

Let us recall some facts about infinitesimal automorphisms of Cartan geometries.

DEFINITION 3.2. Suppose $(p: \mathcal{G} \to M, \omega)$ is a Cartan geometry. A vector field $\tilde{X} \in \Gamma(T\mathcal{G})$ is called an *infinitesimal automorphism* of $(p: \mathcal{G} \to M, \omega)$, if $\tilde{X}$ is right-invariant for the principal right action on $\mathcal{G}$ and $\mathcal{L}_{\tilde{X}} \omega = 0$.

A Cartan connection ω on $p\colon \mathcal{G} \to M$ induces a bijection between right-invariant vector fields $\tilde X \in \Gamma(T\mathcal{G})$ and equivariant functions $\omega(\tilde X)\colon \mathcal{G} \to \mathfrak{g}$. Hence, right-invariant vector fields on $\mathcal{G}$ are in bijection with sections of the adjoint tractor bundle $\mathcal{A}M$. A section s of $\mathcal{A}M$ corresponds to an infinitesimal automorphism of the Cartan geometry if and only if s is parallel for the linear connection

$$\nabla^{\mathcal{A}} s + \kappa(\Pi(s), \cdot), \tag{85}$$

where $\nabla^{\mathcal{A}}$ is the adjoint tractor connection, $\Pi\colon \mathcal{A}M \to TM$ the natural projection, and $\kappa \in \Omega^2(M, \mathcal{A}M)$ the curvature of the Cartan geometry; see [**29, 36**].

The equivalence of categories established in Theorem 2.8 implies that any infinitesimal automorphism $X \in \Gamma(TM)$ of an almost c-projective manifold can be lifted uniquely to an infinitesimal automorphism of its normal Cartan geometry and conversely, any infinitesimal automorphism of the Cartan geometry projects to an infinitesimal automorphism of the underlying almost c-projective manifold. This implies, in particular, that Π induces a bijection between sections of the adjoint tractor bundle of the almost c-projective manifold that are parallel for the connection (85) and infinitesimal automorphisms of the almost c-projective manifold.

For the convenience of the reader let us explicitly compute the modified adjoint tractor connection (85). For these purposes let us identify the adjoint tractor bundle with the $(1,0)$-part of its complexification. As such it is filtered as

$$\mathcal{A}M = \mathcal{A}^{-1}M \supset \mathcal{A}^0 M \supset \mathcal{A}^1 M,$$

where $\mathcal{A}^{-1}M / \mathcal{A}^0 M \cong T^{1,0}M$, $\mathcal{A}^0 M / \mathcal{A}^1 M \cong \mathfrak{gl}(T^{1,0}M, \mathbb{C})$ and $\mathcal{A}^1 M \cong \wedge^{1,0}$. Hence, for any choice of connection $\nabla \in [\nabla]$, we can identify an element of $\mathcal{A}M$ with a triple

$$\begin{pmatrix} X^b \\ \phi_b{}^c \\ \mu_b \end{pmatrix}, \quad \text{where} \quad \begin{cases} X^b \in T^{1,0}M \\ \phi_b{}^c \in \mathfrak{gl}(T^{1,0}M, \mathbb{C}) \\ \mu_b \in \wedge^{1,0}. \end{cases}$$

Note that $\phi_b{}^c$ may be decomposed further into its trace-free and trace parts according to the decomposition (18) $\mathfrak{gl}(T^{1,0}M, \mathbb{C}) \cong \mathfrak{sl}(T^{1,0}M, \mathbb{C}) \oplus \mathcal{E}(0,0)$. However, we shall not make use of this decomposition. From the formulae in (52) defining the tractor connection on the standard complex tractor bundle $\mathcal{T}$ one easily deduces that tractor connection on $\mathcal{A}M = \mathfrak{sl}(\mathcal{T})$ is given by

$$\nabla^{\mathcal{A}}_a \begin{pmatrix} X^b \\ \phi_b{}^c \\ \mu_b \end{pmatrix} = \begin{pmatrix} \nabla_a X^b - \phi_a{}^b \\ \nabla_a \phi_b{}^c + \delta_a{}^c \mu_b + \mathrm{P}_{ab} X^c + (\mu_a + \mathrm{P}_{ad} X^d)\delta_b{}^c \\ \nabla_a \mu_b - \mathrm{P}_{ac}\phi_b{}^c \end{pmatrix} \tag{86}$$

$$\nabla^{\mathcal{A}}_{\bar a} \begin{pmatrix} X^b \\ \phi_b{}^c \\ \mu_b \end{pmatrix} = \begin{pmatrix} \nabla_{\bar a} X^b \\ \nabla_{\bar a}\phi_b{}^c + \mathrm{P}_{\bar a b} X^c + \mathrm{P}_{\bar a d} X^d \delta_b{}^c \\ \nabla_{\bar a} \mu_b - \mathrm{P}_{\bar a c}\phi_b{}^c \end{pmatrix}. \tag{87}$$

From (85) we deduce that:

PROPOSITION 3.9. *Suppose $(M, J, [\nabla])$ is an almost c-projective manifold of dimension $2n \geq 6$. Then the projection $\Pi\colon \mathcal{A}M \to \mathcal{A}M/\mathcal{A}^0 M \cong T^{1,0}M$ induces a*

3.4. ADJOINT TRACTORS AND INFINITESIMAL AUTOMORPHISMS

bijection between sections of $\mathcal{A}M$ that are parallel for

$$\nabla_a^{\mathcal{A}} \begin{pmatrix} X^b \\ \phi_b{}^c \\ \mu_b \end{pmatrix} + \begin{pmatrix} 0 \\ W_{da}{}^c{}_b X^d + W_{\bar{d}a}{}^c{}_b X^{\bar{d}} + T_{af}{}^{\bar{e}} T_{\bar{e}\bar{d}}{}^f X^{\bar{d}} \delta_b{}^c \\ C_{cab} X^c + C_{\bar{c}ab} X^{\bar{c}} \end{pmatrix}$$

$$\nabla_{\bar{a}}^{\mathcal{A}} \begin{pmatrix} X^b \\ \phi_b{}^c \\ \mu_b \end{pmatrix} + \begin{pmatrix} T_{\bar{c}\bar{a}}{}^b X^{\bar{c}} \\ W_{d\bar{a}}{}^c{}_b X^d + W_{\bar{d}\bar{a}}{}^c{}_b X^{\bar{d}} + (\nabla_e T_{\bar{d}\bar{a}}{}^e X^{\bar{d}} - T_{df}{}^{\bar{e}} T_{\bar{e}\bar{a}}{}^f X^d) \delta_b{}^c \\ C_{c\bar{a}b} X^c + C_{\bar{c}\bar{a}b} X^{\bar{c}} \end{pmatrix}$$

and infinitesimal automorphisms of the almost c-projective manifold.

Proposition 3.9 can, of course, also be obtained directly by prolonging cleverly the equations of Proposition 3.7. Note that the form of the equations in Proposition 3.7 immediately shows that Π maps parallel sections for the connection in Proposition 3.9 to c-projective vector fields. To see the converse, one may verify that that for a c-projective vector field X^b and for any choice of $\nabla \in [\nabla]$ the section

$$\begin{pmatrix} X^b \\ \phi_b{}^c \\ \mu_b \end{pmatrix} = \begin{pmatrix} X^b \\ \nabla_b X^c \\ -\frac{1}{n+1}(\nabla_a \nabla_b X^a + 2\mathsf{P}_{(ab)} X^b) \end{pmatrix}$$

is parallel for the connection given in Proposition 3.9 and observe that this differential operator indeed defines the inverse to the claimed bijection.

REMARK 3.3. If the dimension of the almost c-projective manifold is $2n = 4$, then Proposition 3.9 still holds taking into account that $W_{ab}{}^c{}_d \equiv 0$.

REMARK 3.4. Note that the differential operator

$$D^{\mathcal{A}} \colon T^{1,0}M \to (\wedge^{1,0} \otimes T^{0,1}M) \oplus (S^2 \wedge^{1,0} \otimes T^{1,0}M)_\circ$$
$$X^c \mapsto (\nabla_{\bar{b}} X^c, (\nabla_{(b} \nabla_{d)} X^c + X^c \mathsf{P}_{(bd)})_\circ)$$

is also invariant. It is the first operator in the BGG sequence of the adjoint tractor bundle. As for projective structures, this operator differs from $\mathcal{D}^{\text{aut}}$, the operator that controls the infinitesimal automorphisms of the almost c-projective manifold. For a discussion of this phenomena in the context of general parabolic geometries see [**29**].

CHAPTER 4

Metrisability of almost c-projective structures

On any (pseudo-)Kähler manifold (M, J, g) one may consider the c-projective structure that is induced by the Levi-Civita connection of g. The c-projective manifolds that arise in this way from a (pseudo-)Kähler metric are the most extensively studied c-projective manifolds; see [41, 56, 82, 91] and, more recently, [44, 78, 81]. A natural but difficult problem in this context is to characterise the c-projective structures that arise from (pseudo-)Kähler metrics or, more generally, the almost c-projective structures that arise from $(2, 1)$-symplectic (also called quasi-Kähler) metrics. In the following sections we shall show that, suitably interpreted, this problem is controlled by an invariant linear overdetermined system of PDE and we shall explicitly prolong this system. Under the assumptions that J is integrable and the c-projective manifold $(M, J, [\nabla^g])$ arose via the Levi-Civita connection ∇^g of a Kähler metric g, a prolongation of the system of PDE governing the Kähler metrics that are c-projectively equivalent to g was first given in [41, 82] and rediscovered in the setting of Hamiltonian 2-forms on Kähler manifolds in [2].

4.1. Almost Hermitian manifolds

We begin by recalling some basic facts.

DEFINITION 4.1. Suppose (M, J) is an almost complex manifold of dimension $2n \geq 4$. A *Hermitian metric* on (M, J) is a (pseudo-)Riemannian metric $g_{\alpha\beta} \in \Gamma(S^2 T^* M)$ that is J-invariant:
$$J_\alpha{}^\gamma J_\beta{}^\delta g_{\gamma\delta} = g_{\alpha\beta}.$$

We call such a triple (M, J, g) an *almost Hermitian manifold*, or, if J is integrable, a *Hermitian manifold*. Note that we drop the awkward (pseudo-)prefix.

To an almost Hermitian manifold (M, J, g) one can associate a nondegenerate J-invariant 2-form $\Omega \in \Gamma(\wedge^2 T^* M)$ given by

(88) $$\Omega_{\alpha\beta} := J_\alpha{}^\gamma g_{\gamma\beta}.$$

It is called the *fundamental 2-form* or *Kähler form* of (M, J, g). If Ω is closed ($d\Omega = 0$), we say (M, J, g) is *almost Kähler* or *almost pseudo-Kähler* accordingly as g is Riemannian or pseudo-Riemannian; the "almost" prefix is dropped if J is integrable.

We write $g^{\alpha\beta}$ for the inverse of the metric $g_{\alpha\beta}$:
$$g_{\alpha\gamma} g^{\gamma\beta} = \delta_\alpha{}^\beta.$$

We raise and lower indices of tensors on an almost Hermitian manifold (M, J, g) with the metric and its inverse. The *Poisson tensor* on M is $\Omega^{\alpha\beta} = J_\gamma{}^\beta g^{\alpha\gamma}$, with

(89) $$\Omega_{\alpha\beta} \Omega^{\beta\gamma} = -\delta_\alpha{}^\gamma.$$

Viewing $\wedge^{1,0} \otimes \wedge^{0,1}$ as a complex vector bundle equipped with the real structure given by swapping its factors, a Hermitian metric can, by definition, also be seen as a real nondegenerate section $g_{a\bar{b}}$ of $\wedge^{1,0}\otimes\wedge^{0,1}$. We denote by $g^{\bar{a}b} \in \Gamma(T^{0,1}M\otimes T^{1,0}M)$ its inverse, characterised by

$$g_{a\bar{b}}g^{\bar{b}c} = \delta_a{}^c \quad \text{and} \quad g_{a\bar{b}}g^{\bar{c}a} = \delta_{\bar{b}}{}^{\bar{c}}.$$

Let us denote by ∇^g the Levi-Civita connection of a Hermitian metric g. Differentiating the identity $J_\alpha{}^\gamma J_\gamma{}^\beta = -\delta_\alpha{}^\beta$ shows that

(90) $$(\nabla^g_\alpha J_\beta{}^\epsilon) J_\epsilon{}^\gamma + J_\beta{}^\epsilon \nabla^g_\alpha J_\epsilon{}^\gamma = 0.$$

Since $\nabla^g_\alpha \Omega_{\beta\gamma} = g_{\gamma\epsilon} \nabla^g_\alpha J_\beta{}^\epsilon$, it follows immediately from (90) that

(91) $$\nabla^g_\alpha \Omega_{\beta\gamma} + J_\beta{}^\epsilon J_\gamma{}^\zeta \nabla^g_\alpha \Omega_{\epsilon\zeta} = 0.$$

Viewing $\nabla^g_\alpha \Omega_{\beta\gamma}$ as 2-form with values in T^*M, equation (91) says that the part of type $(1,1)$ vanishes identically. On the other hand, the vector bundle map

$$\wedge^2 T^*M \otimes T^*M \to \wedge^3 T^*M$$
$$\Psi_{\alpha\beta\gamma} \mapsto \Psi_{[\alpha\beta\gamma]}$$

induces an isomorphism between 2-forms with values in T^*M of type $(0,2)$ and 3-forms on M of type $(2,1) + (1,2)$, i.e. real sections of $\wedge^{2,1} \oplus \wedge^{1,2}$. Since

(92) $$\nabla^g_{[\alpha} \Omega_{\beta\gamma]} = \tfrac{1}{3}(d\Omega)_{\alpha\beta\gamma},$$

the identity

(93) $$2\nabla^g_\alpha \Omega_{\beta\gamma} = (d\Omega)_{\alpha\beta\gamma} - J_\beta{}^\epsilon J_\gamma{}^\zeta (d\Omega)_{\alpha\epsilon\zeta} - N^J_{\beta\gamma}{}^\epsilon \Omega_{\epsilon\alpha}$$

shows that type $(0,2)$ component of $\nabla^g_\alpha \Omega_{\beta\gamma}$ is identified with the $(2,1) + (1,2)$ component of $d\Omega$ [48]. The type $(3,0) + (0,3)$ component of $d\Omega$ is determined by the Nijenhuis tensor N^J, hence so is the type $(2,0)$ part of $\nabla^g_\alpha \Omega_{\beta\gamma}$ (which has type $(0,2)$ when viewed as a 2-form with values in TM using g).

If M has dimension $2n \geq 6$, $\nabla^g_\alpha \Omega_{\beta\gamma}$ can be decomposed into 4 components, which correspond to 4 real irreducible $\mathrm{U}(p,q)$-submodules in $\wedge^2 \mathbb{C}^n \otimes \mathbb{C}^n$, where $\mathrm{U}(p,q)$ denotes the (pseudo-)unitary group of signature (p,q) with $p + q = n$, the signature of $g_{\alpha\bar\beta}$. If $2n = 4$, then $\nabla^g_\alpha \Omega_{\beta\gamma}$ has only two components. The different possibilities of a subset of these invariants vanishing leads to the Gray–Hervella classification of almost Hermitian manifolds into 16, respectively 4, classes in dimension $2n \geq 6$, respectively $2n = 4$, see [51]. In the following, we shall be interested in the class of almost Hermitian manifolds which in the literature (at least in the case of metrics of definite signature) are referred to as *quasi-Kähler* or *(2,1)-symplectic*, see [48, 51]. We extend this terminology to indefinite signature, as we have done for Hermitian metrics in general.

DEFINITION 4.2. Suppose (M, J, g) is an almost Hermitian manifold of dimension $2n \geq 4$. Then (M, J, g) is called a *quasi-Kähler* or *(2,1)-symplectic* manifold, if

(94) $$\nabla^g_\alpha \Omega_{\beta\gamma} + J_\alpha{}^\epsilon J_\beta{}^\zeta \nabla^g_\epsilon \Omega_{\zeta\gamma} = 0,$$

which is the case if and only if

(95) $$\nabla^g_\alpha J_\beta{}^\gamma = -J_\alpha{}^\epsilon J_\beta{}^\zeta \nabla^g_\epsilon J_\zeta{}^\gamma.$$

4.1. ALMOST HERMITIAN MANIFOLDS

Since $\nabla_\alpha \Omega_{\beta\gamma}$, as a 2-form with values in T^*M, has no component of type $(1,1)$, (94) means, equivalently, that $\nabla_\alpha \Omega_{\beta\gamma}$ has type $(2,0)$, i.e. has no $(0,2)$ part; equivalently $d\Omega$ has no component of type $(2,1)+(1,2)$, i.e. it has type $(3,0)+(0,3)$, which explains the "$(2,1)$-symplectic" terminology. The class of $(2,1)$-symplectic manifolds of dimension $2n \geq 6$ contains as a subclass the *almost (pseudo-)Kähler* manifolds, which are symplectic, and the subclass of *nearly Kähler* manifolds, i.e. those almost Hermitian manifolds that satisfy $\nabla^g_\alpha \Omega_{\beta\gamma} = -\nabla^g_\beta \Omega_{\alpha\gamma}$, which is manifestly equivalent to $3\nabla^g_\alpha \Omega_{\beta\gamma} = (d\Omega)_{\alpha\beta\gamma}$. Since in dimension $2n = 4$ any 3-form has type $(2,1)+(1,2)$, the condition for an almost Hermitian manifold of dimension 4 to be $(2,1)$-symplectic is equivalent to the condition to be almost (pseudo-)Kähler. If J is integrable, i.e. (M,J,g) a Hermitian manifold, then (M,J,g) is $(2,1)$-symplectic if and only if $d\Omega = 0$, i.e. (M,J,g) is (pseudo-)Kähler.

DEFINITION 4.3. Let (M,J,g) be an almost Hermitian manifold of dimension $2n \geq 4$. Then a *Hermitian connection* on M is an affine connection ∇ with $\nabla J = 0$ and $\nabla g = 0$.

Such Hermitian connections exist and are uniquely determined by their torsion. A discussion of Hermitian connections and of the freedom in prescribing their torsion can for instance be found in [48] (see also [69]). The following proposition shows that $(2,1)$-symplectic manifolds can be characterised as those almost Hermitian manifolds which admit a minimal Hermitian connection; for a proof see [48].

PROPOSITION 4.1. *Suppose (M,J,g) is an almost Hermitian manifold of dimension $2n \geq 4$. Then (M,J,g) admits a (unique) Hermitian connection whose torsion T is of type $(0,2)$ as 2-form with values in TM, equivalently $T = -\frac{1}{4}N^J$, if and only if it is $(2,1)$-symplectic.*

For a a $(2,1)$-symplectic manifold (M,J,g) we refer to the unique Hermitian connection ∇ of Proposition 4.1 as the *canonical connection* of (M,J,g). In terms of the Levi-Civita connection ∇^g of g it is given by

$$\nabla_\alpha X^\beta = \nabla^g_\alpha X^\beta + \tfrac{1}{2}(\nabla^g_\alpha J_\gamma{}^\beta)J_\epsilon{}^\gamma X^\epsilon. \tag{96}$$

For the convenience of the reader let us check that this connection has the desired properties. For an arbitrary almost Hermitian manifold (M,J,g) the formula (96) is obviously a complex connection, since

$$\nabla_\alpha(J_\gamma{}^\beta X^\gamma) = \tfrac{1}{2}(\nabla^g_\alpha(J_\gamma{}^\beta X^\gamma) + J_\gamma{}^\beta \nabla^g_\alpha X^\gamma)$$
$$J_\gamma{}^\beta \nabla_\alpha X^\gamma = \tfrac{1}{2}(\nabla^g_\alpha(J_\gamma{}^\beta X^\gamma) + J_\gamma{}^\beta \nabla^g_\alpha X^\gamma),$$

which implies $(\nabla_\alpha J_\beta{}^\gamma)X^\beta = \nabla_\alpha(J_\gamma{}^\beta X^\gamma) - J_\gamma{}^\beta \nabla_\alpha X^\gamma = 0$ for all vector fields X^α. Since ∇^g is a metric connection, the connection given by (96) is a metric connection if and only if $(\nabla^g_\alpha J_\zeta{}^\epsilon)J_\beta{}^\zeta g_{\gamma\epsilon} = (\nabla^g_\alpha \Omega_{\zeta\gamma})J_\beta{}^\zeta$ is skew in γ and β, which follows immediately from (93). Hence, on any almost Hermitian manifold formula (96) defines a Hermitian connection. Moreover, since ∇^g is torsion free, the torsion T of (96) satisfies

$$T_{\alpha\beta}{}^\gamma = \tfrac{1}{2}((\nabla^g_\alpha J_\epsilon{}^\gamma)J_\beta{}^\epsilon - (\nabla^g_\beta J_\epsilon{}^\gamma)J_\alpha{}^\epsilon). \tag{97}$$

Recall that the Nijenhuis tensor can be expressed in terms of ∇^g (actually in terms of any torsion free connection) as

$$N^J_{\alpha\beta}{}^\gamma = -(\nabla^g_\alpha J_\epsilon{}^\gamma)J_\beta{}^\epsilon + (\nabla^g_\beta J_\epsilon{}^\gamma)J_\alpha{}^\epsilon - J_\alpha{}^\epsilon \nabla_\epsilon J_\beta{}^\gamma + J_\beta{}^\epsilon \nabla_\epsilon J_\alpha{}^\gamma, \tag{98}$$

which by (95) reduces in the case of a $(2,1)$-symplectic manifold to the equation
$$(99) \qquad N_{\alpha\beta}^J{}^\gamma = -2((\nabla^g_\alpha J_\epsilon{}^\gamma)J_\beta{}^\epsilon - (\nabla^g_\beta J_\epsilon{}^\gamma)J_\alpha{}^\epsilon).$$
Comparing (97) with (99) shows that on a $(2,1)$-symplectic manifold the torsion T of (96) satisfies $T = -\frac{1}{4}N^J$ as required. Note that, if the Levi-Civita connection ∇^g of (M, J, g) is a complex connection, then also $\nabla^g_\alpha \Omega_{\beta\gamma} = 0$, which by (92) implies that $\Omega_{\alpha\beta}$ is closed. Moreover, the identity (93) shows that J is necessarily integrable in this case. Conversely, the same identity shows that, if J is integrable and the fundamental 2-form closed, then the Levi-Civita connection is a complex connection, cf. Corollary 1.3. Hence, the connection in (96) coincides with the Levi-Civita connection of on an almost Hermitian manifold if and only if (M, J, g) is (pseudo-)Kähler.

REMARK 4.1. We have already observed that formula (96) defines a Hermitian connection on any almost Hermitian manifold, which is usually referred to as the *first canonical connection* following [69]. In the case of a $(2,1)$-symplectic manifold the first canonical connection coincides also with *the second canonical connection* of [69], which is also called *Chern connection*; see [48].

Let (M, J, g) be a $(2,1)$-symplectic manifold and denote by R the curvature of its canonical connection ∇. Since ∇ is Hermitian, we have $R \in \Omega^2(M, \mathfrak{u}(TM))$, where $\mathfrak{u}(TM) \subset T^*M \otimes TM$ denotes the subbundle of unitary bundle endomorphisms of (TM, J, g). Setting $R_{\alpha\beta\gamma\delta} \equiv R_{ab}{}^\epsilon{}_\delta g_{\epsilon\gamma}$, the property $R \in \Omega^2(M, \mathfrak{u}(TM))$ of the curvature of a $(2,1)$-symplectic manifold can be expressed as
$$(100) \qquad R_{\alpha\beta\gamma\delta} = R_{[\alpha\beta][\gamma\delta]} \quad \text{and} \quad R_{\alpha\beta\gamma[\delta}J_{\epsilon]}{}^\gamma = 0.$$
Moreover, recall that for any linear connection the Bianchi symmetry holds. Hence, R satisfies
$$(101) \qquad R_{[\alpha\beta}{}^\gamma{}_{\delta]} = \nabla_{[\alpha}T_{\beta\delta]}{}^\gamma + T_{\epsilon[\alpha}{}^\gamma T_{\beta\delta]}{}^\epsilon,$$
where $T_{\alpha\beta}{}^\gamma = -\frac{1}{4}N^J_{\alpha\beta}{}^\gamma$ is the torsion of ∇. Note that (101) for a minimal connection is of course precisely equivalent to the already established identities (41) and (42). Since ∇ is a complex connection, R decomposes as a 2-form with values in the complex endomorphism of TM into three components according to type as explained in Chapter 2.4. In barred and unbarred indices R can therefore be encoded by the three tensors
$$R_{ab}{}^c{}_d \qquad R_{a\bar{b}}{}^c{}_d \qquad R_{\bar{a}\bar{b}}{}^c{}_d,$$
or equivalently by their complex conjugates, where $R_{ab}{}^c{}_d \equiv \Pi_a^\alpha \Pi_b^\beta \Pi_\gamma^c \Pi_d^\delta R_{\alpha\beta}{}^\gamma{}_\delta$ and so on. Since ∇ preserves in addition a (pseudo-)Riemannian metric, the additional symmetry $R_{\alpha\beta\gamma\delta} = -R_{\alpha\beta\delta\gamma}$ implies

$R_{ab\bar{c}d} \equiv R_{ab}{}^e{}_d g_{e\bar{c}} = -R_{ab}{}^{\bar{e}}{}_{\bar{c}} g_{\bar{e}d} \equiv -R_{abd\bar{c}} \qquad R_{\bar{a}b\bar{c}d} \equiv R_{\bar{a}b}{}^e{}_d g_{e\bar{c}} = -R_{\bar{a}b}{}^{\bar{e}}{}_{\bar{c}} g_{\bar{e}d} \equiv -R_{\bar{a}bd\bar{c}}$

$R_{a\bar{b}\bar{c}d} \equiv R_{a\bar{b}}{}^e{}_d g_{e\bar{c}} = -R_{a\bar{b}}{}^{\bar{e}}{}_{\bar{c}} g_{\bar{e}d} \equiv -R_{a\bar{b}d\bar{c}} \qquad R_{\bar{a}\bar{b}\bar{c}d} \equiv R_{\bar{a}\bar{b}}{}^e{}_d g_{e\bar{c}} = -R_{\bar{a}\bar{b}}{}^{\bar{e}}{}_{\bar{c}} g_{\bar{e}d} \equiv -R_{\bar{a}\bar{b}d\bar{c}}.$

Note that the first two identities (which are conjugates of each other) show that for the canonical connection of a $(2,1)$-symplectic manifold (in contrast to a general minimal complex connection) the curvature components $R_{ab}{}^c{}_d$ and $R_{ab}{}^{\bar{c}}{}_{\bar{d}} = \overline{R_{\bar{a}\bar{b}}{}^c{}_d}$ are not independent of each other, since they are related by g. Hence, the curvature R of the canonical connection of a $(2,1)$-symplectic manifold can be encoded by the two tensors
$$R_{ab\bar{c}d} = R_{[ab]\bar{c}d} \quad \text{and} \quad R_{a\bar{b}\bar{c}d}$$

(or their complex conjugates). By (101), (100) and the fact that the torsion of ∇ has type $(0,2)$ one deduces straightforwardly that the curvature and the torsion of a $(2,1)$-symplectic manifold satisfy the symmetries

(102) $\quad R_{ab\bar{c}d} = -\nabla_{\bar{c}} T_{abd} \qquad R_{ab\bar{c}d} + R_{bd\bar{c}a} + R_{da\bar{c}b} = 0$

(103) $\quad \nabla_{[a} T_{bc]d} = 0$

(104) $\quad R_{a\bar{b}\bar{c}d} - R_{d\bar{b}\bar{c}a} = -T_{ad}{}^{\bar{e}} T_{\bar{e}\bar{b}\bar{c}} \qquad R_{a\bar{b}\bar{c}d} - R_{a\bar{c}\bar{b}d} = -T_{\bar{b}\bar{c}}{}^{e} T_{ead}$

(105) $\quad R_{a\bar{b}\bar{c}d} - R_{\bar{c}d a\bar{b}} = T_{\bar{a}\bar{b}\bar{c}} T_{da}{}^{\bar{e}} + T_{fda} T_{\bar{c}\bar{b}}{}^{f},$

where $T_{abc} = T_{ab}{}^{d} g_{c\bar{d}}$ and $T_{\bar{a}\bar{b}\bar{c}} = T_{\bar{a}\bar{b}}{}^{\bar{d}} g_{d\bar{c}}$ (cf. also (41) and (42)). Now let us consider the Ricci tensor Ric of the canonical connection of a $(2,1)$-symplectic manifold. By definition we have

$$\mathrm{Ric}_{ab} = R_{ca}{}^{c}{}_{b} \qquad \mathrm{Ric}_{\bar{a}\bar{b}} = R_{\bar{c}\bar{a}}{}^{\bar{c}}{}_{\bar{b}} \qquad \mathrm{Ric}_{a\bar{b}} = R_{\bar{c}a}{}^{\bar{c}}{}_{\bar{b}} \qquad \mathrm{Ric}_{\bar{a}b} = R_{c\bar{a}}{}^{c}{}_{b}.$$

From the identities (102) we we conclude that

(106) $\quad \begin{aligned} \mathrm{Ric}_{ab} &= -\nabla_{\bar{c}} T^{\bar{c}}{}_{ab} & \mathrm{Ric}_{[ab]} &= \tfrac{1}{2} \nabla_{\bar{c}} T_{ab}{}^{\bar{c}} \\ \mathrm{Ric}_{\bar{a}\bar{b}} &= -\nabla_{c} T^{c}{}_{\bar{a}\bar{b}} & \mathrm{Ric}_{[\bar{a}\bar{b}]} &= \tfrac{1}{2} \nabla_{c} T_{\bar{a}\bar{b}}{}^{c}. \end{aligned}$

Moreover, taking a Ricci type contraction in (105) shows immediately that the J-invariant part of the Ricci tensor of the canonical connection of a $(2,1)$-symplectic manifold is symmetric:

(107) $\qquad\qquad\qquad \mathrm{Ric}_{a\bar{b}} = \mathrm{Ric}_{\bar{b}a}.$

The canonical connection of a $(2,1)$-symplectic manifold is special in the sense of Section 2.5, since it preserves the volume form of g; hence (106) and (107) confirm in particular what we deduced there for the Ricci curvature of special connections.

We already observed that $(2,1)$-symplectic is equivalent to (pseudo-)Kähler when J is integrable. Hence, in this case, the canonical connection simply coincides with the Levi-Civita connection. Suppose (M, J, g) is now a (pseudo-)Kähler manifold. Then the identities (102)–(105) imply that R is determined by any of the following tensors

$$R_{a\bar{b}c\bar{d}} \qquad R_{\bar{a}bc\bar{d}} \qquad R_{a\bar{b}\bar{c}d} \qquad R_{\bar{a}b\bar{c}d}$$

which are now subject to the following symmetries

(108) $\quad R_{a\bar{b}c\bar{d}} = -R_{\bar{b}ac\bar{d}} \qquad R_{a\bar{b}c\bar{d}} = -R_{a\bar{b}\bar{d}c} \qquad R_{a\bar{b}c\bar{d}} = R_{\bar{c}ba\bar{d}} \qquad R_{a\bar{b}c\bar{d}} = R_{ad\bar{c}\bar{b}}$

as well as

$$\overline{R}_{\bar{a}b\bar{c}d} \equiv \overline{R_{a\bar{b}c\bar{d}}} = R_{b\bar{a}d\bar{c}}.$$

REMARK 4.2. Let us remark that for the curvature R of a (pseudo-)Kähler manifold, we have $R_{[\alpha\beta}{}^{\gamma}{}_{\delta]} = 0$ which, together with the symmetries (100), implies

(109) $\qquad R_{\alpha\beta\gamma\delta} = R_{\gamma\delta\alpha\beta} \qquad \text{and} \qquad J_{\alpha}{}^{\epsilon} J_{\beta}{}^{\zeta} R_{\epsilon\zeta}{}^{\gamma}{}_{\delta} = R_{\alpha\beta}{}^{\gamma}{}_{\delta}.$

The symmetries of (100) and (109) are precisely the ones in (108) expressed in real indices. Note also that (109) shows immediately that the Ricci tensor $\mathrm{Ric}_{\alpha\beta} = R_{\epsilon\alpha}{}^{\epsilon}{}_{\beta}$ is symmetric and J-invariant, which is consistent with (107).

56   4. METRISABILITY OF ALMOST C-PROJECTIVE STRUCTURES

Moreover, note that it is immediate that the rank of the bundle of (pseudo-)Kähler curvatures is $(n(n+1)/2)^2$ and that this bundle further decomposes under U(n) as

$$S^2 \wedge^{1,0} \otimes S^2 \wedge^{0,1} = (S^2 \wedge^{1,0} \otimes_\circ S^2 \wedge^{0,1}) \oplus (\wedge^{1,0} \otimes_\circ \wedge^{0,1}) \oplus \mathbb{R},$$

where the subscript ∘ means trace-free part and $\mathbb{R}$ stands for the trivial bundle. Under this decomposition, the (pseudo-)Kähler curvature splits as

$$(110) \quad R_{a\bar{b}c\bar{d}} = U_{a\bar{b}c\bar{d}} - 2(\Xi_{a\bar{b}}g_{c\bar{d}} + \Xi_{c\bar{d}}g_{a\bar{b}} + \Xi_{a\bar{d}}g_{c\bar{b}} + \Xi_{c\bar{b}}g_{a\bar{d}}) - 2\Lambda(g_{a\bar{b}}g_{c\bar{d}} + g_{a\bar{d}}g_{c\bar{b}}),$$

where

$$U_{a\bar{b}c\bar{d}} = U_{c\bar{b}a\bar{d}} = U_{a\bar{d}c\bar{b}} \qquad g^{\bar{b}c}U_{a\bar{b}c\bar{d}} = 0 \qquad g^{\bar{b}a}\Xi_{a\bar{b}} = 0.$$

This is a Kähler analogue of the usual decomposition of Riemannian curvature into the conformal Weyl tensor, the trace-free Ricci tensor, and the scalar curvature. The tensor $U_{a\bar{b}c\bar{d}}$ is called the *Bochner curvature* (or *tensor*) and is the orthogonal projection of the conformal Weyl curvature onto the intersection of the space of Kähler curvatures with the space of conformal Weyl tensors [2]. The analogue of constant curvature in (pseudo-)Kähler geometry is to insist that $R_{a\bar{b}d\bar{c}} = \Lambda(g_{a\bar{b}}g_{c\bar{d}} + g_{c\bar{b}}g_{a\bar{d}})$, where the (a priori) smooth function Λ is constant by the Bianchi identity. This is called *constant holomorphic sectional curvature* and (for $\Lambda > 0$) locally characterises $\mathbb{CP}^n$ and its Fubini–Study metric as in Chapter 2.6 (where the normalisation is such that $\Lambda = 1$).

4.2. Other curvature decompositions

It will be useful, both in this article and elsewhere, to decompose the (pseudo-)Kähler curvature tensor from various different viewpoints, some of which ignore the complex structure. Without a complex structure, barred and unbarred indices are unavailable so firstly we should rewrite the irreducible decomposition (110) using only real indices. We recall that

$$(111) \qquad R_{\alpha\beta\gamma\delta} = R_{[\alpha\beta][\gamma\delta]} \qquad R_{[\alpha\beta\gamma]\delta} = 0 \qquad R_{\alpha\beta\gamma[\delta}J_{\epsilon]}{}^\gamma = 0$$

and the real version of (110) applies to any tensor satisfying these identities. Recalling that $\Omega_{\alpha\beta} = J_{\alpha\beta} = J_\alpha{}^\gamma g_{\beta\gamma}$, we obtain

$$(112)$$
$$R_{\alpha\beta\gamma\delta} = U_{\alpha\beta\gamma\delta}$$
$$+ g_{\alpha\gamma}\Xi_{\beta\delta} - g_{\beta\gamma}\Xi_{\alpha\delta} - g_{\alpha\delta}\Xi_{\beta\gamma} + g_{\beta\delta}\Xi_{\alpha\gamma}$$
$$+ \Omega_{\alpha\gamma}\Sigma_{\beta\delta} - \Omega_{\beta\gamma}\Sigma_{\alpha\delta} - \Omega_{\alpha\delta}\Sigma_{\beta\gamma} + \Omega_{\beta\delta}\Sigma_{\alpha\gamma} + 2\Omega_{\alpha\beta}\Sigma_{\gamma\delta} + 2\Omega_{\gamma\delta}\Sigma_{\alpha\beta}$$
$$+ \Lambda(g_{\alpha\gamma}g_{\beta\delta} - g_{\beta\gamma}g_{\alpha\delta} + \Omega_{\alpha\gamma}\Omega_{\beta\delta} - \Omega_{\beta\gamma}\Omega_{\alpha\delta} + 2\Omega_{\alpha\beta}\Omega_{\gamma\delta}),$$

where

- $U_{\alpha\beta\gamma\delta}$ is totally trace-free with respect to $g^{\alpha\beta}$ and $\Omega^{\alpha\beta}$
- $\Sigma_{\alpha\beta} \equiv J_\alpha{}^\gamma \Xi_{\beta\gamma}$ whilst $\Xi_{\alpha\beta}$ is symmetric, trace-free, and of type $(1,1)$:

$$\Xi_{\alpha\beta} = \Xi_{(\alpha\beta)} \qquad \Xi_\alpha{}^\alpha = 0 \qquad \Sigma_{\alpha\beta} = \Sigma_{[\alpha\beta]}.$$

A simple way to see this is to check that all parts of this decomposition satisfy (111) as they should and then apply $\Pi_a^\alpha \overline{\Pi}_{\bar{b}}^\beta \Pi_c^\gamma \overline{\Pi}_{\bar{d}}^\delta$, using the various identities from Chapter 1.1 including (4), to recover (110). One can also read off from (112)

the corresponding decomposition of the Ricci tensor in (pseudo-)Kähler geometry. Specifically,
$$\mathrm{Ric}_{\beta\delta} = 2(n+2)\Xi_{\beta\delta} + 2(n+1)\Lambda g_{\beta\delta} \qquad \mathrm{Scal} = 4n(n+1)\Lambda$$
and, conversely,
$$\Lambda = \tfrac{1}{4n(n+1)}\mathrm{Scal} \qquad \Xi_{\alpha\beta} = \tfrac{1}{2(n+2)}\left(\mathrm{Ric}_{\alpha\beta} - \tfrac{1}{2n}\mathrm{Scal}\, g_{\alpha\beta}\right).$$

Other natural realms in which one may view (pseudo-)Kähler geometry are
- projective
- conformal
- c-projective
- symplectic

and in each case decompose the curvature accordingly. The projective Weyl curvature tensor [**42**] on a Riemannian manifold of dimension m is given by
$$R_{\alpha\beta\gamma\delta} - \tfrac{1}{m-1}g_{\alpha\gamma}\mathrm{Ric}_{\beta\delta} + \tfrac{1}{m-1}g_{\beta\gamma}\mathrm{Ric}_{\alpha\delta}.$$

If this vanishes, then, in conjunction with the interchange symmetry $R_{\alpha\beta\gamma\delta} = R_{\gamma\delta\alpha\beta}$, we deduce that $R_{\alpha\beta\gamma\delta} = \lambda(g_{\alpha\gamma}g_{\beta\delta} - g_{\beta\gamma}g_{\alpha\delta})$ where, if $m \geq 3$, the (a priori) smooth function λ is constant by the Bianchi identity. This is Beltrami's Theorem that the only projectively flat (pseudo-)Riemannian geometries are constant curvature (when $m = 2$ one instead uses that the projective Cotton–York tensor vanishes). In any case, comparison with (112) shows that for $n \geq 2$ the only projectively flat (pseudo-)Kähler manifolds are flat. The conformal Weyl curvature is given by
$$R_{\alpha\beta\gamma\delta} - g_{\alpha\gamma}Q_{\beta\delta} + g_{\beta\gamma}Q_{\alpha\delta} - g_{\beta\delta}Q_{\alpha\gamma} + g_{\alpha\delta}Q_{\beta\gamma},$$
where $Q_{\alpha\beta}$ is the Riemannian Schouten tensor
$$Q_{\alpha\beta} = \frac{1}{m-2}\left(\mathrm{Ric}_{\alpha\beta} - \frac{1}{2(m-1)}\mathrm{Scal}\, g_{\alpha\beta}\right).$$

Thus, if the conformal Weyl curvature vanishes on a (pseudo-)Kähler manifold, then
$$2R_\alpha{}^\delta{}_{\gamma[\delta}J_{\epsilon]}{}^\gamma = J_{\epsilon\alpha}Q_\beta{}^\beta + 2(n-2)J_\epsilon{}^\gamma Q_{\alpha\gamma}.$$

From (111), we see that for $n \geq 3$ the only conformally flat (pseudo-)Kähler manifolds are flat. For $n = 2$ it follows only that the geometry is scalar flat and, in fact, Tanno [**97**] showed that 4-dimensional conformally flat Kähler manifolds are locally of the form $\mathbb{CP}^1 \times \Sigma$ where $\mathbb{CP}^1$ has the Fubini–Study metric up to constant scale and the complex surface Σ has a constant negative scalar curvature of equal magnitude but opposite sign to that on $\mathbb{CP}^1$.

From the c-projective viewpoint, if we compare the decomposition (112) with (24), then we conclude, firstly that $W_{\alpha\beta}{}^\gamma{}_\delta = H_{\alpha\beta}{}^\gamma{}_\delta$ (see the proof of Proposition 4.4 for a barred/unbarred index proof of this), and then that

(113)
$$H_{\alpha\beta\gamma\delta} = U_{\alpha\beta\gamma\delta} - g_{\alpha\delta}\Xi_{\beta\gamma} + g_{\beta\delta}\Xi_{\alpha\gamma} + \tfrac{1}{n+1}(g_{\beta\gamma}\Xi_{\alpha\delta} - g_{\alpha\gamma}\Xi_{\beta\delta})$$
$$+ 2\Omega_{\alpha\beta}\Sigma_{\gamma\delta} - \Omega_{\alpha\delta}\Sigma_{\beta\gamma} + \Omega_{\beta\delta}\Sigma_{\alpha\gamma} - \tfrac{1}{n+1}(2\Omega_{\gamma\delta}\Sigma_{\alpha\beta} - \Omega_{\beta\gamma}\Sigma_{\alpha\delta} + \Omega_{\alpha\gamma}\Sigma_{\beta\delta}).$$

Notice, in particular, that

(114) $$H_{\alpha\beta\gamma}{}^\beta = 2\tfrac{n(n+1)}{n+1}\Xi_{\alpha\gamma}$$

from which we can deduce the following c-projective counterpart to Beltrami's Theorem.

THEOREM 4.2. *Suppose a (pseudo-)Kähler metric is c-projectively flat. Then it has constant holomorphic sectional curvature.*

PROOF. To be c-projectively flat, the harmonic curvature tensor $H_{\alpha\beta}{}^\gamma{}_\delta$ must vanish. Then from (114) we find that $\Xi_{\alpha\beta} = 0$ and from (113) that also $U_{\alpha\beta\gamma\delta} = 0$. According to (112) we find that $R_{\alpha\beta\gamma\delta}$ is of the required form. □

Finally, we may view (pseudo-)Kähler geometry from the purely symplectic viewpoint as follows. For any torsion-free connection preserving $\Omega_{\alpha\beta}$, the tensor $R_{\alpha\beta}{}^\epsilon{}_\delta \Omega_{\epsilon\gamma}$ is symmetric in $\gamma\delta$ and may be decomposed into irreducible pieces under $\mathrm{Sp}(2n, \mathbb{R})$:

$$(115) \quad R_{\alpha\beta}{}^\epsilon{}_\delta \Omega_{\epsilon\gamma} = V_{\alpha\beta\gamma\delta} + \Omega_{\alpha\gamma}\Phi_{\beta\delta} - \Omega_{\beta\gamma}\Phi_{\alpha\delta} + \Omega_{\alpha\delta}\Phi_{\beta\gamma} - \Omega_{\beta\delta}\Phi_{\alpha\gamma} + 2\Omega_{\alpha\beta}\Phi_{\gamma\delta},$$

where

$$V_{\alpha\beta\gamma\delta} = V_{[\alpha\beta](\gamma\delta)} \qquad V_{[\alpha\beta\gamma]\delta} = 0 \qquad \Omega^{\alpha\beta} V_{\alpha\beta\gamma\delta} = 0 \qquad \Phi_{\alpha\beta} = \Phi_{(\alpha\beta)}.$$

PROPOSITION 4.3. *On a (pseudo-)Kähler manifold, if the tensor $V_{\alpha\beta\gamma\delta}$ vanishes, then the metric has constant holomorphic sectional curvature.*

PROOF. From (115), we find that

$$\Omega^{\alpha\beta} R_{\alpha\beta}{}^\epsilon{}_\delta \Omega_{\epsilon\gamma} = \Omega^{\alpha\beta} \left[\Omega_{\alpha\gamma}\Phi_{\beta\delta} - \Omega_{\beta\gamma}\Phi_{\alpha\delta} + \Omega_{\alpha\delta}\Phi_{\beta\gamma} - \Omega_{\beta\delta}\Phi_{\alpha\gamma} + 2\Omega_{\alpha\beta}\Phi_{\gamma\delta} \right]$$
$$= 4(n+1)\Phi_{\gamma\delta}$$

whereas computing according to (112) leads to

$$\Omega^{\alpha\beta} R_{\alpha\beta}{}^\epsilon{}_\delta \Omega_{\epsilon\gamma} = 4(n+2)\Xi_{\gamma\delta} + 4(n+1)\Lambda g_{\gamma\delta}.$$

Equating these expressions and contracting (115) with $g^{\alpha\gamma}$ yields

$$R_{\alpha\beta}{}^\epsilon{}_\delta J_\epsilon{}^\alpha = -2(n+1)\Lambda\Omega_{\beta\delta} - 2\tfrac{n+2}{n+1}\Sigma_{\beta\delta}.$$

However, from (112) we find that

$$R_{\alpha\beta}{}^\epsilon{}_\delta J_\epsilon{}^\alpha = -\Omega^{\alpha\gamma} R_{\alpha\beta\gamma\delta} = -2(n+1)\Lambda\Omega_{\beta\delta} - 2(n+1)\Sigma_{\beta\delta}.$$

We conclude that $\Xi_{\beta\delta} = 0$ and (112) now has the required form. □

4.3. Metrisability of almost c-projective manifolds

Suppose $(M, J, [\nabla])$ is an almost c-projective manifold. It is natural to ask whether $[\nabla]$ contains the canonical connection of a $(2,1)$-symplectic metric on (M, J).

DEFINITION 4.4. On an almost c-projective manifold $(M, J, [\nabla])$ a $(2,1)$-symplectic metric $g \in \Gamma(S^2 T^*M)$ on (M, J) is *compatible* with the c-projective class $[\nabla]$ if and only if its canonical connection is contained in $[\nabla]$. The almost c-projective structure on M is said to be *metrisable* or *(2,1)-symplectic* or *quasi-Kähler* (or *Kähler* or *pseudo-Kähler* when J is integrable) if it admits a compatible $(2,1)$-symplectic metric g (respectively a Kähler or pseudo-Kähler metric g, if J is integrable).

4.3. METRISABILITY OF ALMOST C-PROJECTIVE MANIFOLDS

The volume form $\mathrm{vol}(g)$ of g is a positive section of $\wedge^{2n}T^*M$, which we view as a c-projective density of weight $(-(n+1),-(n+1))$ under the identification of oriented real line bundles $\wedge^{2n}T^*M = \mathcal{E}_\mathbb{R}(-(n+1),-(n+1))$ determined by

$$\varepsilon_{ab\cdots c}\bar{\varepsilon}_{\bar{d}\bar{e}\cdots\bar{f}} \in \Gamma(\wedge^{2n}T^*(n+1,n+1)),$$

where $\varepsilon_{ab\cdots c} \in \Gamma(\wedge^{n,0}(n+1,0))$ is the tautological form from Chapter 2.1. We now write $\mathrm{vol}(g) = \tau_g^{-(n+1)}$ uniquely to determine a positive section τ_g of $\mathcal{E}_\mathbb{R}(1,1)$. The canonical connection ∇ of g is a special connection in the c-projective class, and for all $\ell \in \mathbb{Z}$, $\tau_g^\ell = \mathrm{vol}(g)^{-\ell/(n+1)} \in \Gamma(\mathcal{E}_\mathbb{R}(\ell,\ell))$ is a ∇-parallel trivialisation of $\mathcal{E}_\mathbb{R}(\ell,\ell)$.

In the integrable case, the metrisability of a c-projective structure gives easily the following constraints on the harmonic curvature.

PROPOSITION 4.4. *Let $(M, J, [\nabla])$ be a c-projective manifold of dimension $2n \geq 4$. If $[\nabla]$ is induced by the Levi-Civita of a (pseudo-)Kähler metric on (M, J), then the harmonic curvature only consists of the $(1,1)$-part*

$$W_{a\bar{b}}{}^c{}_d = H_{a\bar{b}}{}^c{}_d$$

of the (c-projective) Weyl curvature.

PROOF. Suppose first that $2n \geq 6$. Then we have to show that $W_{ab}{}^c{}_d$ vanishes. Recall that, by construction, $W_{ab}{}^c{}_d$ is the connection-independent part of the $(2,0)$-component of the curvature of any connection in the c-projective class. Hence, if $[\nabla]$ is induced from the Levi-Civita connection of a (pseudo-)Kähler metric on (M,J), then $W_{ab}{}^c{}_d$ vanishes identically, since the curvature of a (pseudo-)Kähler metric is J-invariant. If $2n = 4$, then $W_{ab}{}^c{}_d$ is always identically zero and the $(2,0)$-part C_{abc} of the Cotton–York tensor is independent of the choice of connection in the c-projective class. Since the Ricci tensor $\mathrm{Ric}_{\alpha\beta}$ of a (pseudo-)Kähler metric g is J-invariant (109), we have $\mathsf{P}_{a\bar{b}} = \frac{1}{n+1}\mathrm{Ric}_{a\bar{b}}$ and $\mathrm{Ric}_{ab} = \mathsf{P}_{ab} = 0$. Hence, if $2n = 4$ and the c-projective structure is metrisable, then $C_{abc} = \nabla_a \mathsf{P}_{bc} - \nabla_b \mathsf{P}_{ac}$ vanishes identically, which proves the claim. □

We now link compatible metrics to solutions of the first BGG operator associated to a real analogue $\mathcal{V}$ of the standard complex tractor bundle $\mathcal{T}$. Any almost c-projective manifold $(M, J, [\nabla])$ admits a complex vector bundle

$$\mathcal{V}_\mathbb{C} = \mathcal{T} \otimes \overline{\mathcal{T}}.$$

Although the construction of $\mathcal{T}$ and $\overline{\mathcal{T}}$ requires the existence and choice of an $(n+1)^{\mathrm{st}}$ root $\mathcal{E}(1,0)$ of $\wedge^n T^{1,0}M$, the vector bundle $\mathcal{T} \otimes \overline{\mathcal{T}}$ is defined independently of such a choice. Moreover, swapping the two factors defines a real structure on $\mathcal{T} \otimes \overline{\mathcal{T}}$ and hence $\mathcal{V}_\mathbb{C}$ is the complexification of a real vector bundle $\mathcal{V}$ over M corresponding to that real structure. The filtration (49) of $\mathcal{T}$ induces filtrations on $\mathcal{V}$ and $\mathcal{V}_\mathbb{C}$ given by

$$\mathcal{V}_\mathbb{C} = \mathcal{V}_\mathbb{C}^{-1} \supset \mathcal{V}_\mathbb{C}^0 \supset \mathcal{V}_\mathbb{C}^1,$$

where

$$\mathcal{V}_\mathbb{C}^{-1}/\mathcal{V}_\mathbb{C}^0 \cong T^{0,1}M \otimes T^{1,0}M(-1,-1)$$
$$\mathcal{V}_\mathbb{C}^0/\mathcal{V}_\mathbb{C}^1 \cong (T^{1,0}M \oplus T^{0,1}M)(-1,-1)$$
$$\mathcal{V}_\mathbb{C}^1 \cong \mathcal{E}(-1,-1).$$

60   4. METRISABILITY OF ALMOST C-PROJECTIVE STRUCTURES

For any choice of connection $\nabla \in [\nabla]$ we can therefore identify an element of $\mathcal{V}_{\mathbb{C}}$ with a quadruple

$$\begin{pmatrix} \eta^{\bar{b}c} \\ X^b \mid Y^{\bar{b}} \\ \rho \end{pmatrix}, \text{ where } \begin{cases} \eta^{\bar{b}c} \in T^{0,1}M \otimes T^{1,0}M(-1,-1), \\ X^b \in T^{1,0}M(-1,-1), \quad Y^{\bar{b}} \in T^{0,1}M(-1,-1), \\ \rho \in \mathcal{E}(-1,-1), \end{cases}$$

and elements of $\mathcal{V}$ can be identified with the real elements of $\mathcal{V}_{\mathbb{C}}$:

(116) $$\overline{\eta^{\bar{c}b}} = \eta^{\bar{b}c}, \quad \overline{X^b} = Y^{\bar{b}} \quad \text{and} \quad \bar{\rho} = \rho.$$

The formulae in (52) for the tractor connection on $\mathcal{T}$ immediately imply that the tractor connection on $\mathcal{V}_{\mathbb{C}} = \mathcal{T} \otimes \overline{\mathcal{T}}$ is given by

(117) $$\nabla_a^{\mathcal{V}_{\mathbb{C}}} \begin{pmatrix} \eta^{\bar{b}c} \\ X^b \mid Y^{\bar{b}} \\ \rho \end{pmatrix} = \begin{pmatrix} \nabla_a \eta^{\bar{b}c} + \delta_a{}^c Y^{\bar{b}} \\ \nabla_a X^b + \rho \delta_a{}^b - \mathsf{P}_{a\bar{c}}\eta^{\bar{c}b} \mid \nabla_a Y^{\bar{b}} - \mathsf{P}_{ac}\eta^{\bar{b}c} \\ \nabla_a \rho - \mathsf{P}_{a\bar{b}}Y^{\bar{b}} - \mathsf{P}_{ab}X^b \end{pmatrix}$$

(118) $$\nabla_{\bar{a}}^{\mathcal{V}_{\mathbb{C}}} \begin{pmatrix} \eta^{\bar{b}c} \\ X^b \mid Y^{\bar{b}} \\ \rho \end{pmatrix} = \begin{pmatrix} \nabla_{\bar{a}} \eta^{\bar{b}c} + \delta_{\bar{a}}{}^{\bar{b}} X^c \\ \nabla_{\bar{a}} X^b - \mathsf{P}_{\bar{a}\bar{c}}\eta^{\bar{c}b} \mid \nabla_{\bar{a}} Y^{\bar{b}} + \rho \delta_{\bar{a}}{}^{\bar{b}} - \mathsf{P}_{\bar{a}c}\eta^{\bar{b}c} \\ \nabla_{\bar{a}} \rho - \mathsf{P}_{\bar{a}\bar{b}}Y^{\bar{b}} - \mathsf{P}_{\bar{a}b}X^b \end{pmatrix}.$$

Note that the real structure on $\mathcal{V}_{\mathbb{C}}$ is parallel for this connection and that, consequently, the tractor connection on $\mathcal{V}$ is the restriction of (117) and (118) to real sections (116).

Now consider, for a section $\eta^{\bar{b}c}$ of $T^{0,1}M \otimes T^{1,0}M(-1,-1)$, the system of equations

(119) $$\nabla_a \eta^{\bar{b}c} + \delta_a{}^c Y^{\bar{b}} = 0, \quad \nabla_{\bar{a}} \eta^{\bar{b}c} + \delta_{\bar{a}}{}^{\bar{b}} X^c = 0$$

for some sections X^c of $T^{1,0}M(-1,-1)$ and $Y^{\bar{b}}$ of $T^{0,1}M(-1,-1)$. It follows immediately from the invariance of (59) that the system (119) is c-projectively invariant. In fact, if $\eta^{\bar{b}c} \in \Gamma(T^{0,1}M \otimes T^{1,0}M(-1,-1))$ satisfies (119) for some connection $\nabla \in [\nabla]$, for some $X^c \in \Gamma(T^{1,0}M(-1,-1))$, and for some $Y^{\bar{b}} \in \Gamma(T^{1,0}M(-1,-1))$, then $\eta^{\bar{b}c}$ satisfies (119) for $\hat{\nabla} \in [\nabla]$ with

(120) $$\hat{X}^c = X^c - \Upsilon_{\bar{b}}\eta^{\bar{b}c} \quad \text{and} \quad \hat{Y}^{\bar{b}} = Y^{\bar{b}} - \Upsilon_c \eta^{\bar{b}c}.$$

Moreover, if (119) is satisfied, one must have $Y^{\bar{b}} = -\frac{1}{n}\nabla_a \eta^{\bar{b}a}$ and $X^c = -\frac{1}{n}\nabla_{\bar{a}}\eta^{\bar{a}c}$. If $\eta^{\bar{b}c}$ is a real section, then the first equation in (119) is satisfied if and only if the second equation of (119) holds, in which case $\overline{X^b} = Y^{\bar{b}}$. We can reformulate these observations as follows. There is an invariant differential operator

(121) $$D_{\mathbb{C}}^{\mathcal{V}} \colon T^{0,1}M \otimes T^{1,0}M(-1,-1) \to \begin{matrix} (\wedge^{1,0} \otimes T^{0,1}M \otimes T^{1,0}M(-1,-1))_\circ \\ \oplus \\ (\wedge^{0,1} \otimes T^{0,1}M \otimes T^{1,0}M(-1,-1))_\circ \end{matrix}$$

given by $\eta^{\bar{b}c} \mapsto (\nabla_a \eta^{\bar{b}c} - \frac{1}{n}\delta_a{}^c \nabla_d \eta^{\bar{b}d}, \nabla_{\bar{a}}\eta^{\bar{b}c} - \frac{1}{n}\delta_{\bar{a}}{}^{\bar{b}}\nabla_{\bar{d}}\eta^{\bar{d}c})$. Restricting $D^{\mathcal{V}_{\mathbb{C}}}$ to real sections $\eta^{\bar{b}c} = \overline{\eta^{\bar{c}b}}$ gives an invariant differential operator $D^{\mathcal{V}}$. It is the first operator in the BGG sequence corresponding to the tractor bundle $\mathcal{V}$ and $D^{\mathcal{V}_{\mathbb{C}}}$ is its complexification.

PROPOSITION 4.5. *Let $(M, J, [\nabla])$ be an almost c-projective manifold of dimension $2n \geq 4$. Then, when n is even, the map sending a Hermitian metric $g_{b\bar{c}}$ to the real section $\eta^{\bar{a}b} = g^{\bar{a}b}\tau_g^{-1}$ of $T^{0,1}M \otimes T^{1,0}M(-1,-1)$ restricts to a bijection*

4.3. METRISABILITY OF ALMOST C-PROJECTIVE MANIFOLDS

between compatible $(2,1)$-symplectic Hermitian metrics on $(M, J, [\nabla])$ and nondegenerate sections in the kernel of $D^{\mathcal{V}}$. The inverse map sends $\eta^{\bar{a}b}$ to the Hermitian metric $g_{b\bar{c}}$ with $g^{\bar{a}b} = (\det \eta)\eta^{\bar{a}b}$, where

$$\det \eta := \tfrac{1}{n!}\,\bar{\varepsilon}_{\bar{a}\bar{c}\cdots\bar{e}}\,\varepsilon_{bd\cdots f}\,\eta^{\bar{a}b}\eta^{\bar{c}d}\cdots\eta^{\bar{e}f} \in \Gamma(\mathcal{E}_{\mathbb{R}}(1,1)) \tag{122}$$

and $\varepsilon_{ab\cdots c}$ denotes the tautological section of $\wedge^{n,0}(n+1,0)$. When n is odd, the mapping $\eta^{\bar{a}b} \mapsto g^{\bar{a}b} := (\det \eta)\eta^{\bar{a}b}$ is 2-1 and, conversely, the mapping $g^{\bar{a}b} \mapsto \eta^{\bar{a}b} := \tau_g^{-1} g^{\bar{a}b}$ picks a preferred sign for $\eta^{\bar{a}b}$ but, otherwise, the same conclusions hold.

PROOF. Assume first that $g_{b\bar{c}}$ is a compatible $(2,1)$-symplectic Hermitian metric, i.e. its canonical connection ∇ is contained in $[\nabla]$. Then $\eta^{\bar{a}b} = g^{\bar{a}b}\tau_g^{-1}$ is a real section of $T^{0,1}M \otimes T^{1,0}M(-1,-1)$, which satisfies (119) for ∇ with $X^c = 0$ and $Y^{\bar{c}} = \overline{X^c} = 0$. Hence, $\eta^{\bar{b}c}$ is in the kernel of $D^{\mathcal{V}}$, and $\det \eta = \tau_g^{n+1}\tau_g^{-n} = \tau_g$.

Conversely, suppose that $\eta^{\bar{b}c} \in \Gamma(T^{0,1}M \otimes T^{1,0}M(-1,-1))$ is a real nondegenerate section satisfying (119) for some connection $\nabla \in [\nabla]$ with $X^b \in \Gamma(T^{1,0}M(-1,-1))$ and $Y^{\bar{b}} = X^{\bar{b}} \in \Gamma(T^{0,1}M(-1,-1))$. Since $\eta^{\bar{a}b}$ is nondegenerate, there is a unique 1-form Υ_b such that $\eta^{\bar{a}b}\Upsilon_b = X^{\bar{a}}$. Let us denote by $\hat{\nabla} \in [\nabla]$ the connection obtained by c-projectively changing ∇ via Υ_b. Then we deduce form (120) that $\hat{\nabla}_a \eta^{\bar{b}c} = \hat{\nabla}_{\bar{a}} \eta^{\bar{b}c} = 0$. Since $\varepsilon_{ab\cdots c}$ is parallel for any connection in the c-projective class, $\det \eta$ is parallel for $\hat{\nabla}$. Hence, $g^{\bar{b}c} = \eta^{\bar{b}c} \det \eta$ is a real nondegenerate section of $T^{0,1}M \otimes T^{1,0}M$ that is parallel for $\hat{\nabla}$, i.e. its inverse $g_{b\bar{c}}$ is a $(2,1)$-symplectic Hermitian metric whose canonical connection is $\hat{\nabla} \in [\nabla]$. □

The real vector bundle $\mathcal{V}$ can be realised naturally in two alternative ways as follows. First, let us view $\mathcal{T}$ as a real vector bundle $\mathcal{T}_{\mathbb{R}}$ equipped with a complex structure $J^{\mathcal{T}}$ (thus, equivalently, $\mathcal{T}_{\mathbb{R}} \otimes \mathbb{C} \cong \mathcal{T} \oplus \overline{\mathcal{T}}$). Then we can identify $\mathcal{V}$ as the $J^{\mathcal{T}}$-invariant elements in $S^2 \mathcal{T}_{\mathbb{R}}$. However, since $J^{\mathcal{T}}$ induces an isomorphism between $J^{\mathcal{T}}$-invariant elements in $S^2 \mathcal{T}_{\mathbb{R}}$ and such elements in $\wedge^2 \mathcal{T}_{\mathbb{R}}$, cf. (88), we may, secondly, realise $\mathcal{V}$ as the latter. Realised as the bundle of $J^{\mathcal{T}}$-invariant elements in $S^2 \mathcal{T}_{\mathbb{R}}$ we can, for any choice of connection in the c-projective class, identify an element of $\mathcal{V}$ with a triple

$$\begin{pmatrix} \eta^{\beta\gamma} \\ X^{\beta} \\ \rho \end{pmatrix}, \text{ where } \begin{cases} \eta^{\beta\gamma} \in S^2 TM \otimes \mathcal{E}_{\mathbb{R}}(-1,-1) \text{ with } J_{\delta}{}^{\beta} J_{\epsilon}{}^{\gamma} \eta^{\delta\epsilon} = \eta^{\beta\gamma} \\ X^{\beta} \in TM \otimes \mathcal{E}_{\mathbb{R}}(-1,-1) \\ \rho \in \mathcal{E}_{\mathbb{R}}(-1,-1). \end{cases}$$

In this picture the tractor connection becomes

$$\nabla_{\alpha}^{\mathcal{V}} \begin{pmatrix} \eta^{\beta\gamma} \\ X^{\beta} \\ \rho \end{pmatrix} = \begin{pmatrix} \nabla_{\alpha}\eta^{\beta\gamma} + \delta_{\alpha}{}^{(\beta} X^{\gamma)} + J_{\alpha}{}^{(\beta} J_{\epsilon}{}^{\gamma)} X^{\epsilon} \\ \nabla_{\alpha} X^{\beta} + \rho \delta_{\alpha}{}^{\beta} - \mathsf{P}_{\alpha\gamma}\eta^{\beta\gamma} \\ \nabla_{\alpha}\rho - \mathsf{P}_{\alpha\beta} X^{\beta} \end{pmatrix}. \tag{123}$$

The formulae (117) and (118) may be recovered from (123) by natural projection:

$$\nabla_a^{\mathcal{V}_{\mathbb{C}}} = \Pi_a^{\alpha} \nabla_{\alpha}^{\mathcal{V}}, \quad \nabla_{\bar{a}}^{\mathcal{V}_{\mathbb{C}}} = \overline{\Pi}_{\bar{a}}^{\alpha} \nabla_{\alpha}^{\mathcal{V}}, \quad \eta^{\bar{b}c} = \overline{\Pi}_{\beta}^{\bar{b}} \Pi_{\gamma}^{c} \eta^{\beta\gamma}, \quad X^b = \Pi_{\beta}^{b} X^{\beta}, \quad Y^{\bar{b}} = \overline{\Pi}_{\beta}^{\bar{b}} X^{\beta},$$

so that, for example,

$$\Pi_a^{\alpha} \Pi_b^{\beta}(\nabla_{\alpha} X^{\beta} + \rho \delta_{\alpha}{}^{\beta} - \mathsf{P}_{\alpha\gamma}\eta^{\beta\gamma}) = \nabla_a X^b + \rho \delta_a{}^b - \mathsf{P}_{a\gamma}\eta^{\gamma b}$$
$$= \nabla_a X^b + \rho \delta_a{}^b - \mathsf{P}_{a\bar{c}}\eta^{\bar{c}b},$$

as in (117). To pass explicitly to the second (skew) viewpoint on $\mathcal{V}$ described above, one can write $\Phi^{\beta\gamma} = J_{\alpha}{}^{\gamma}\eta^{\alpha\beta}$ and $Y^{\beta} = J_{\alpha}{}^{\beta} X^{\alpha}$. Then, for any choice of connection in the c-projective class, an element of $\mathcal{V}$ may alternatively be identified with a triple

$$\begin{pmatrix} \Phi^{\beta\gamma} \\ Y^\beta \\ \rho \end{pmatrix}, \text{ where } \begin{cases} \Phi^{\beta\gamma} \in \wedge^2 TM \otimes \mathcal{E}_\mathbb{R}(-1,-1) \text{ with } J_\delta{}^\beta J_\epsilon{}^\gamma \Phi^{\delta\epsilon} = \Phi^{\beta\gamma} \\ Y^\beta \in TM \otimes \mathcal{E}_\mathbb{R}(-1,-1) \\ \rho \in \mathcal{E}_\mathbb{R}(-1,-1). \end{cases}$$

The tractor connection becomes

$$(124) \qquad \nabla_\alpha^\mathcal{V} \begin{pmatrix} \Phi^{\beta\gamma} \\ Y^\beta \\ \rho \end{pmatrix} = \begin{pmatrix} \nabla_\alpha \Phi^{\beta\gamma} + \delta_\alpha{}^{[\beta}Y^{\gamma]} + J_\alpha{}^{[\beta}J_\epsilon{}^{\gamma]}Y^\epsilon \\ \nabla_\alpha Y^\beta + \rho J_\alpha{}^\beta + \mathsf{P}_{\alpha\gamma}\Phi^{\beta\gamma} \\ \nabla_\alpha \rho + \mathsf{P}_{\alpha\beta}J_\gamma{}^\beta Y^\gamma \end{pmatrix}.$$

The formulae (117) and (118) are again projections of (124):

$$\eta^{\bar{b}c} = i\overline{\Pi}_\beta^{\bar{b}}\Pi_\gamma^c \Phi^{\beta\gamma} \qquad X^b = -i\Pi_\beta^b Y^\beta \qquad Y^{\bar{b}} = i\overline{\Pi}_\beta^{\bar{b}}Y^\beta.$$

4.4. The metrisability equation and mobility

Let $(M, J, [\nabla])$ be an almost c-projective manifold. By Proposition 4.5, solutions to the metrisability problem on M, i.e. compatible $(2,1)$-symplectic metrics up to sign, correspond bijectively to nondegenerate solutions η of the equation $D^\mathcal{V}\eta = 0$. We refer to this equation as the *metrisability equation* on $(M, J, [\nabla])$. It may be written explicitly in several ways.

First, viewing $\mathcal{V}$ as the real part of $\mathcal{V}_\mathbb{C}$, $\eta^{\bar{b}c}$ satisfies, by (119), the conjugate equations:

$$(125) \qquad \nabla_a \eta^{\bar{b}c} + \delta_a{}^c X^{\bar{b}} = 0 \qquad \text{and} \qquad \nabla_{\bar{a}} \eta^{\bar{b}c} + \delta_{\bar{a}}{}^{\bar{b}} X^c = 0$$

for some (and hence any) connection $\nabla \in [\nabla]$ and some section X^a of $T^{1,0}M \otimes \mathcal{E}_\mathbb{R}(-1,-1)$ with conjugate $X^{\bar{a}}$. In the alternative realisation (123) of $\mathcal{V}$, the metrisability equation for J-invariant sections $\eta^{\alpha\beta}$ of $S^2 TM \otimes \mathcal{E}_\mathbb{R}(-1,-1)$ is

$$(126) \qquad \nabla_\alpha \eta^{\beta\gamma} + \delta_\alpha{}^{(\beta}X^{\gamma)} + J_\alpha{}^{(\beta}J_\epsilon{}^{\gamma)}X^\epsilon = 0$$

for some section X^α of $TM \otimes \mathcal{E}_\mathbb{R}(-1,-1)$. Similarly, using the realisation (124) of $\mathcal{V}$, the metrisability equation for J-invariant sections $\Phi^{\alpha\beta}$ of $\wedge^2 TM \otimes \mathcal{E}_\mathbb{R}(-1,-1)$ is

$$(127) \qquad \nabla_\alpha \Phi^{\beta\gamma} + \delta_\alpha{}^{[\beta}Y^{\gamma]} + J_\alpha{}^{[\beta}J_\epsilon{}^{\gamma]}Y^\epsilon = 0$$

for some section Y^α of $TM \otimes \mathcal{E}_\mathbb{R}(-1,-1)$.

DEFINITION 4.5. The (*degree of*) *mobility* of an almost c-projective manifold, is the dimension of the space

$$\mathfrak{m}_c[\nabla] := \ker D^\mathcal{V} = \left\{ \eta^{\alpha\beta} \;\middle|\; \begin{array}{l} J_\gamma{}^\alpha J_\epsilon{}^\beta \eta^{\gamma\epsilon} = \eta^{\alpha\beta} \\ \nabla_\alpha \eta^{\beta\gamma} + \delta_\alpha{}^{(\beta}X^{\gamma)} + J_\alpha{}^{(\beta}J_\epsilon{}^{\gamma)}X^\epsilon = 0 \text{ for some } X^\alpha \end{array} \right\}$$

of solutions to the metrisability equation.

In the sequel, the notion of mobility will only be of interest to us when the metrisability equation has a nondegenerate solution. Then $(M, J, [\nabla])$ has mobility ≥ 1, and the mobility is the dimension of the space of compatible $(2,1)$-symplectic metrics. For any $(2,1)$-symplectic Hermitian metric g on a complex manifold (M, J), the mobility of the c-projective class $[\nabla]$ of its canonical connection ∇ is ≥ 1, and will be called the *mobility of* g. If such a metric g has mobility one, i.e. the constant multiples of g are the only metrics compatible with c-projective class $[\nabla]$, then most natural questions about the geometry of the c-projective manifold $(M, J, [\nabla])$ can be reformulated as questions about the Hermitian manifold (M, J, g). For example the c-projective vector fields of $(M, J, [\nabla])$ are Killing or homothetic vector fields for g. Hence, roughly speaking, there is essentially no difference between the geometry of the Hermitian manifold (M, J, g) and the geometry of the c-projective manifold $(M, J, [\nabla])$.

We shall therefore typically assume in the sequel that (M, J, g), or rather, its c-projective class $(M, J, [\nabla])$, has mobility ≥ 2, and hence admits compatible metrics $\tilde{g}$ that are not proportional to g; we then say g and $\tilde{g}$ are *c-projectively equivalent*. Although all metrics in a given c-projective class are on the same footing, it is often convenient to fix a *background metric* g, corresponding to a nondegenerate solution η of (125). Then any section $\tilde{\eta}$ of $T^{0,1}M \otimes T^{1,0}M(-1,-1)$ may be written

$$\tilde{\eta}^{\bar{a}c} = \eta^{\bar{a}b} A_b{}^c$$

for uniquely determined $A_b{}^c$—explicitly, we have:

$$A^{\bar{a}b} = (\det \eta)\tilde{\eta}^{\bar{a}b} \quad \text{and} \quad A_a{}^b = g_{a\bar{c}}A^{\bar{c}b}.$$

Since η and $\tilde{\eta}$ are real, $A_b{}^c$ is g-Hermitian (i.e. the isomorphism $T^{0,1}M \to \Omega^{1,0}$ induced by g intertwines the transpose of $A_a{}^b$ with its conjugate):

$$\overline{A_a{}^b} = A^{\bar{b}}{}_{\bar{a}} := g^{\bar{b}d} A_d{}^c g_{c\bar{a}}.$$

Using the canonical connection ∇ of g, the metrisability equation (125) for $\tilde{\eta}$ may be rewritten as an equation for $A_a{}^b$, which we call the *mobility equation*:

(128) $\quad \nabla_a A_b{}^c = -\delta_a{}^c \Lambda_b, \quad \text{or} \quad \text{(equivalently)} \quad \nabla_{\bar{c}} A_a{}^b = -g_{a\bar{c}} \Lambda^b,$

where $\Lambda^b = \Pi^b_\beta \Lambda^\beta$ with Λ^β real, and $\Lambda_b = \Pi^\beta_b \Lambda_\beta = \Pi^\beta_b g_{\beta\alpha}\Lambda^\alpha = g_{b\bar{a}}\Lambda^{\bar{a}}$ with $\Lambda^{\bar{a}} = \Pi^{\bar{a}}_\alpha \Lambda^\alpha$. Taking a trace gives $\Lambda_c = \nabla_c \lambda$ and $\Lambda_{\bar{c}} = \nabla_{\bar{c}}\lambda$, with $\lambda = -A_a{}^a = -A_{\bar{a}}{}^{\bar{a}}$ real. The metric g itself corresponds to the solution $A_a{}^b = \delta_a{}^b$ of (128), with $\Lambda_c = 0$.

Since the background metric g trivialises the bundles $\mathcal{E}(\ell,\ell)$ by ∇-parallel sections $\tau_g^\ell = (\det \eta)^\ell$, we often assume these bundles are trivial. We may also raise and lower indices using g to obtain further equivalent forms of the mobility equations:

(129) $\quad \nabla_a A^{\bar{b}c} = -\delta_a{}^c \Lambda^{\bar{b}} \quad \text{or} \quad \nabla_{\bar{a}} A^{\bar{b}c} = -\delta_{\bar{a}}{}^{\bar{b}} \Lambda^c,$

(130) $\quad \nabla_a A_{b\bar{c}} = -g_{a\bar{c}} \Lambda_b \quad \text{or} \quad \nabla_{\bar{a}} A_{b\bar{c}} = -g_{b\bar{a}} \Lambda_{\bar{c}}.$

Like the metrisability equation, the mobility equation can be rewritten in explicitly real terms. If we let $\tilde{\eta}^{\alpha\gamma} = \eta^{\alpha\beta} A_\beta{}^\gamma$ and raise indices using g, then the metrisability equation (126) maybe rewritten as a mobility equation for the unweighted tensor $A^{\alpha\beta} \in \Gamma(S^2_J(TM))$:

(131) $\quad \nabla_\alpha A^{\beta\gamma} = -\delta_\alpha{}^{(\beta} \Lambda^{\gamma)} - J_\alpha{}^{(\beta} J_\delta{}^{\gamma)} \Lambda^\delta.$

We thus have that

(132) $\quad \Lambda_\alpha = \nabla_\alpha \lambda \quad \text{where} \quad \lambda = -\tfrac{1}{2} A_\beta{}^\beta.$

Tracing back through the identifications, note that

(133) $\quad A^{\alpha\beta} = \left(\frac{\mathrm{vol}(\tilde{g})}{\mathrm{vol}(g)}\right)^{1/(n+1)} \tilde{g}^{\alpha\beta},$

where $\tilde{g}^{\alpha\beta} = (\det \tilde{\eta})\tilde{\eta}^{\alpha\beta}$ is the inverse metric induced by $\tilde{\eta}^{\alpha\beta}$.

We may, of course, also lower indices to obtain:

(134) $$\nabla_\alpha A_{\beta\gamma} = -g_{\alpha(\beta}\Lambda_{\gamma)} + \Omega_{\alpha(\beta}J_{\gamma)}{}^\delta \Lambda_\delta.$$

This is the form of the mobility equation used in [**41,95**] and [**44**, Equation (3)] to study c-projectively equivalent Kähler metrics. This is a special case of Proposition 4.5, in which we suppose that there is a (pseudo-)Kähler metric in our c-projective class and we ask about other (pseudo-)Kähler metrics in the same c-projective class.

Finally, we may rewrite (127) as a mobility equation with respect to a background (2, 1)-symplectic metric g with fundamental 2-form Ω and canonical connection ∇. Trivialising $\mathcal{E}(1,1)$ and lowering indices using g, we obtain

$$\nabla_\alpha \Phi_{\beta\gamma} + g_{\alpha[\beta}Y_{\gamma]} - \Omega_{\alpha[\beta}J_{\gamma]}{}^\delta Y_\delta = 0$$

for a 2-form $\Phi_{\alpha\beta}$. In the integrable case (i.e. when g is (pseudo-)Kähler) this is the equation for *Hamiltonian 2-forms* in the terminology of [**2**]. We extend this terminology to the (2, 1)-symplectic setting and refer to its c-projectively invariant version (127) as the equation for *Hamiltonian 2-vectors* $\Phi^{\alpha\beta}$ on an almost c-projective manifold.

REMARK 4.3. If g is a Kähler metric, then applying the contracted differential Bianchi identity $g^{e\bar{b}}\nabla_{[e}R_{a]\bar{b}c\bar{d}} = 0$ to the Bochner curvature decomposition (110), we deduce that if the Bochner curvature is coclosed, i.e. $g^{e\bar{b}}\nabla_e U_{ab c\bar{d}} = 0$, then $A_{c\bar{d}} := (n+2)\Xi_{c\bar{d}} + \Lambda g_{c\bar{d}}$ satisfies the mobility equation in the form (130). Equivalently, the corresponding J-invariant 2-form, which is a modification of the Ricci form, is a Hamiltonian 2-form. This was one of the motivations for the introduction of Hamiltonian 2-forms in [**2**], and is explored further in [**5**].

REMARK 4.4. Many concepts and results in c-projective geometry have analogues in real projective differential geometry. We recall that on a smooth manifold M of dimension $m \geq 2$, a (real) projective structure is a class $[\nabla]$ of projectively equivalent affine connections, cf. (10). It is shown in [**43**] that the operator

(135) $$\Gamma(M, S^2 TM(-2)) \ni \eta^{\beta\gamma} \mapsto (\nabla_\alpha \eta^{\beta\gamma})_\circ,$$

where $S^2 TM(-2)$ denotes the bundle of contravariant symmetric tensors of projective weight -2 and $\circ$ denotes the trace-free part, is projectively invariant (it is a first BGG operator) and that, when n is even and otherwise up to sign, nondegenerate solutions are in bijection with compatible (pseudo-)Riemannian metrics, i.e. metrics whose Levi-Civita connection is in the projective class $[\nabla]$. We define the *mobility* of $[\nabla]$, or of any compatible (pseudo-)Riemannian metric, to be the dimension of this space

$$\mathfrak{m}[\nabla] := \{\eta^{\beta\gamma} \in \Gamma(M, S^2 TM) \mid \nabla_\alpha \eta^{\beta\gamma} = \delta_\alpha{}^\beta \mu^\gamma + \delta_a{}^\gamma \mu^\beta \text{ for some } \mu^\alpha\}$$

of solutions to this projective *metrisability* or *mobility equation*, where we reserve the latter term for the case that the projective structure admits a compatible metric.

4.5. Prolongation of the metrisability equation

Suppose $(M, J, [\nabla])$ is an almost c-projective manifold and let us prolong the invariant system of differential equations on sections $\eta^{\bar{b}c}$ of $T^{0,1}M \otimes T^{1,0}M(-1,-1)$ given by (119). We have already observed that (119) implies that

(136) $$X^b = -\tfrac{1}{n}\nabla_{\bar{a}}\eta^{\bar{a}b} \qquad \text{and} \qquad Y^{\bar{b}} = -\tfrac{1}{n}\nabla_a \eta^{\bar{b}a}.$$

4.5. PROLONGATION OF THE METRISABILITY EQUATION

Moreover, we immediately deduce from (119) that

(137) $\quad (\nabla_a \nabla_b - \nabla_b \nabla_a)\eta^{\bar{c}d} + T_{ab}{}^e \nabla_e \eta^{\bar{c}d} = 2\delta_{[a}{}^d \nabla_{b]} Y^{\bar{c}} - T_{ab}{}^{\bar{c}} X^d$

(138) $\quad (\nabla_{\bar{a}} \nabla_{\bar{b}} - \nabla_{\bar{b}} \nabla_{\bar{a}})\eta^{\bar{c}d} + T_{\bar{a}\bar{b}}{}^{\bar{e}} \nabla_{\bar{e}} \eta^{\bar{c}d} = 2\delta_{[\bar{a}}{}^{\bar{c}} \nabla_{\bar{b}]} X^d - T_{\bar{a}\bar{b}}{}^d Y^{\bar{c}}.$

The left hand sides of equations (137) and (138) equal

$$R_{ab}{}^d{}_e \eta^{\bar{c}e} + R_{ab}{}^{\bar{c}}{}_{\bar{e}} \eta^{\bar{e}d} + 2\mathsf{P}_{[ab]} \eta^{\bar{c}d} - \tfrac{1}{n+1}(\nabla_{\bar{e}} T_{ab}{}^{\bar{e}})\eta^{\bar{c}d}$$

(139) $\quad = W_{ab}{}^d{}_e \eta^{\bar{c}e} + 2\delta_{[a}{}^d \mathsf{P}_{b]e} \eta^{\bar{c}e} + (\nabla_{\bar{e}} T_{ab}{}^{\bar{c}}) \eta^{\bar{e}d} - \tfrac{1}{n+1}(\nabla_{\bar{e}} T_{ab}{}^{\bar{e}})\eta^{\bar{c}d}$

$$R_{\bar{a}\bar{b}}{}^{\bar{c}}{}_{\bar{e}} \eta^{\bar{e}d} + R_{\bar{a}\bar{b}}{}^d{}_e \eta^{\bar{c}e} + 2\mathsf{P}_{[\bar{a}\bar{b}]} \eta^{\bar{c}d} - \tfrac{1}{n+1}(\nabla_e T_{\bar{a}\bar{b}}{}^e)\eta^{\bar{c}d}$$

(140) $\quad = W_{\bar{a}\bar{b}}{}^{\bar{c}}{}_{\bar{e}} \eta^{\bar{e}d} + 2\delta_{[\bar{a}}{}^{\bar{c}} \mathsf{P}_{\bar{b}]\bar{e}} \eta^{\bar{e}d} + (\nabla_e T_{\bar{a}\bar{b}}{}^d) \eta^{\bar{c}e} - \tfrac{1}{n+1}(\nabla_e T_{\bar{a}\bar{b}}{}^e)\eta^{\bar{c}d},$

where we have used Theorem 2.13 to rewrite the curvature tensors $R_{ab}{}^c{}_d$, $R_{\bar{a}\bar{b}}{}^{\bar{c}}{}_{\bar{d}}$, $R_{\bar{a}\bar{b}}{}^c{}_d$, and $R_{ab}{}^{\bar{c}}{}_{\bar{d}}$. We conclude from (137) and (139), taking a trace with respect to a and d, and from (138) and (140), taking a trace with respect to $\bar{a}$ and $\bar{c}$, that

(141) $\quad \nabla_b Y^{\bar{c}} = \mathsf{P}_{be} \eta^{\bar{c}e} + \tfrac{1}{n} U_b{}^{\bar{c}} \qquad \nabla_{\bar{b}} X^d = \mathsf{P}_{\bar{b}\bar{e}} \eta^{\bar{e}d} + \tfrac{1}{n} V_{\bar{b}}{}^d,$

where

(142) $\quad U_b{}^{\bar{c}} := \tfrac{n}{n-1} T_{ab}{}^{\bar{c}} X^a + \tfrac{n}{n-1}(\nabla_{\bar{e}} T_{ab}{}^{\bar{c}})\eta^{\bar{e}a} - \tfrac{n}{(n+1)(n-1)}(\nabla_{\bar{e}} T_{ab}{}^{\bar{e}})\eta^{\bar{c}a}$

(143) $\quad V_{\bar{b}}{}^d := \tfrac{n}{n-1} T_{\bar{a}\bar{b}}{}^d Y^{\bar{a}} + \tfrac{n}{n-1}(\nabla_e T_{\bar{a}\bar{b}}{}^d)\eta^{\bar{a}e} - \tfrac{n}{(n+1)(n-1)}(\nabla_e T_{\bar{a}\bar{b}}{}^e)\eta^{\bar{a}d},$

depend linearly on $\eta^{\bar{b}c}$ and on X^a respectively $Y^{\bar{a}}$.

REMARK 4.5. Suppose J is integrable. Then the equations (141) imply $\nabla_b Y^{\bar{c}} = \mathsf{P}_{be} \eta^{\bar{c}e}$ and $\nabla_{\bar{b}} X^d = \mathsf{P}_{\bar{b}\bar{e}} \eta^{\bar{e}d}$. Hence, in this case, the equalities between (137) and (139) and between (138) and (140) show that

(144) $\quad W_{ab}{}^d{}_e \eta^{\bar{c}e} \equiv 0 \qquad W_{\bar{a}\bar{b}}{}^{\bar{c}}{}_{\bar{e}} \eta^{\bar{e}d} \equiv 0.$

If $\eta^{\bar{a}b}$ is a nondegenerate solution of (119), then (144) implies that $W_{ab}{}^c{}_d$ and its conjugate are identically zero, which confirms again Proposition 4.4 for $2n \geq 6$.

Now consider

(145) $\quad (\nabla_a \nabla_b - \nabla_b \nabla_a)\eta^{\bar{c}d} - R_{ab}{}^d{}_e \eta^{\bar{c}e} + R_{ab}{}^{\bar{c}}{}_{\bar{e}} \eta^{\bar{e}d} + \mathsf{P}_{ab} \eta^{\bar{c}d} - \mathsf{P}_{ba} \eta^{\bar{c}d}.$

By Equation (119) and Theorem 2.13 we may rewrite (145) as

(146) $\quad -\delta_{\bar{b}}{}^{\bar{c}} \nabla_a X^d + \delta_a{}^d \nabla_{\bar{b}} Y^{\bar{c}} = W_{ab}{}^d{}_e \eta^{\bar{c}e} + W_{ab}{}^{\bar{c}}{}_{\bar{e}} \eta^{\bar{e}d} + \delta_a{}^d \mathsf{P}_{\bar{b}e} \eta^{\bar{c}e} - \delta_{\bar{b}}{}^{\bar{c}} \mathsf{P}_{ae} \eta^{\bar{e}d}.$

Taking the trace in (146) with respect to $\bar{b}$ and $\bar{c}$ shows that

(147) $\quad \nabla_a X^d = \mathsf{P}_{a\bar{e}} \eta^{\bar{e}d} - \tfrac{1}{n} \delta_a{}^d (\mathsf{P}_{\bar{b}\bar{e}} \eta^{\bar{b}e} - \nabla_{\bar{b}} Y^{\bar{b}}) - \tfrac{1}{n} W_{ab}{}^d{}_e \eta^{\bar{b}e}$

and with respect a and d that

(148) $\quad \nabla_{\bar{b}} Y^{\bar{c}} = \mathsf{P}_{\bar{b}e} \eta^{\bar{c}e} - \tfrac{1}{n} \delta_{\bar{b}}{}^{\bar{c}} (\mathsf{P}_{ae} \eta^{\bar{e}a} - \nabla_a X^d) + \tfrac{1}{n} W_{ab}{}^{\bar{c}}{}_{\bar{e}} \eta^{\bar{e}a}.$

As the contraction of (147) with respect to a and d and the contraction of (148) with respect $\bar{b}$ and $\bar{c}$ must lead to the same result, we see that

(149) $\quad \tfrac{1}{n}(\mathsf{P}_{\bar{b}e} \eta^{\bar{b}e} - \nabla_{\bar{b}} Y^{\bar{b}}) = \tfrac{1}{n}(\mathsf{P}_{ae} \eta^{\bar{e}a} - \nabla_a X^a),$

which we denote by $\rho \in \Gamma(\mathcal{E}(-1,-1))$. Inserting (136) into (149) therefore implies that

(150) $\quad \rho = \tfrac{1}{n^2}(\nabla_{\bar{a}} \nabla_b \eta^{\bar{a}b} + n \mathsf{P}_{\bar{a}b} \eta^{\bar{a}b}) = \tfrac{1}{n^2}(\nabla_b \nabla_{\bar{a}} \eta^{\bar{a}b} + n \mathsf{P}_{a\bar{b}} \eta^{\bar{b}a}).$

By Theorem 2.13 we have

$$(\nabla_a \nabla_{\bar{b}} - \nabla_{\bar{b}} \nabla_a) X^c = R_{a\bar{b}}{}^c{}_d X^d + \mathsf{P}_{a\bar{b}} X^c - \mathsf{P}_{\bar{b}a} X^c \qquad (151)$$
$$= W_{a\bar{b}}{}^c{}_d X^d + \delta_a{}^c \mathsf{P}_{\bar{b}d} X^d + \mathsf{P}_{a\bar{b}} X^c.$$

Inserting the second equation of (141) and (147) into the left hand side of (151) one computes that

$$(\nabla_a \nabla_{\bar{b}} - \nabla_{\bar{b}} \nabla_a) X^a = n \nabla_{\bar{b}} \rho + \mathsf{P}_{a\bar{b}} X^a - n \mathsf{P}_{\bar{b}\bar{e}} Y^{\bar{e}} + C_{a\bar{b}\bar{e}} \eta^{\bar{e}a} + Z_{\bar{b}}, \qquad (152)$$

with

$$Z_{\bar{b}} := \tfrac{n}{(n+1)(n-1)}(\nabla_a T_{\bar{e}\bar{b}}{}^a) Y^{\bar{e}} + \tfrac{1}{n-1} T_{\bar{e}\bar{b}}{}^a \mathsf{P}_{ad} \eta^{\bar{e}d} + \tfrac{1}{n(n-1)} T_{\bar{e}\bar{b}}{}^a U_a{}^{\bar{e}}$$
$$- \tfrac{1}{(n+1)(n-1)}(\nabla_a \nabla_d T_{\bar{e}\bar{b}}{}^d) \eta^{\bar{e}a} + \tfrac{1}{(n-1)}(\nabla_a \nabla_d T_{\bar{e}\bar{b}}{}^a) \eta^{\bar{e}d}, \qquad (153)$$

where we have used (119), (142) and that $W_{a\bar{b}}{}^a{}_d$ is zero. Note again that $Z_{\bar{b}}$ depends linearly on $\eta^{\bar{a}b}$, X^a and $Y^{\bar{a}}$. From (151), the expression (152) must be equal to $n \mathsf{P}_{\bar{b}d} X^d + \mathsf{P}_{a\bar{b}} X^a$, which implies that

$$\nabla_{\bar{b}} \rho = \mathsf{P}_{\bar{b}a} X^a + \mathsf{P}_{\bar{b}\bar{e}} Y^{\bar{e}} - \tfrac{1}{n} C_{a\bar{b}\bar{e}} \eta^{\bar{e}a} - \tfrac{1}{n} Z_{\bar{b}}. \qquad (154)$$

Rewriting $(\nabla_a \nabla_{\bar{b}} - \nabla_{\bar{b}} \nabla_a) Y^{\bar{c}}$ analogously shows immediately that

$$\nabla_a \rho = \mathsf{P}_{ad} X^d + \mathsf{P}_{a\bar{e}} Y^{\bar{e}} + \tfrac{1}{n} C_{a\bar{b}d} \eta^{\bar{b}d} + \tfrac{1}{n} Q_a, \qquad (155)$$

where

$$Q_a := \tfrac{n}{(n+1)(n-1)}(\nabla_{\bar{e}} T_{da}{}^{\bar{e}}) X^d + \tfrac{1}{n-1} T_{da}{}^{\bar{b}} \mathsf{P}_{\bar{b}\bar{e}} \eta^{\bar{e}d} + \tfrac{1}{n(n-1)} T_{da}{}^{\bar{b}} V_{\bar{b}}{}^d$$
$$- \tfrac{1}{(n+1)(n-1)}(\nabla_{\bar{b}} \nabla_{\bar{e}} T_{da}{}^{\bar{e}}) \eta^{\bar{b}d} + \tfrac{1}{(n-1)}(\nabla_{\bar{b}} \nabla_{\bar{e}} T_{da}{}^{\bar{b}}) \eta^{\bar{e}d}, \qquad (156)$$

depends linearly on $\eta^{\bar{a}b}$, X^a and $Y^{\bar{a}}$. In summary, we have proved the following.

THEOREM 4.6. *Suppose $(M, J, [\nabla])$ is an almost c-projective manifold. The canonical projection $\pi \colon \mathcal{V}_{\mathbb{C}} := \mathcal{T} \otimes \overline{\mathcal{T}} \to T^{0,1} M \otimes T^{1,0} M(-1,-1)$ induces a bijection between sections of $\mathcal{V}_{\mathbb{C}}$ that are parallel for the linear connection*

$$\nabla_a^{\mathcal{V}_{\mathbb{C}}} \begin{pmatrix} \eta^{\bar{b}c} \\ X^b \mid Y^{\bar{b}} \\ \rho \end{pmatrix} + \frac{1}{n} \begin{pmatrix} 0 \\ W_{ad}{}^b{}_c \eta^{\bar{d}c} \mid -U_a{}^{\bar{b}} \\ -C_{a\bar{b}c} \eta^{\bar{b}c} - Q_a \end{pmatrix} \qquad (157)$$

and

$$\nabla_{\bar{a}}^{\mathcal{V}_{\mathbb{C}}} \begin{pmatrix} \eta^{\bar{b}c} \\ X^b \mid Y^{\bar{b}} \\ \rho \end{pmatrix} + \frac{1}{n} \begin{pmatrix} 0 \\ -V_{\bar{a}}{}^b \mid W_{\bar{a}c}{}^{\bar{b}}{}_{\bar{d}} \eta^{\bar{d}c} \\ -C_{\bar{a}c\bar{b}} \eta^{\bar{b}c} - Z_a \end{pmatrix} \qquad (158)$$

and elements in the kernel of $D^{\mathcal{V}_{\mathbb{C}}}$, where $U_a{}^{\bar{b}}$, $V_{\bar{a}}{}^b$, Z_a and Q_a are defined as in (142), (143), (153) and (156). The inverse of this bijection is induced by a differential operator $L \colon T^{0,1} M \otimes T^{1,0} M(-1,-1) \to \mathcal{V}_{\mathbb{C}}$, which for a choice of connection $\nabla \in [\nabla]$ can be written as

$$L \colon \eta^{\bar{b}c} \mapsto \begin{pmatrix} \eta^{\bar{b}c} \\ -\tfrac{1}{n} \nabla_{\bar{a}} \eta^{\bar{a}b} \mid -\tfrac{1}{n} \nabla_a \eta^{\bar{b}a} \\ \tfrac{1}{n^2}(\nabla_{\bar{a}} \nabla_b \eta^{\bar{a}b} + n \mathsf{P}_{\bar{a}b} \eta^{\bar{a}b}) \end{pmatrix}.$$

If J is integrable, $W_{a\bar{b}}{}^c{}_d = H_{a\bar{b}}{}^c{}_d$ and $W_{\bar{a}b}{}^{\bar{c}}{}_{\bar{d}} = H_{\bar{a}b}{}^{\bar{c}}{}_{\bar{d}}$ (by Theorem 2.13) and $U_a{}^{\bar{b}}$, $V_{\bar{a}}{}^b$, Q_a and $Z_{\bar{a}}$ are identically zero.

4.5. PROLONGATION OF THE METRISABILITY EQUATION

Let now $\mathcal{V}$ be the real form of the vector bundle $\mathcal{V}_{\mathbb{C}}$, as defined in the previous section. Obviously, the connection in Theorem 4.6 preserves $\mathcal{V}$ and therefore Proposition 4.5 and Theorem 4.6 imply that:

COROLLARY 4.7. *Suppose $(M, J, [\nabla])$ is an almost c-projective manifold of dimension $2n \geq 4$. Then, up to sign, there exists a bijection between compatible $(2, 1)$-symplectic Hermitian metrics and sections s of $\mathcal{V}$ that satisfy*

- $\pi(s) \equiv \eta^{\bar{b}c}$ *is nondegenerate*
- s *is parallel for the connection given by* (157) *and* (158).

Note, that since s is a real section, it is covariant constant for (157) *if and only if it is covariant constant for* (158).

Suppose s is a section of $\mathcal{V}$ that is parallel for the tractor connection. Then $\pi(s) \equiv \eta^{\bar{b}c}$ is still in the kernel of $D^{\mathcal{V}}$ and hence Theorem 4.6 implies that s is also parallel for the connection given by (157) and (158), i.e. $\pi(s) \equiv \eta^{\bar{a}b}$ must satisfy $W_{a\bar{d}}{}^{b}{}_{c}\eta^{\bar{d}c} = 0$, $W_{\bar{a}c}{}^{b}{}_{\bar{d}}\eta^{\bar{d}c} = 0$, $U_a{}^b = 0$, $V_{\bar{a}}{}^b = 0$, $C_{a\bar{b}c}\eta^{\bar{b}c} + Q_a = 0$ and $C_{\bar{a}b\bar{c}}\eta^{\bar{c}b} + Z_{\bar{a}} = 0$. The following proposition gives a geometric interpretation of parallel sections of the tractor connection and hence of so-called *normal* solutions of the first BGG operator $D^{\mathcal{V}}$ in the terminology of [**33**].

PROPOSITION 4.8. *Suppose $(M, J, [\nabla])$ is an almost c-projective manifold of dimension $2n \geq 4$. Then, if n is even, there is a bijection between sections s of $\mathcal{V}$ such that*

- $\pi(s) \equiv \eta^{\bar{b}c}$ *is nondegenerate*
- s *is parallel for the tractor connection $\nabla^{\mathcal{V}}$ on $\mathcal{V}$*

and compatible $(2,1)$-symplectic metrics g satisfying the generalised Einstein condition:

$$(159) \qquad \mathrm{Ric}_{ab} = 0 \quad \text{and} \quad \mathrm{Ric}_{a\bar{b}} = k g_{a\bar{b}} \text{ for some constant } k \in \mathbb{R},$$

where Ric is the Ricci tensor of the canonical connection of g. If J is integrable, then (159) *simply characterises (pseudo-)Kähler–Einstein metrics. If n is odd, the same conclusions are valid up to sign.*

PROOF. Suppose $s \in \Gamma(\mathcal{V})$ is parallel for the tractor connection $\nabla^{\mathcal{V}}$ and that $\pi(s) \equiv \eta^{\bar{b}c} \in \ker D^{\mathcal{V}}$ is nondegenerate. Then Proposition 4.5 implies that the inverse of $g^{\bar{a}b} \equiv \eta^{\bar{a}b} \det \eta$ is a compatible $(2,1)$-symplectic Hermitian metric. Now let $\nabla \in [\nabla]$ be the canonical connection of $g_{a\bar{b}}$. With respect to the splitting of $\mathcal{V}$ determined by ∇ the section s corresponds to the section

$$(160) \qquad \begin{pmatrix} \eta^{\bar{b}c} \\ X^b \mid X^{\bar{b}} \\ \rho \end{pmatrix} = \begin{pmatrix} \eta^{\bar{b}c} \\ 0 \mid 0 \\ \frac{1}{n}\mathsf{P}_{a\bar{b}}\eta^{\bar{b}a} \end{pmatrix}.$$

From $\nabla^{\mathcal{V}} s = 0$ it therefore follows on the one hand that $\mathsf{P}_{ac}\eta^{\bar{b}c} = 0$, which implies $\mathsf{P}_{ac} = 0$ by the nondegeneracy of $\eta^{\bar{b}c}$. Since $\mathrm{Ric}_{ab} = (n-1)\mathsf{P}_{ab} + 2\mathsf{P}_{[ab]}$, we see that the first condition of (159) holds for g. On the other hand, we deduce from $\nabla^{\mathcal{V}} s = 0$ that $\mathsf{P}_{a\bar{c}}\eta^{\bar{c}b} = \rho \delta_a{}^b$ and $\nabla_a \rho = \nabla_{\bar{a}} \rho = 0$. Since $\mathsf{P}_{a\bar{c}} = \frac{1}{n+1}\mathrm{Ric}_{a\bar{b}}$, we conclude that

$$\mathrm{Ric}_{a\bar{c}} g^{\bar{c}b} = (n+1)\rho(\det \eta)\delta_a{}^b.$$

Hence $g_{a\bar{b}}$ satisfies also the second condition of (159).

68   4. METRISABILITY OF ALMOST C-PROJECTIVE STRUCTURES

Conversely, suppose $g_{a\bar{b}}$ is a compatible $(2,1)$-symplectic Hermitian metric satisfying (159). Let us write $s \in \Gamma(\mathcal{V})$ for the corresponding parallel section of the prolongation connection given by (157) and (158). With respect to the splitting of $\mathcal{V}$ determined by the canonical connection ∇ of $g_{a\bar{b}}$, the section s is again given by (160), where $\eta^{\bar{a}b}\det(\eta) = g^{\bar{a}b}$. By assumption we have

(161) $$\mathrm{Ric}_{ab} = 0,$$

which is equivalent to $\mathsf{P}_{ab} = 0$, and also that

(162) $$\mathsf{P}_{a\bar{b}} = \tfrac{1}{n+1}\mathrm{Ric}_{a\bar{b}} = \tfrac{k}{n+1}g_{a\bar{b}}$$

for some constant k. Moreover, (161) yields

(163) $$0 = \mathrm{Ric}_{ab} = -g^{\bar{c}d}\nabla_{\bar{c}}T_{da}{}^{\bar{f}}g_{b\bar{f}} \qquad 0 = \mathrm{Ric}_{[ab]} = \tfrac{1}{2}\nabla_{\bar{c}}T_{ab}{}^{\bar{c}}$$

which shows immediately that (with respect to ∇) in (157) and (158) we have

$$U_a{}^{\bar{b}} = 0 \qquad V_{\bar{a}}{}^{b} = \overline{U_a{}^{\bar{b}}} = 0 \qquad Q_a = 0 \qquad Z_{\bar{a}} = \overline{Q_a} = 0.$$

Hence, to prove that s is parallel for $\nabla^{\mathcal{V}}$ it remains to show that $W_{a\bar{d}}{}^{b}{}_{c}g^{\bar{d}c}$ and $C_{a\bar{b}c}g^{\bar{b}c}$ (or equivalently their conjugates) are identically zero. From Theorem 2.13 and (162) we obtain

(164) $$W_{a\bar{d}}{}^{b}{}_{c}g^{\bar{d}c} = R_{a\bar{d}}{}^{b}{}_{c}g^{\bar{d}c} - \delta_a{}^{b}\mathsf{P}_{\bar{d}c}g^{\bar{d}c} - \mathsf{P}_{\bar{d}a}g^{\bar{d}b} = R_{a\bar{d}}{}^{b}{}_{c}g^{\bar{d}c} - k\delta_a{}^{b}.$$

Therefore, if we lower the b index in (164) with the metric, we obtain

$$W_{a\bar{d}bc}g^{\bar{d}c} = R_{a\bar{d}bc}g^{\bar{d}c} - kg_{a\bar{b}}.$$

Since ∇ preserves g, the tensors $R_{a\bar{d}bc} = -R_{\bar{d}a bc}$ and $-R_{a\bar{d}c\bar{b}}$ coincide. Hence, $R_{a\bar{d}bc}g^{\bar{d}c} = R_{\bar{d}ac\bar{b}}g^{\bar{d}c} = \mathrm{Ric}_{a\bar{b}} = kg_{a\bar{b}}$, which shows that (164) vanishes identically. From (161) and (162) it follows immediately that $C_{a\bar{b}c} = \nabla_a\mathsf{P}_{\bar{b}c} - \nabla_{\bar{b}}\mathsf{P}_{ac}$ vanishes identically, which completes the proof. $\square$

REMARK 4.6. As observed in Chapter 4.3, $\mathcal{V}_{\mathbb{C}} = \mathcal{T} \otimes \bar{\mathcal{T}}$, and sections of $\mathcal{V}$ may be viewed as Hermitian forms on $\mathcal{T}^*$. This has an interpretation in terms of the construction of the complex affine cone $\pi_{\mathcal{C}}\colon \mathcal{C} \to M$ described in Chapter 3.2: by Lemma 3.4, a Hermitian form on $\mathcal{T}^*$ pulls back to a Hermitian form on $T^*\mathcal{C}$. If this form is nondegenerate, its inverse defines a Hermitian metric on $\mathcal{C}$. Further, if the section of $\mathcal{V}$ is parallel with respect to a connection on $\mathcal{V}$ induced by a connection on $\mathcal{T}$, then the latter connection induces a metric connection on $\mathcal{C}$.

In particular, if we have a compatible metric satisfying the generalised Einstein condition of Proposition 4.8, then it generically induces a metric on $\mathcal{C}$ which is parallel for the connection $\nabla^{\mathcal{C}}$ induced by the tractor connection on $\mathcal{T}$.

4.6. The c-projective Hessian

Let us consider the dual $\mathcal{W}$ of the tractor bundle $\mathcal{V}$ of an almost c-projective manifold. Its complexification is given by $\mathcal{W}_{\mathbb{C}} = \mathcal{T}^* \otimes \bar{\mathcal{T}}^*$, which admits a filtration

$$\mathcal{W}_{\mathbb{C}} = \mathcal{W}_{\mathbb{C}}^{-1} \supset \mathcal{W}_{\mathbb{C}}^{0} \supset \mathcal{W}_{\mathbb{C}}^{1},$$

such that for any connection $\nabla \in [\nabla]$ we can write an element of $\mathcal{W}_{\mathbb{C}}$ as

$$\begin{pmatrix} \sigma \\ \mu_b \mid \lambda_{\bar{b}} \\ \zeta_{b\bar{c}} \end{pmatrix}, \text{ where } \begin{cases} \sigma \in \mathcal{E}(1,1), \\ \mu_b \in \wedge^{1,0}M(1,1), \quad \lambda_{\bar{b}} \in \wedge^{0,1}M(1,1), \\ \zeta_{b\bar{c}} \in \wedge^{1,1}M(1,1), \end{cases}$$

and the tractor connection as

(165) $$\nabla_a^{\mathcal{W}_{\mathbb{C}}} \begin{pmatrix} \sigma \\ \mu_b \mid \lambda_{\bar{b}} \\ \zeta_{b\bar{c}} \end{pmatrix} = \begin{pmatrix} \nabla_a \sigma - \mu_a \\ \nabla_a \mu_b + \mathsf{P}_{ab}\sigma \mid \nabla_a \lambda_{\bar{b}} + \mathsf{P}_{a\bar{b}}\sigma - \zeta_{a\bar{b}} \\ \nabla_a \zeta_{b\bar{c}} + \mathsf{P}_{a\bar{c}}\mu_b + \mathsf{P}_{ab}\lambda_{\bar{c}} \end{pmatrix}$$

and

(166) $$\nabla_{\bar{a}}^{\mathcal{W}_{\mathbb{C}}} \begin{pmatrix} \sigma \\ \mu_b \mid \lambda_{\bar{b}} \\ \zeta_{b\bar{c}} \end{pmatrix} = \begin{pmatrix} \nabla_{\bar{a}}\sigma - \lambda_{\bar{a}} \\ \nabla_{\bar{a}}\mu_b + \mathsf{P}_{\bar{a}b}\sigma - \zeta_{b\bar{a}} \mid \nabla_{\bar{a}}\lambda_{\bar{b}} + \mathsf{P}_{\bar{a}\bar{b}}\sigma \\ \nabla_{\bar{a}}\zeta_{b\bar{c}} + \mathsf{P}_{\bar{a}\bar{c}}\mu_b + \mathsf{P}_{\bar{a}b}\lambda_{\bar{c}} \end{pmatrix}.$$

The first BGG operator associated to $\mathcal{W}_{\mathbb{C}}$ or $\mathcal{W}$ is a c-projectively invariant operator of order two, which we call the *c-projective Hessian*. It can be written as

(167) $$D^{\mathcal{W}} \colon \mathcal{E}(1,1) \to S^2 \wedge^{1,0} M(1,1) \oplus S^2 \wedge^{0,1} M(1,1)$$
$$D^{\mathcal{W}}\sigma = (\nabla_{(a}\nabla_{b)}\sigma + \mathsf{P}_{(ab)}\sigma,\ \nabla_{(\bar{a}}\nabla_{\bar{b})}\sigma + \mathsf{P}_{(\bar{a}\bar{b})}\sigma),$$

or alternatively as

(168) $$D^{\mathcal{W}}\sigma = \nabla_{(\alpha}\nabla_{\beta)}\sigma + \mathsf{P}_{(\alpha\beta)}\sigma - J_{(\alpha}{}^{\gamma} J_{\beta)}{}^{\delta}(\nabla_{\gamma}\nabla_{\delta}\sigma + \mathsf{P}_{\gamma\delta}\sigma),$$

for any connection $\nabla \in [\nabla]$. The reader might easily verify the c-projective invariance of $D^{\mathcal{W}}$ directly using Proposition 2.5, the identities (16), and the formulae for the change of Rho tensor in Corollary 2.12. The following Proposition gives a geometric interpretation of nonvanishing real solutions $\sigma = \bar{\sigma} \in \Gamma(\mathcal{E}(1,1))$ of the invariant overdetermined system $D^{\mathcal{W}}\sigma = 0$.

PROPOSITION 4.9. *Let $(M, J, [\nabla])$ be an almost c-projective manifold and $\sigma \in \Gamma(\mathcal{E}(1,1))$ a real nowhere vanishing section. Then $D^{\mathcal{W}}\sigma = 0$ if and only if the Ricci tensor of the special connection $\nabla^{\sigma} \in [\nabla]$ associated to σ satisfies $\mathrm{Ric}_{(ab)} = 0$. In particular, if J is integrable, then $D^{\mathcal{W}}\sigma = 0$ if and only if the Ricci tensor of ∇^{σ} satisfies $\mathrm{Ric}_{ab} = 0$, i.e. the Ricci tensor is symmetric and J-invariant.*

PROOF. Let $\sigma = \bar{\sigma} \in \Gamma(\mathcal{E}(1,1))$ be nowhere vanishing. Recall that the Ricci tensor of the special connection ∇^{σ} associated to σ satisfies

$$\mathrm{Ric}_{\bar{a}b} = \mathrm{Ric}_{b\bar{a}} \qquad \mathrm{Ric}_{[ab]} = \tfrac{1}{2}\nabla^{\sigma}_{\bar{c}}T_{ab}{}^{\bar{c}}.$$

With respect to ∇^{σ} the equation $D^{\mathcal{W}_{\mathbb{C}}}\sigma = 0$ reduces to

$$\mathsf{P}_{(ab)}\sigma = 0 \qquad \mathsf{P}_{(\bar{a}\bar{b})}\sigma = 0,$$

i.e. to $\mathrm{Ric}_{(ab)} = \tfrac{1}{n-1}\mathsf{P}_{(ab)} = 0$ and $\mathrm{Ric}_{(\bar{a}\bar{b})} = \tfrac{1}{n-1}\mathsf{P}_{(\bar{a}\bar{b})} = 0$, since σ is nonvanishing. □

It follows immediately that if a c-projective manifold $(M, J, [\nabla])$ admits a compatible (pseudo-)Kähler metric g, then $\tau_g = \mathrm{vol}(g)^{-1/(n+1)} \in \Gamma(\mathcal{E}(1,1))$ satisfies $D^{\mathcal{W}}\tau_g = 0$. By Proposition 4.5, $\tau_g = \det \eta$, where η is the nondegenerate solution of the metrisability equation corresponding to g. This observation continues to hold without the nondegeneracy assumption.

PROPOSITION 4.10. *Let $(M, J, [\nabla])$ be a c-projective manifold and suppose that $\eta^{\bar{b}c} \in \Gamma(T^{0,1}M \otimes T^{1,0}M(-1,-1))$ is a real section satisfying (125). Then $\sigma \equiv \det \eta \in \Gamma(\mathcal{E}(1,1))$ is a real section in the kernel of the c-projective Hessian (which might be identically zero).*

PROOF. Let $U \subset M$ be the open subset (possibly empty), where σ is nowhere vanishing or equivalently where $\eta^{\bar b c}$ is invertible. By Proposition 4.5 the section $\eta^{\bar b c}(\det \eta) \in \Gamma(T^{0,1}M \otimes T^{1,0}M)$ defines the inverse of a compatible (pseudo-)Kähler metric on U and its Levi-Civita connection on U is ∇^σ. Since the Ricci tensor of a (pseudo-)Kähler metric is J-invariant (109), i.e. $\operatorname{Ric}_{ab} = \operatorname{Ric}_{\bar a \bar b} = 0$, we deduce from Proposition 4.9 that σ satisfies $D^{\mathcal{W}}\sigma = 0$ on U whence, by continuity, on $\overline{U}$. Since σ vanishes identically on the open set $M \setminus \overline{U}$, we obtain that $D^{\mathcal{W}}\sigma$ is identically zero on all of M. □

REMARK 4.7. For an almost c-projective manifold admitting a compatible $(2,1)$-symplectic metric g, the section $\tau_g \in \Gamma(\mathcal{E}(1,1))$ is in the kernel of the c-projective Hessian, if the Ricci tensor of the canonical connection ∇ of g satisfies

$$\operatorname{Ric}_{(ab)} = -\nabla_{\bar c} T^{\bar c}{}_{(ab)} = \tfrac{1}{4} \nabla_{\bar c} N^{\bar c}{}_{(ab)} = 0,$$

where we use g to raise and lower indices. It is well known that nearly Kähler manifolds can be characterised as $(2,1)$-symplectic manifolds such that T_{abc} is totally skew (see e.g. [63]). It then follows straightforwardly from the identities (102)–(103) that the canonical connection of a nearly Kähler manifold preserves its torsion, i.e. $\nabla T = -\tfrac{1}{4}\nabla N = 0$ (see [58, 88]), and $R_{ab\bar c d}$ vanishes identically. Hence, Proposition 4.10 extends to the nearly Kähler setting.

4.7. Prolongation of the c-projective Hessian

The c-projective Hessian will play a crucial role in the sequel. We therefore prolong the associated equation. Suppose $\sigma \in \Gamma(\mathcal{E}(1,1))$ is in the kernel of the c-projective Hessian:

(169) $\qquad \nabla_{(a}\nabla_{b)}\sigma + \mathsf{P}_{(ab)}\sigma = 0 \qquad \nabla_{(\bar a}\nabla_{\bar b)}\sigma + \mathsf{P}_{(\bar a \bar b)}\sigma = 0,$

Then we deduce from (46) that (169) is equivalent to

(170) $\qquad \nabla_a \nabla_b \sigma + \mathsf{P}_{ab}\sigma = \nabla_{[a}\nabla_{b]}\sigma + \mathsf{P}_{[ab]}\sigma = \tfrac{1}{2(n+1)}(\nabla_{\bar c} T_{ab}{}^{\bar c})\sigma - \tfrac{1}{2}T_{ab}{}^{\bar c}\nabla_{\bar c}\sigma$

(171) $\qquad \nabla_{\bar a} \nabla_{\bar b} \sigma + \mathsf{P}_{\bar a \bar b}\sigma = \nabla_{[\bar a}\nabla_{\bar b]}\sigma + \mathsf{P}_{[\bar a \bar b]}\sigma = \tfrac{1}{2(n+1)}(\nabla_c T_{\bar a \bar b}{}^c)\sigma - \tfrac{1}{2}T_{\bar a \bar b}{}^c \nabla_c \sigma,$

where we abbreviate the left-hand sides by Φ_{ab} respectively $\Psi_{\bar a \bar b}$, which depend linearly on σ and on $\lambda_{\bar a} := \nabla_{\bar a}\sigma$ respectively $\mu_a := \nabla_a \sigma$. From (45) we moreover deduce that

$$\nabla_a \lambda_{\bar b} + \mathsf{P}_{a\bar b}\sigma = \nabla_a \nabla_{\bar b}\sigma + \mathsf{P}_{a\bar b}\sigma = \nabla_{\bar b}\nabla_a \sigma + \mathsf{P}_{\bar b a}\sigma = \nabla_{\bar b}\mu_a + \mathsf{P}_{\bar b a}\sigma,$$

which we denote by $\zeta_{a\bar b} \in \wedge^{1,1}M(1,1)$. Consequently, we have

(172) $\qquad \nabla_a \nabla_{\bar c} \mu_b - \nabla_{\bar c}\nabla_a \mu_b =$
$\qquad \qquad \nabla_a \zeta_{b\bar c} - (\nabla_a \mathsf{P}_{\bar c b})\sigma - \mathsf{P}_{\bar c b}\mu_a + (\nabla_{\bar c}\mathsf{P}_{ab})\sigma + \mathsf{P}_{ab}\lambda_{\bar c} + \Sigma_{ab\bar c},$

where

(173) $\qquad \Sigma_{ab\bar c} := -\tfrac{1}{2(n+1)}((\nabla_{\bar c}\nabla_{\bar d}T_{ab}{}^{\bar d})\sigma + (\nabla_{\bar d}T_{ab}{}^{\bar d})\lambda_{\bar c})$
$\qquad \qquad + \tfrac{1}{2}((\nabla_{\bar c}T_{ab}{}^{\bar d})\lambda_{\bar d} - T_{ab}{}^{\bar d}\mathsf{P}_{\bar c \bar d} + T_{ab}{}^{\bar d}\Psi_{\bar c \bar d})$

depends linearly on σ, μ_a and $\lambda_{\bar a}$. From Proposition 2.13 and the identity (45) we obtain that the expression (172) must be also equal to

(174) $\qquad \nabla_a \nabla_{\bar c}\mu_b - \nabla_{\bar c}\nabla_a \mu_b = -W_{a\bar c}{}^d{}_b \mu_d - \mathsf{P}_{\bar c b}\mu_a - \mathsf{P}_{a\bar c}\mu_b,$

which shows that

(175) $$\nabla_a \zeta_{b\bar{c}} = -\mathsf{P}_{ab}\lambda_{\bar{c}} - \mathsf{P}_{a\bar{c}}\mu_b - W_{a\bar{c}}{}^d{}_b\mu_d + C_{a\bar{c}b}\sigma - \Sigma_{ab\bar{c}}.$$

Similarly, one shows that

(176) $$\nabla_{\bar{a}} \zeta_{b\bar{c}} = -\mathsf{P}_{\bar{a}b}\lambda_{\bar{c}} - \mathsf{P}_{\bar{a}\bar{c}}\mu_b - W_{\bar{a}b}{}^{\bar{d}}{}_{\bar{c}}\lambda_{\bar{d}} + C_{\bar{a}b\bar{c}}\sigma - \Xi_{\bar{a}b\bar{c}},$$

where

(177) $$\Xi_{\bar{a}b\bar{c}} := -\tfrac{1}{2(n+1)}((\nabla_b \nabla_d T_{\bar{a}\bar{c}}{}^d)\sigma + (\nabla_d T_{\bar{a}\bar{c}}{}^d)\mu_b)$$
$$+ \tfrac{1}{2}((\nabla_b T_{\bar{a}\bar{c}}{}^d)\mu_d - T_{\bar{a}\bar{c}}{}^d \mathsf{P}_{bd}\sigma + T_{\bar{a}\bar{c}}{}^d \Phi_{bd})$$

depends linearly on σ, μ_a and $\lambda_{\bar{a}}$. In summary, we have shown the following theorem:

THEOREM 4.11. *Suppose $(M, J, [\nabla])$ is a c-projective manifold. Then the canonical projection $\pi\colon \mathcal{W}_\mathbb{C} \to \mathcal{E}(1,1)$ induces a bijection between sections of $\mathcal{W}_\mathbb{C}$ that are parallel for the linear connection*

(178) $$\nabla_a^{\mathcal{W}_\mathbb{C}} \begin{pmatrix} \sigma \\ \mu_b \mid \lambda_{\bar{b}} \\ \zeta_{b\bar{c}} \end{pmatrix} + \begin{pmatrix} 0 \\ -\Phi_{ab} \mid 0 \\ W_{a\bar{c}}{}^d{}_b\mu_d - C_{a\bar{c}b}\sigma + \Sigma_{ab\bar{c}} \end{pmatrix}$$

(179) $$\nabla_{\bar{a}}^{\mathcal{W}_\mathbb{C}} \begin{pmatrix} \sigma \\ \mu_b \mid \lambda_{\bar{b}} \\ \zeta_{b\bar{c}} \end{pmatrix} + \begin{pmatrix} 0 \\ 0 \mid -\Psi_{\bar{a}\bar{b}} \\ W_{\bar{a}b}{}^{\bar{d}}{}_{\bar{c}}\lambda_{\bar{d}} - C_{\bar{a}b\bar{c}}\sigma + \Xi_{\bar{a}b\bar{c}} \end{pmatrix}.$$

and sections $\sigma \in \Gamma(\mathcal{E}(1,1))$ in the kernel of the c-projective Hessian, where Φ_{ab}, $\Psi_{\bar{a}\bar{b}}$, $\Sigma_{ab\bar{c}}$ and $\Xi_{\bar{a}b\bar{c}}$ are defined as in (170), (171), (173) and (177). The inverse of this bijection is induced by a linear differential operator L, which, for a choice of connection $\nabla \in [\nabla]$, can be written as

$$L\colon \mathcal{E}(1,1) \to \mathcal{W}_\mathbb{C}$$

$$L(\sigma) = \begin{pmatrix} \sigma \\ \nabla_a \sigma \mid \nabla_{\bar{a}} \sigma \\ \nabla_a \nabla_{\bar{b}} \sigma + \mathsf{P}_{a\bar{b}} \sigma \end{pmatrix}.$$

The following Proposition characterises normal solutions of $D^{\mathcal{W}}(\sigma) = 0$, i.e. real sections $\sigma = \bar{\sigma} \in \Gamma(\mathcal{E}(1,1))$ in the kernel of the c-projective Hessian that in addition satisfy:

(180) $$\Phi_{ab} = 0 \qquad\qquad \Psi_{\bar{a}\bar{b}} = 0$$

(181) $$W_{a\bar{c}}{}^d{}_b \nabla_d \sigma - C_{a\bar{c}b}\sigma + \Sigma_{ab\bar{c}} = 0 \qquad W_{\bar{a}b}{}^{\bar{d}}{}_{\bar{c}}\nabla_{\bar{d}} \sigma - C_{\bar{a}b\bar{c}}\sigma + \Xi_{\bar{a}b\bar{c}} = 0,$$

where Φ, Ψ, Σ and Ξ depend linearly on σ and $\nabla \sigma$.

PROPOSITION 4.12. *Let $(M, J, [\nabla])$ be an almost c-projective manifold and suppose that $\sigma \in \Gamma(\mathcal{E}(1,1))$ is a real nowhere vanishing section in the kernel of the c-projective Hessian. Then σ satisfies (180) if and only if the Ricci tensor $\mathrm{Ric}_{\alpha\beta}$ of the special connection $\nabla^\sigma \in [\nabla]$ corresponding to σ satisfies*

$$\mathrm{Ric}_{ab} = 0 \qquad \text{and} \qquad \nabla_a^\sigma \mathrm{Ric}_{b\bar{c}} = 0 = \nabla_{\bar{a}}^\sigma \mathrm{Ric}_{b\bar{c}}.$$

If the Ricci tensor $\mathrm{Ric}_{b\bar{c}} = \overline{\mathrm{Ric}_{\bar{c}b}}$ is, in addition, nondegenerate, then it defines a $(2,1)$-symplectic Hermitian metric satisfying the generalised Einstein condition (159) with canonical connection ∇^σ.

PROOF. Let $\sigma = \bar\sigma \in \Gamma(\mathcal{E}(1,1))$ be a real nowhere vanishing section in the kernel of (167). With respect to the special connection $\nabla^\sigma \in [\nabla]$ corresponding to σ, the equations (180) reduce to
$$0 = \tfrac{1}{2(n+1)}\nabla^\sigma_{\bar c} T_{ab}{}^{\bar c} = \tfrac{1}{n+1}\mathrm{Ric}_{[ab]} = \mathsf{P}_{[ab]}$$
$$0 = \tfrac{1}{2(n+1)}\nabla^\sigma_{c} T_{\bar a b}{}^{c} = \tfrac{1}{n+1}\mathrm{Ric}_{[\bar a b]} = \mathsf{P}_{[\bar a b]},$$
which, since σ is in the kernel of the c-projective Hessian, is equivalent to $\mathrm{Ric}_{ab} = 0 = \mathrm{Ric}_{\bar a \bar b}$. If these equations are satisfied, is follows immediately that also $\Sigma_{ab\bar c}$ and $\Xi_{\bar a b \bar c}$ are identically zero (with respect to ∇^σ) and that the equations (181) reduces to
$$C_{a\bar c b}\sigma = (\nabla^\sigma_a \mathsf{P}_{\bar c b})\sigma = (\nabla^\sigma_a \mathsf{P}_{b\bar c})\sigma = \tfrac{1}{n+1}(\nabla^\sigma_a \mathrm{Ric}_{b\bar c})\sigma = 0$$
$$C_{\bar a b \bar c}\sigma = (\nabla^\sigma_{\bar a}\mathsf{P}_{b\bar c})\sigma = \tfrac{1}{n+1}(\nabla^\sigma_{\bar a}\mathrm{Ric}_{b\bar c})\sigma = 0$$
which proves the claim, since σ is nowhere vanishing. $\square$

CHAPTER 5

Metrisability, conserved quantities and integrability

In this section we investigate the implications of mobility ≥ 2 for the geodesic flow of a (pseudo-)Kähler manifold (M, J, g): we show that any metric $\tilde{g}$ c-projectively equivalent, but not homothetic, to g gives rise to families of commuting linear and quadratic integrals for the geodesic flow of g, and characterise when this implies integrability of the flow.

5.1. Conserved quantities for the geodesic flow

For any smooth manifold M, the total space of its cotangent bundle $p\colon T^*M \to M$ has a canonical exact symplectic structure $d\Theta$, where $\Theta\colon TT^*M \to \mathbb{R}$ is the *tautological 1-form* defined by $\Theta_\alpha(X) = \alpha(Tp(X))$. The Poisson bracket of smooth functions on T^*M preserves the subalgebra

$$C^\infty_{\mathrm{pol}}(T^*M, \mathbb{R}) \cong \bigoplus_{k \geq 0} C^\infty(M, S^k TM)$$

of functions which are polynomial on the fibres of p, where a symmetric tensor Q of valence $(k, 0)$, i.e. a section of $S^k TM$, is identified with the function $\alpha \mapsto Q(\alpha, \ldots, \alpha)$ on T^*M (which is homogeneous of degree k on each fibre of p). The induced bracket

$$\{\cdot, \cdot\}\colon C^\infty(M, S^j TM) \times C^\infty(M, S^k TM) \to C^\infty(S^{j+k-1} TM)$$

on symmetric multivectors is sometimes called the (symmetric) *Schouten–Nijenhuis* bracket or sometimes the *Nijenhuis concomitant*. As discussed in [**105**, §4.5], it may be computed using any torsion-free connection ∇ on TM as

(182) $$\{Q, R\}^{\alpha \cdots \epsilon} = j\, Q^{\zeta(\alpha \cdots \beta} \nabla_\zeta R^{\gamma\delta \cdots \epsilon)} - k\, R^{\zeta(\delta \cdots \epsilon} \nabla_\zeta Q^{\alpha \cdots \beta\gamma)}.$$

When $j = 1$ and Q is a vector field, $\{Q, R\}$ is just the Lie derivative $\mathcal{L}_Q R$.

Now suppose g is a (pseudo-)Riemannian metric on M. Then the inverse metric $g^{\alpha\beta}$ induces a function on T^*M which is quadratic on each fibre. The flow of the corresponding Hamiltonian vector field on T^*M is the image of the geodesic flow on TM under the vector bundle isomorphism $TM \to T^*M$ defined by g.

DEFINITION 5.1. A smooth function $I\colon TM \to \mathbb{R}$ on a (pseudo-)Riemannian manifold (M, g) is called an *integral of the geodesic flow* (or an *integral*) of g, if for any affinely parametrised geodesic γ, the function $s \mapsto I(\dot{\gamma}(s))$ is constant.

The interpretation of the geodesic flow as a Hamiltonian flow on T^*M allows us to describe integrals as functions on T^*M.

PROPOSITION 5.1. $Q\colon T^*M \to \mathbb{R}$ defines an integral I of the geodesic flow of g if and only if it is a conserved quantity for $g^{\alpha\beta}$ i.e. has vanishing Poisson bracket with $g^{\alpha\beta}$.

We shall only consider integrals defined by $Q \in C^\infty_{\mathrm{pol}}(T^*M, \mathbb{R})$. Without loss of generality, we may assume such an integral is homogeneous, hence given by a symmetric tensor $Q^{\alpha\cdots\gamma} \in C^\infty(M, S^k TM)$. Using the Levi-Civita connection of g to compute the Schouten–Nijenhuis bracket, we obtain
$$\{g, Q\}^{\beta\gamma\cdots\epsilon} = 2\, g^{\alpha(\beta} \nabla_\alpha Q^{\gamma\cdots\epsilon)},$$
which is obtained from $\nabla_{(\alpha} Q_{\gamma\cdots\epsilon)}$ by raising all indices (using g) and multiplying by 2. When $k = 1$, $\{g, Q\} = 0$ if and only if Q^α is a Killing vector field. Thus we recover Clairaut's Theorem, that Killing vector fields define integrals of the geodesic flow. More generally, a *symmetric Killing tensor* of valence $(0, \ell)$ on a (pseudo-)Riemannian manifold (M, g) is a tensor $H_{\alpha\beta\cdots\delta} \in S^\ell T^*M$ that satisfies
$$\nabla_{(\alpha} H_{\beta\gamma\cdots\epsilon)} = 0, \tag{183}$$
where $\ell \geq 1$ can be any integer and ∇ is the Levi-Civita connection of g.

COROLLARY 5.2. $Q^{\alpha\cdots\gamma} \in C^\infty(M, S^k TM)$ defines an integral of the geodesic flow of g if and only if $Q_{\alpha\cdots\gamma}$ is a symmetric Killing tensor of g.

5.2. Holomorphic Killing fields

Let (M, J, g) be a (pseudo-)Kähler manifold with Levi-Civita connection ∇ and Kähler form $\Omega_{\alpha\beta} = J_\alpha{}^\gamma g_{\gamma\beta}$.

DEFINITION 5.2. A vector field X on (M, J, g) is called a *holomorphic Killing field* if it preserves the complex structure J and the metric g, i.e. $\mathcal{L}_X J = 0$ and $\mathcal{L}_X g = 0$.

In terms of the Levi-Civita connection ∇ the defining properties of a holomorphic Killing field can be rewritten as:
$$\nabla_\alpha X^\beta = -J_\alpha{}^\gamma J_\delta{}^\beta \nabla_\gamma X^\delta \quad \text{and} \quad \nabla_\alpha X_\beta + \nabla_\beta X_\alpha = 0. \tag{184}$$
It follows immediately from the definition of a holomorphic Killing field X that X also preserves the Kähler form, which means that $\mathcal{L}_X \Omega = d(i_X \Omega) = 0$ or equivalently
$$\nabla_\alpha(\Omega_{\gamma\beta} X^\gamma) - \nabla_\beta(\Omega_{\gamma\alpha} X^\gamma) = 0. \tag{185}$$
In particular, this equation is satisfied if there exists a smooth function $f\colon M \to \mathbb{R}$ such that $-i_X \Omega = df$, i.e. $\Omega_{\alpha\gamma} X^\gamma = \nabla_\alpha f$, or, using the Poisson structure $\Omega^{\alpha\beta}$,
$$X^\beta = \Omega^{\alpha\beta} \nabla_\alpha f = J_\alpha{}^\beta \nabla^\alpha f, \tag{186}$$
in which case X is said to be the *symplectic gradient* of f.

PROPOSITION 5.3. *If X and Y are symplectic gradients of functions f and h, then $\mathcal{L}_X h = 0$ if and only if $\mathcal{L}_Y f = 0$ if and only if $\Omega^{\alpha\beta}(\nabla_\alpha f)(\nabla_\beta h) = 0$ if and only if $\Omega_{\alpha\beta} X^\alpha Y^\beta = 0$. These equivalent conditions imply that X and Y commute: $[X, Y] = 0$.*

PROOF. $i_X dh = -i_X(i_Y \Omega) = i_Y(i_X \Omega) = -i_Y df$ and so the equivalences are trivial. Now $\mathcal{L}_X h = 0$ implies $0 = \mathcal{L}_X dh = -\mathcal{L}_X(i_Y \Omega) = -i_{[X,Y]} \Omega$, since $\mathcal{L}_X \Omega = 0$. Hence $[X, Y] = 0$, since Ω is nondegenerate. □

5.2. HOLOMORPHIC KILLING FIELDS

In this situation, X and Y have *isotropic* span with respect to Ω, and they are said to *Poisson commute*, since f and h have vanishing Poisson bracket.

We now return to holomorphic Killing fields.

PROPOSITION 5.4. *Let $f\colon M \to \mathbb{R}$ be a smooth function. Then the symplectic gradient $X^\beta = \Omega^{\alpha\beta}\nabla_\alpha f$ is a holomorphic Killing field if and only if the Hessian $\nabla^2 f$ is J-invariant, i.e.*

$$\nabla_a \nabla_b f = 0 = \nabla_{\bar{a}} \nabla_{\bar{b}} f. \tag{187}$$

PROOF. Since any two equations of (184) and (185) imply the third, we deduce that a vector field of the form $X^\beta = \Omega^{\alpha\beta}\nabla_\alpha f$ is a holomorphic Killing field if and only if

$$\nabla_\alpha J_\beta{}^\gamma \nabla_\gamma f + \nabla_\beta J_\alpha{}^\gamma \nabla_\gamma f = 0 \tag{188}$$

or equivalently

$$\nabla_\alpha \nabla_\beta f = J_\alpha{}^\gamma J_\beta{}^\delta \nabla_\gamma \nabla_\delta f, \tag{189}$$

which is equivalent to (187). $\square$

We call f in this case a *Killing potential* or a *Hamiltonian* for the holomorphic Killing field X. Note that a holomorphic Killing field always admits such a potential locally (and on any open subset U with $H^1(U, \mathbb{R}) = 0$).

Suppose now that g is a compatible (pseudo-)Kähler metric on a c-projective manifold $(M, J, [\nabla])$. Then we may write any real section $\sigma \in \Gamma(\mathcal{E}(1, 1))$ as $\sigma = h\tau_g$ for some function $h\colon M \to \mathbb{R}$, where τ_g is the trivialisation of $\mathcal{E}(1, 1)$ determined by g.

PROPOSITION 5.5. *Let $(M, J, [\nabla])$ be a c-projective manifold and $h \in C^\infty(M, \mathbb{R})$.*
  (1) *If τ_g is the (real) trivialisation of $\mathcal{E}(1, 1)$ corresponding to a compatible metric g, then $\sigma = h\tau_g$ is in the kernel of the c-projective Hessian $D^{\mathcal{W}}\sigma = 0$ if and only if h is a Killing potential with respect to (g, J).*
  (2) *If g and $\tilde{g}$ are compatible metrics whose corresponding trivialisations of $\mathcal{E}(1, 1)$ are related by $\tau_{\tilde{g}} = e^{-f}\tau_g$, then h is a Killing potential with respect to (g, J) if and only if $e^f h$ is a Killing potential with respect to $(\tilde{g}, J)$.*

PROOF. For the first part, compute $D^{\mathcal{W}}\sigma$ using the Levi-Civita connection ∇^g. Since τ_g is parallel, and the Ricci tensor of g is J-invariant, $D^{\mathcal{W}}\sigma = 0$ if and only if the J-invariant part of the Hessian of h is zero, and Proposition 5.4 applies. The second part follows from the first. $\square$

These observations may be generalised to (possible degenerate) solutions η of the metrisability equation. Given any J-invariant section $\eta^{\alpha\beta}$ of $S^2 TM \otimes \mathcal{E}_\mathbb{R}(-1, -1)$ and any section σ of $\mathcal{E}_\mathbb{R}(1, 1)$, we define vector fields $\Lambda(\eta, \sigma)$ and $K(\eta, \sigma)$ by

$$\Lambda^\gamma(\eta, \sigma) = \eta^{\alpha\gamma}\nabla_\alpha \sigma - \tfrac{1}{n}\sigma \nabla_\alpha \eta^{\alpha\gamma} \tag{190}$$

$$K^\beta(\eta, \sigma) = J_\gamma{}^\beta \Lambda^\gamma(\eta, \sigma) = \Phi^{\alpha\beta}\nabla_\alpha \sigma - \tfrac{1}{n}\sigma \nabla_\alpha \Phi^{\alpha\beta}, \tag{191}$$

where $\Phi^{\alpha\beta} = J_\gamma{}^\beta \eta^{\alpha\gamma}$.

PROPOSITION 5.6. *$\Lambda(\eta, \sigma)$ and $K(\eta, \sigma)$ are c-projectively invariant, and if η is a nondegenerate solution of the metrisability equation corresponding to a metric g and*

$\sigma = h \det \eta$ is in the kernel of the c-projective Hessian, then $\Lambda(\eta,\sigma)$ is holomorphic, and $K(\eta,\sigma)$ is the holomorphic Killing field of g with Killing potential h.

PROOF. For a c-projectively equivalent connection $\hat{\nabla} \in [\nabla]$, we have
$$\eta^{\alpha\gamma}\hat{\nabla}_\alpha \sigma - \tfrac{1}{n}\sigma\hat{\nabla}_\alpha \eta^{\alpha\gamma} = \eta^{\alpha\gamma}\nabla_\alpha \sigma + \eta^{\alpha\gamma}\Upsilon_\alpha \sigma - \tfrac{1}{n}\sigma\nabla_\alpha \eta^{\alpha\gamma} - \eta^{\alpha\gamma}\Upsilon_\alpha \sigma$$
and the Υ terms cancel, showing that $\Lambda(\eta,\sigma)$—and hence also $K(\eta,\sigma)$—is independent of the choice of $\nabla \in [\nabla]$.

Now if η is nondegenerate, corresponding to a compatible metric g with $\tau_g = \det \eta$, we use ∇^g to compute
$$K^\beta(\eta,\sigma) = \Phi^{\alpha\beta}\nabla^g_\alpha(h\tau_g) = \Omega^{\alpha\beta}\nabla_\alpha h,$$
which is the holomorphic Killing field associated to h. $\square$

REMARK 5.1. Suppose that $(M, J, [\nabla])$ is an almost c-projective manifold and consider the tensor product
$$\mathcal{V}_\mathbb{C} \otimes \mathcal{W}_\mathbb{C} = \mathcal{T}^* \otimes \overline{\mathcal{T}}^* \otimes \mathcal{T} \otimes \overline{\mathcal{T}}.$$
Since $\mathcal{T}^* \otimes \mathcal{T} = \mathcal{A}M \oplus \mathcal{E}(0,0)$, there is a natural projection

(192) $$\Pi \colon \mathcal{V}_\mathbb{C} \otimes \mathcal{W}_\mathbb{C} \to \begin{array}{c} \mathcal{A}M \oplus \mathcal{E}(0,0) \\ \oplus \\ \overline{\mathcal{A}M} \oplus \mathcal{E}(0,0) \end{array}$$

or equivalently a natural projection

(193) $$\Pi \colon \mathcal{V} \otimes \mathcal{W} \to \mathcal{A}M \oplus \mathcal{E}(0,0).$$

Hence, the results in [25] imply that there are two invariant bilinear differential operators

(194) $\Lambda \colon T^{0,1}M \otimes T^{1,0}M(-1,-1) \times \mathcal{E}(1,1) \to T^{1,0}M \oplus T^{0,1}M$
(195) $c \colon T^{0,1}M \otimes T^{1,0}M(-1,-1) \times \mathcal{E}(1,1) \to \mathcal{E}(0,0) \oplus \mathcal{E}(0,0),$

which are constructed as follows. Consider the two differential operators $L \colon T^{0,1}M \otimes T^{1,0}M(-1,-1)) \to \mathcal{V}_\mathbb{C}$ and $L \colon \mathcal{E}(1,1)) \to \mathcal{W}_\mathbb{C}$ from Theorem 4.6 respectively 4.11. Recall that in terms of a connection $\nabla \in [\nabla]$ they can be written as
$$L(\eta^{\bar{b}c}) = \begin{pmatrix} \eta^{\bar{b}c} \\ -\tfrac{1}{n}\nabla_{\bar{a}}\eta^{\bar{a}b} \mid -\tfrac{1}{n}\nabla_a \eta^{\bar{b}a} \\ \tfrac{1}{n}(\nabla_{\bar{a}}\nabla_b \eta^{\bar{a}b} + \mathsf{P}_{\bar{a}b}\eta^{\bar{a}b}) \end{pmatrix} \qquad L(\sigma) = \begin{pmatrix} \sigma \\ \nabla_a \sigma \mid \nabla_{\bar{a}}\sigma \\ \nabla_a \nabla_{\bar{b}}\sigma + \mathsf{P}_{a\bar{b}}\sigma \end{pmatrix}.$$

Then Λ and c are defined as the projections to $T_\mathbb{C}M = \mathcal{A}_\mathbb{C}M/\mathcal{A}^0_\mathbb{C}M$ respectively to $\mathcal{E}(0,0) \oplus \mathcal{E}(0,0)$ of $\Pi(L(\eta^{\bar{b}c}) \otimes L(\sigma))$.

In particular, for a choice of connection $\nabla \in [\nabla]$, the invariant differential operator Λ is given by

(196) $$\Lambda(\eta,\sigma) = \left(\eta^{\bar{b}c}\nabla_{\bar{b}}\sigma - \tfrac{1}{n}\sigma\nabla_{\bar{b}}\eta^{\bar{b}c} \mid \eta^{\bar{b}c}\nabla_c \sigma - \tfrac{1}{n}\sigma\nabla_c \eta^{\bar{b}c}\right).$$

Note that if $\eta^{\bar{b}c}$ and σ are real sections, the two components of (196) are conjugate to each other. In this case we may identify $\eta^{\bar{b}c}$ with a J-invariant section $\eta^{\beta\gamma}$ of $S^2 TM$.

5.3. Hermitian symmetric Killing tensors

Suppose (M, J) is an almost complex manifold and $k \geq 1$. Then we call a symmetric tensor $H_{\alpha\beta\cdots\epsilon} \in \Gamma(S^{2k}T^*M)$ *Hermitian*, if it satisfies

$$(197) \qquad J_{(\alpha}{}^{\beta} H_{\beta\gamma\cdots\epsilon)} = 0.$$

Since, by definition, $H_{\beta\gamma\cdots\epsilon} = H_{(\beta\gamma\cdots\epsilon)}$, equation (197) is equivalent to

$$(198) \qquad J_\alpha{}^\beta H_{\beta\gamma\cdots\epsilon} + J_\gamma^\beta H_{\alpha\beta\cdots\epsilon} + \cdots + J_\epsilon^\beta H_{\alpha\gamma\cdots\beta} = 0.$$

Viewing a symmetric tensor H of valence $(0, 2k)$ as an element in $S^{2k}T^*M \otimes \mathbb{C} = S^{2k}\wedge^1$ via complexification, we can use the projectors from Chapter 1 to decompose H into components according to the decomposition of $S^{2k}\wedge^1$ into irreducible vector bundles:

$$(199) \qquad S^{2k}\wedge^1 = \bigoplus_{j=0}^{2k} S^{2k-j}\wedge^{1,0} \otimes S^j \wedge^{0,1}.$$

Since this decomposition is in particular invariant under the action of J, all the components of a tensor $H_{\alpha\beta\cdots\gamma\delta} \in S^{2k}\wedge^1$ that satisfies (198) must independently satisfy (198). If $H_{ab\cdots d\bar{e}\bar{f}\cdots\bar{h}}$ is a section of $S^{2k-j}\wedge^{1,0} \otimes S^j\wedge^{0,1}$ that satisfies (198), then this equation says that $2(k-j)iH = 0$, which implies that $H \equiv 0$ unless $j = k$. We conclude that Hermitian symmetric tensors of valence $(0, 2k)$ can be viewed as real sections of the vector bundle

$$S^k \wedge^{1,0} \otimes S^k \wedge^{0,1},$$

which is the complexification of the vector bundle that consists of those elements in $S^{2k}T^*M$ that satisfy (197).

REMARK 5.2. Note that, if H is a symmetric tensor of valence $(0, 2k+1)$ satisfying (197), then the above reasoning immediately implies that $H \equiv 0$. The same arguments apply, mutatis mutandis, to symmetric tensors $Q^{\alpha\beta\cdots\epsilon}$ of valence $(2k, 0)$, and to weighted tensors of valence $(0, 2k)$ and $(2k, 0)$.

Suppose now (M, J, g) is a (pseudo-)Kähler manifold and $\ell = 2k$ is even, then we can restrict equation (183) for symmetric Killing tensors of valence $(0, 2k)$ to Hermitian tensors. If we complexify (183), we obtain the following system of differential equations on tensors $H \in \Gamma(S^k \wedge^{1,0} \otimes S^k\wedge^{0,1})$:

$$(200) \qquad \nabla_{(a} H_{bc\cdots d)\bar{e}\bar{f}\cdots\bar{h}} = 0 \quad \text{and} \quad \nabla_{(\bar{a}} H_{|bc\cdots d|\bar{e}\bar{f}\cdots\bar{g})} = 0,$$

where $|\cdots|$ means that one does not symmetrise over these indices. Real solutions of (200) thereby correspond to Hermitian symmetric Killing tensors of valence $(0, 2k)$ and obviously for real solutions the two equations of (200) are conjugates of each other. The following proposition shows that (suitably interpreted) the Killing equation for Hermitian symmetric tensors of valence $(0, 2k)$ is c-projectively invariant.

PROPOSITION 5.7. *Suppose $(M, J, [\nabla])$ is an almost c-projective manifold of dimension $2n \geq 4$. If $H_{ab\cdots d\bar{e}\bar{f}\cdots\bar{h}} \in \Gamma(S^k \wedge^{1,0} \otimes S^k \wedge^{0,1}(2k, 2k))$ satisfies*

$$(201) \qquad \nabla_{(a} H_{bc\cdots d)\bar{e}\bar{f}\cdots\bar{h}} = 0 \qquad \nabla_{(\bar{a}} H_{|bc\cdots d|\bar{e}\bar{f}\cdots\bar{h})} = 0,$$

for some connection in $\nabla \in [\nabla]$, then it does so for any other connection in the c-projective class.

PROOF. Suppose $H \in \Gamma(S^k \wedge^{1,0} \otimes S^k \wedge^{0,1}(2k,2k))$ satisfies (201) for some connection $\nabla \in [\nabla]$ and let $\hat\nabla \in [\nabla]$ be another connection in the c-projective class. Then it follows from Proposition 2.5 and Corollary 2.4 that

$$\hat\nabla_a H_{b\cdots d\bar e\cdots\bar h} = \nabla_a H_{b\cdots d\bar e\cdots\bar h} - k\Upsilon_a H_{b\cdots d\bar e\cdots\bar h} - \Upsilon_b H_{a\cdots d\bar e\cdots\bar h} - \cdots - \Upsilon_d H_{b\cdots a\bar e\cdots\bar h}$$
$$\qquad + 2k\Upsilon_a H_{b\cdots d\bar e\cdots\bar h}$$
$$= \nabla_a H_{b\cdots d\bar e\cdots\bar h} + k\Upsilon_a H_{b\cdots d\bar e\cdots\bar h} - \Upsilon_b H_{a\cdots d\bar e\cdots\bar h} - \cdots - \Upsilon_d H_{b\cdots a\bar e\cdots\bar h}.$$

Since $\nabla_{(a} H_{b\cdots d)\bar e\cdots\bar h} = 0$ by assumption and

$$\Upsilon_{(a} H_{b\cdots d)\bar e\cdots\bar h} = \tfrac{1}{k+1}(\Upsilon_a H_{b\cdots d\bar e\cdots\bar h} + \Upsilon_b H_{a\cdots d\bar e\cdots\bar h} + \cdots + \Upsilon_d H_{b\cdots a\bar e\cdots\bar h}),$$

we conclude that the symmetrisation over the unbarred indices on the right hand side is zero, which proves that the first equation of (201) is independent of the connection. Analogous reasoning shows that this is also true for the second equation of (201). □

We refer to solutions of the c-projectively invariant equation (201) as *c-projective Hermitian symmetric Killing tensors of valence* $(0,2k)$.

COROLLARY 5.8. *Suppose $(M, J, [\nabla])$ is a metrisable c-projective manifold with compatible (pseudo-)Kähler metric g. Then a real section $H \in \Gamma(S^k \wedge^{1,0} \otimes S^k \wedge^{0,1})$ is a Hermitian symmetric Killing tensor of g (i.e. a solution of (200) with respect to ∇^g) if and only if $\tau_g^{2k} H$ is a c-projective Hermitian symmetric Killing tensor. In particular, in this case, if $\tilde g$ is another compatible (pseudo-)Kähler metric, then $e^{2kf} H$ is a Hermitian symmetric Killing tensor of $\tilde g$, where f is given by $\tau_{\tilde g} = e^{-f}\tau_g$.*

The differential equation (201) gives rise to a c-projectively invariant operator, which is the first BGG operator

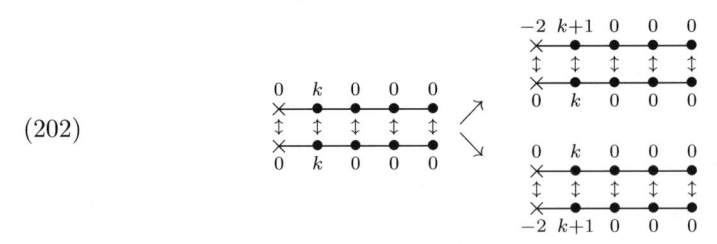

(202)

corresponding to the tractor bundle $\mathcal{W}$, where $\mathcal{W}$ is the Cartan product of k copies of $\wedge^2 \mathcal{T}^*$ and k copies of $\wedge^2 \overline{\mathcal{T}}^*$. As for the BGG operators discussed in previous sections, this implies (see [**20, 52, 89**]), that there is a linear connection on $\mathcal{W}$ whose parallel sections are in bijection to solution of (201). Hence, the dimension of the solution space is bounded by the rank of $\mathcal{W}$.

PROPOSITION 5.9. *Suppose (M, J, g) is a (pseudo-)Kähler manifold of dimension $2n \geq 4$ and let $k \geq 1$ be an integer. Then the space of Hermitian symmetric Killing tensors of valence $(0, 2k)$ of (M, J, g) has dimension at most*

$$\text{(203)} \qquad \left(\frac{(k+1)(k+2)^2\cdots(k+(n-1))^2(k+n)}{(n-1)!n!}\right)^2.$$

In the sequel, we shall be interested in the case $k = 1$, where any compatible metric g defines a c-projective Hermitian symmetric Killing tensor $H_{b\bar a} = \tau_g^2 g_{b\bar a}$ by Corollary 5.8. This has the following c-projectively invariant formulation.

PROPOSITION 5.10. *Let $(M, J, [\nabla])$ be an almost c-projective manifold and let $\eta^{\bar{a}b}$ be a real section of $T^{0,1}M \otimes T^{1,0}M(-1,-1)$. Then*

$$H_{b\bar{a}} := \tfrac{1}{(n-1)!}\,\bar{\varepsilon}_{\bar{a}\bar{c}\cdots\bar{e}}\,\varepsilon_{bd\cdots f}\,\eta^{\bar{c}d}\cdots\eta^{\bar{e}f} \tag{204}$$

is a real section of $\wedge^{1,0} \otimes \wedge^{0,1}(2,2)$ with $H_{b\bar{a}}\eta^{\bar{a}c} = \sigma\delta_b{}^c$, where $\sigma = \det\eta$. If $\eta^{\bar{a}b}$ satisfies (119) for some $X^d, Y^{\bar{c}}$ (depending on ∇) then $H_{b\bar{a}}$ is a c-projective Hermitian symmetric Killing tensor and

$$\eta^{\bar{c}d}\nabla_d H_{b\bar{a}} = Y^{\bar{e}}H_{b\bar{e}}\delta_{\bar{a}}{}^{\bar{c}} - Y^{\bar{c}}H_{b\bar{a}}, \qquad \eta^{\bar{c}d}\nabla_{\bar{c}}H_{b\bar{a}} = X^e H_{e\bar{a}}\delta_b{}^d - X^d H_{b\bar{a}}. \tag{205}$$

PROOF. The first statement is straightforward. For the rest, suppose first that $\eta^{\bar{a}b}$ is nondegenerate, hence parallel with respect to some connection $\hat{\nabla}$ in $[\nabla]$, related to ∇ by Υ with $\Upsilon_b = H_{b\bar{a}}Y^{\bar{a}}$ and $\Upsilon_{\bar{a}} = H_{b\bar{a}}X^b$. Then $H_{b\bar{a}}$ is parallel with respect to $\hat{\nabla}$, hence a Hermitian symmetric Killing tensor, and equation (205) follows by rewriting this condition in terms of ∇. At each point, these are statements about the 1-jet of H, which depends polynomially on the 1-jet of η. They hold when the 0-jet of η is invertible (at a given point, hence in a neighbourhood of that point), hence in general by continuity. □

5.4. Metrisability pencils, Killing fields and Killing tensors

Suppose we have two (real) linearly independent solutions $\eta^{\bar{a}b}$ and $\tilde{\eta}^{\bar{a}b}$ of the metrisability equation (125). Since the metrisability equation is linear, the one parameter family

$$\tilde{\eta}^{\bar{a}b}(t) := \tilde{\eta}^{\bar{a}b} - t\eta^{\bar{a}b} \tag{206}$$

also satisfies (125), and we refer to such a family as a *pencil* of solutions of the metrisability equation, or *metrisability pencil* for short.

By Proposition 4.10, the determinant

$$\tilde{\sigma}(t) := \det\tilde{\eta}(t) \tag{207}$$

of the pencil (206) lies in the kernel of the c-projective Hessian for all $t \in \mathbb{R}$ (as does $\sigma := \det\eta$). If $\tilde{\eta}(t)$ is degenerate for all t, then $\tilde{\sigma}(t)$ is identically zero. Otherwise, we may assume, at least locally:

CONDITION 5.1. η is nondegenerate, i.e. $\sigma = \det\eta$ is nonvanishing, and hence $g^{\alpha\beta} = (\det\eta)\eta^{\alpha\beta}$ is inverse to a compatible metric g.

Assuming Condition 5.1, we may write $\tilde{\eta}^{\bar{a}c} = \eta^{\bar{a}b}A_b{}^c$ as in Chapter 4.4, where the (g, J)-Hermitian metric A satisfies (128). Setting $A_a{}^b(t) := A_a{}^b - t\delta_a{}^b$, we have

$$\tilde{\eta}^{\bar{a}c}(t) = \eta^{\bar{a}b}A_b{}^c(t) \quad \text{and} \quad \tilde{\sigma}(t) = (\det\eta)(\det A(t)).$$

Thus $\tilde{\sigma}(t)$ is essentially the characteristic polynomial $\det A(t)$ of $A_a{}^b$, regarded as a complex linear endomorphism of the complex bundle $T^{1,0}M$.

REMARK 5.3. A pencil is another name for a projective line: if we make a projective change $s = (at+b)/(ct+d)$ of parameter (with $ad - bc \neq 0$) then the pencil may be rewritten, up to overall scale, as $a\tilde{\eta} + b\eta - s(c\tilde{\eta} + d\eta) = (\tilde{\eta} - t\eta)(ad - bc)/(ct + d)$. Assuming that $c\tilde{\eta} + d\eta$ is nondegenerate, the rescaled and reparameterized pencil is thus $(c\tilde{\eta}+d\eta)(\tilde{A}-s\operatorname{Id})$, where $\tilde{A} = (cA+d\operatorname{Id})^{-1}(aA+b\operatorname{Id})$.

We next set $H_{b\bar{a}} := \frac{1}{(n-1)!} \bar{\varepsilon}_{\bar{a}\bar{c}\cdots\bar{e}}\, \varepsilon_{bd\cdots f}\, \eta^{\bar{c}d}\cdots\eta^{\bar{e}f}$ as in (204) and introduce

(208) $\qquad \widetilde{H}_{b\bar{a}}(t) := \frac{1}{(n-1)!} \bar{\varepsilon}_{\bar{a}\bar{c}\cdots\bar{e}}\, \varepsilon_{bd\cdots f}\, \tilde{\eta}^{\bar{c}d}(t)\cdots\tilde{\eta}^{\bar{e}f}(t) = (\mathrm{adj}\,A(t))_b{}^c H_{c\bar{a}},$

where $\mathrm{adj}\,B$ denotes the endomorphism adjugate to B, with $B\,\mathrm{adj}\,B = (\det B)I$.

Proposition 5.10 implies that for all $t \in \mathbb{R}$, $\widetilde{H}_{b\bar{a}}(t)$ is a c-projective Hermitian symmetric Killing tensor of $(M, J, [\nabla])$. Hence for any $s \in \mathbb{R}$ with $\tilde{\eta}(s)$ nondegenerate, $\tilde{\sigma}(s)^{-2}\widetilde{H}_{b\bar{a}}(t)$ defines a family of Hermitian symmetric Killing tensors for the corresponding metric.

Similarly, Proposition 5.6 implies that if $\tilde{\eta}(s)$ is nondegenerate (for $s \in \mathbb{R}$), then for all $t \in \mathbb{R}$, $K(\tilde{\eta}(s), \tilde{\sigma}(t))$ is a holomorphic Killing field with respect to the corresponding metric (hence an *inessential* c-projective vector field). Now observe that

$$K(\tilde{\eta}(s), \tilde{\sigma}(t)) = K(\tilde{\eta}(t) + (t-s)\eta, \tilde{\sigma}(t)) = (t-s)K(\eta, \tilde{\sigma}(t)),$$

since K is bilinear and $K(\tilde{\eta}(t), \tilde{\sigma}(t)) = 0$. By continuity, the vector fields

(209) $\qquad \widetilde{K}(t) := K(\eta, \tilde{\sigma}(t)), \qquad \text{i.e. } \widetilde{K}^\beta(t) = \Omega^{\alpha\beta}\nabla_\alpha \det A(t),$

which are holomorphic Killing fields with respect to g, preserve $\tilde{\eta}(s)$ for all $s, t \in \mathbb{R}$, i.e. $\mathcal{L}_{\widetilde{K}(t)}\tilde{\eta}(s) = 0$, and hence also $\mathcal{L}_{\widetilde{K}(t)}\widetilde{H}(s) = 0 = \mathcal{L}_{\widetilde{K}(t)}\tilde{\sigma}(s)$. Thus $\widetilde{K}(t)$ preserves the Killing potential $\det A(s)$ of $\widetilde{K}(s)$ with respect to g, so Proposition 5.3 implies that $\widetilde{K}(s)$ and $\widetilde{K}(t)$ Poisson-commute. We summarise what we have proven as follows.

THEOREM 5.11. *Let $(M, J, [\nabla])$ be a c-projective manifold with metrisability solutions η and $\tilde{\eta}$ corresponding to compatible (pseudo-)Kähler metric metrics g and $\tilde{g}$ that are not homothetic. Let $\tilde{\eta}(t)$ be the corresponding metrisability pencil (206).*

(1) *The vector fields $\widetilde{K}(t) : t \in \mathbb{R}$ defined by (207)–(209) are Poisson-commuting holomorphic Killing fields with respect to g and $\tilde{g}$.*

(2) *The tensors $\widetilde{H}(t) : t \in \mathbb{R}$ defined by (207)–(208) are c-projective Hermitian symmetric Killing tensors, invariant with respect to $\widetilde{K}(s)$ for any $s \in \mathbb{R}$. In particular, by Corollary 5.8, they induce Hermitian symmetric Killing tensors of g respectively $\tilde{g}$ (by tensoring with τ_g^{-2} respectively $\tau_{\tilde{g}}^{-2}$).*

We call the vector fields $\widetilde{K}(t)$ and tensor densities $\widetilde{H}(t)$ the *canonical Killing fields* and *canonical Killing tensors* (respectively) for the pair $(g, \tilde{g})$; the former are Killing vector fields with respect to any nondegenerate metric in the family (206), and the latter give rise to symmetric Killing tensor fields for any such metric by tensoring with the corresponding trivialisation of $\mathcal{E}(-2, -2)$.

Since the canonical Killing fields $\widetilde{K}(t)$ are holomorphic with $[\widetilde{K}(t), \widetilde{K}(s)] = 0$ for all $s, t \in \mathbb{R}$, we also have $[\widetilde{K}(t), J\widetilde{K}(s)] = 0$. Since J is integrable, $J\widetilde{K}(t)$ are also holomorphic vector fields, and $[J\widetilde{K}(t), J\widetilde{K}(s)] = 0$ for all $s, t \in \mathbb{R}$.

The fact that for all $t \in \mathbb{R}$, $\widetilde{K}(t)$ is a holomorphic Killing field means equivalently (by linearity) that the coefficients of $\widetilde{K}(t)$ are holomorphic Killing fields, whose Killing potentials with respect g are the coefficients of the characteristic polynomial $\det A(t)$. Up to scale, the nontrivial coefficients of $\det A(t)$ can be written

(210) $\qquad \tilde{\sigma}_1 := A_b{}^b, \quad \tilde{\sigma}_2 := A_b{}^{[b}A_c{}^{c]}, \quad \ldots \quad \tilde{\sigma}_n := A_b{}^{[b}A_c{}^c \cdots A_d{}^{d]},$

5.4. METRISABILITY PENCILS, KILLING FIELDS AND KILLING TENSORS

which are real-valued because A is Hermitian with respect to g. Raising an index in (128), we have

$$\nabla^{\bar{c}} A_a{}^b = -g^{\bar{c}b} \Lambda_a, \quad \text{or equivalently,} \quad \nabla^c A_a{}^b = -\delta_a{}^c \Lambda^b. \tag{211}$$

Hence, applying the $(1,0)$-gradient operator $\nabla^a = \Pi_\alpha^a g^{\alpha\beta}\nabla_\beta$ to the canonical potentials in (210), we obtain (up to sign) holomorphic vector fields

$$\Lambda^a_{(1)} := \Lambda^a, \quad \Lambda^a_{(2)} := 2\Lambda^{[a} A_b{}^{b]}, \quad \ldots \quad \Lambda^a_{(n)} := n\Lambda^{[a} A_b{}^b A_c{}^c \cdots A_d{}^{d]} \tag{212}$$

whose imaginary parts are (up to scale) the coefficients of $\widetilde{K}(t)$. In particular, $\Pi_a^\alpha \Lambda^a = \tfrac{1}{2}(\Lambda^\alpha - iK^\alpha)$, where the holomorphic Killing field $K^\alpha = J_\beta{}^\alpha \Lambda^\beta$ is the leading coefficient of $\widetilde{K}(t)$. In general, the coefficients satisfy the recursive relation

$$\Lambda^a_{(k+1)} = A_b{}^a \Lambda^b_{(k)} + \tilde{\sigma}_1 \Lambda^a. \tag{213}$$

PROPOSITION 5.12. *Let g and $\tilde{g}$ be compatible metrics on $(M, J, [\nabla])$ related by a (real) solution A of (128). Then there is an integer ℓ, with $0 \le \ell \le n$, such that $\Lambda^a_{(1)}, \ldots, \Lambda^a_{(\ell)}$ are linearly independent on a dense open subset of M, and $\dim \operatorname{span} \widetilde{K}(t) \le \ell$ on M.*

PROOF. Suppose for some $p \in M$ and $1 \le k \le n$, $\Lambda^a_{(k)}$ is a linear combination of $\Lambda^a_{(1)}, \ldots, \Lambda^a_{(k-1)}$ at p. Then $A_b{}^a \Lambda^b_{(k)}$ is a linear combination of $A_b{}^a \Lambda^b_{(1)}, \ldots, A_b{}^a \Lambda^b_{(k-1)}$, hence of $\Lambda^a_{(1)}, \ldots, \Lambda^a_{(k)}$ by (213). Applying (213) once more, we see that $\Lambda^a_{(k+1)}$ is a linear combination of $\Lambda^a_{(1)}, \ldots, \Lambda^a_{(k)}$. Hence at each $p \in M$, $\dim \operatorname{span}\{\Lambda^a_{(1)}, \ldots, \Lambda^a_{(n)}\} = \dim \operatorname{span} \widetilde{K}(t)$ is the largest integer ℓ such that $\Lambda^a_{(1)}, \ldots, \Lambda^a_{(\ell)}$ are linearly independent at p. However, for any integer k, $\Lambda^a_{(1)}, \ldots, \Lambda^a_{(k)}$ are linearly dependent if and only if the holomorphic k-vector $\Lambda^{[a}_{(1)} \Lambda^b_{(2)} \cdots \Lambda^{e]}_{(k)}$ is zero. Hence the set where $\Lambda^a_{(1)}, \ldots, \Lambda^a_{(k)}$ are linearly independent is empty (for $k > \ell$) or dense (for $k \le \ell$). The result follows. □

Following [2], we refer to the integer ℓ of this Proposition as the *order* of the pencil.

PROPOSITION 5.13. *Let g and $\tilde{g}$ be compatible metrics on $(M, J, [\nabla])$ related by a (real) solution A of (128). Then the endomorphisms $\nabla \Lambda$ and A commute, i.e.*

$$A_a{}^c \nabla_c \Lambda^b - A_c{}^b \nabla_a \Lambda^c = 0. \tag{214}$$

PROOF. We first give a proof using Theorem 5.11, which implies that K^α is a holomorphic Killing field with respect to both g and $\tilde{g}$. It follows that $\mathcal{L}_K A = 0$. However, $\nabla_K A = 0$ (since $\Lambda^{\bar{c}} \nabla_{\bar{c}} A_a{}^b = -\Lambda_a \Lambda^b = \Lambda^c \nabla_c A_a{}^b$) and so $[\nabla K, A] = 0$. Equation (214) is obtained by taking $(1,0)$-parts.

We now give a more direct proof of (214), starting from the observation that

$$\nabla_a \Lambda_{\bar{b}} = -\nabla_a \nabla_{\bar{b}} A_c{}^c = -\nabla_{\bar{b}} \nabla_a \bar{A}^{\bar{c}}{}_{\bar{c}} = \nabla_{\bar{b}} \bar{\Lambda}_a,$$

(i.e. $\nabla_a \Lambda_{\bar{b}}$ is real). Now expand $(\nabla_a \nabla_{\bar{b}} - \nabla_{\bar{b}} \nabla_a) A^{\bar{c}d}$ by curvature and also by using (129) to obtain

$$R_{a\bar{b}}{}^{\bar{c}}{}_{\bar{e}} A^{\bar{e}d} + R_{a\bar{b}}{}^d{}_e A^{\bar{c}e} = \delta_a{}^d \nabla_{\bar{b}} \bar{\Lambda}^{\bar{c}} - \delta_{\bar{b}}{}^{\bar{c}} \nabla_a \Lambda^d. \tag{215}$$

Transvect with $A^{\bar{b}}{}_{\bar{c}}$ to conclude that

$$A^{\bar{b}c} R_{a\bar{b}c\bar{e}} A^{\bar{e}}{}_{\bar{d}} + A^{\bar{b}}{}_{\bar{c}} A^{\bar{c}e} R_{a\bar{b}d e} = g_{a\bar{d}} A^{\bar{b}}{}_{\bar{c}} \nabla_{\bar{b}} \bar{\Lambda}^{\bar{c}} - A^{\bar{b}}{}_{\bar{b}} \nabla_a \Lambda_{\bar{d}}$$

and hence that
$$A^{\bar{b}c}R_{a\bar{b}c\bar{e}}A^{\bar{e}}{}_{\bar{d}} - A^{\bar{c}b}R_{b\bar{d}e\bar{c}}A_a{}^e = 0 \tag{216}$$
(i.e. $A^{\bar{b}c}R_{a\bar{b}c\bar{e}}A^{\bar{e}}{}_{\bar{d}}$ is real). Now transvect (215) with $A_f{}^a\delta_{\bar{c}}{}^{\bar{b}}$ to conclude that
$$-A_f{}^a\mathrm{Ric}_{a\bar{e}}A^{\bar{e}}{}_{\bar{d}} + A_f{}^a R_{a\bar{c}d\bar{e}}A^{\bar{c}e} = A_{f\bar{d}}\nabla_{\bar{c}}\bar{\Lambda}^{\bar{c}} - nA_f{}^a\nabla_a\Lambda_{\bar{d}}$$
and hence that
$$A_a{}^b R_{b\bar{c}d\bar{e}}A^{\bar{c}e} - A^{\bar{b}}{}_{\bar{d}}R_{\bar{b}ca\bar{e}}A^{\bar{e}c} = n(A^{\bar{b}}{}_{\bar{d}}\nabla_{\bar{b}}\bar{\Lambda}_a - A_a{}^b\nabla_b\Lambda_{\bar{d}}).$$
But from (216) the left hand side vanishes whence
$$A_a{}^b\nabla_b\Lambda^d = A^{\bar{b}d}\nabla_{\bar{b}}\bar{\Lambda}_a = A^{\bar{b}d}\nabla_a\Lambda_{\bar{b}} = A_b{}^d\nabla_a\Lambda^b,$$
as required. □

Note that (214) can be used to provide an alternative proof that the holomorphic vector fields (212) commute. If V^a and W^b are two such fields, we must show that
$$0 = [V,W]^b = V^a\nabla_a W^b - W^a\nabla_a V^b. \tag{217}$$
Let us take $V^a = \Lambda^a$ and $W^b = 2\Lambda^{[b}A_c{}^{c]}$. Then (128) yields
$$\nabla_a W^b = (\nabla_a\Lambda^b)A_c{}^c + \Lambda^b\nabla_a A_c{}^c - (\nabla_a\Lambda^c)A_c{}^b - \Lambda^c\nabla_a A_c{}^b$$
$$= (\nabla_a\Lambda^b)A_c{}^c - \Lambda^b\bar{\Lambda}_a - (\nabla_a\Lambda^c)A_c{}^b + \delta_a{}^b\Lambda^c\bar{\Lambda}_c$$
whence $V^a\nabla_a W^b = \Lambda^a(\nabla_a\Lambda^b)A_c{}^c - \Lambda^a(\nabla_a\Lambda^c)A_c{}^b$ and
$$V^a\nabla_a W^b - W^a\nabla_a V^b = \Lambda^a(A_a{}^c\nabla_c\Lambda^b - A_c{}^b\nabla_a\Lambda^c).$$
Similar computations show that all the fields in (212) commute.

REMARK 5.4. It is interesting to compare Proposition 5.13 with what happens in the real projective setting. The mobility equations (128) are replaced by
$$\nabla^\alpha A_\beta{}^\gamma = -\delta_\beta{}^\alpha\Lambda^\gamma - g^{\alpha\gamma}\Lambda_\beta$$
and the development runs in parallel. These equations control the existence of another metric in the projective class other than the assumed background metric and, from the coefficients of the characteristic polynomial of $A_\alpha{}^\beta$, a solution gives rise to n canonically defined potentials for n canonically defined vector fields. These are counterparts to the fields (212) and, as such, need not be Killing. Nevertheless, they commute (cf. also [**16**, Lemma 4.5]) and to see this it is necessary to employ the alternative reasoning that we encountered near the end of the proof just given. The key observation, like (214), is that the endomorphisms $A_\alpha{}^\beta$ and $\nabla_\alpha\Lambda^\beta$ commute and its proof follows exactly the course just given.

5.5. Conserved quantities on c-projective manifolds

On a c-projective manifold $(M, J, [\nabla])$, the construction of an integral from a compatible metric and Killing tensor has a c-projectively invariant formulation: in particular, given a nondegenerate solution $\eta^{\alpha\beta}$ of the metrisability equation, and a c-projective Hermitian symmetric Killing tensor $H_{\gamma\delta}$ of valence $(0,2)$, the Hermitian $(2,0)$-tensor $\eta^{\alpha\gamma}\eta^{\beta\delta}H_{\gamma\delta}$ defines an integral of the geodesic flow of the metric corresponding to η; if $H_{\gamma\delta}$ is associated to $\eta^{\alpha\beta}$ by Proposition 5.10, then this integral is the Hamiltonian associated to the inverse metric $g^{\alpha\beta} = (\det\eta)\,\eta^{\alpha\beta}$.

DEFINITION 5.3. Let $(M, J, [\nabla])$ be a c-projective manifold, and let $\tilde{\eta}^{\alpha\beta}(t) := \tilde{\eta}^{\alpha\beta} - t\eta^{\alpha\beta}$ be a metrisability pencil satisfying Condition 5.1, so that $g^{\alpha\beta} = (\det \eta)\eta^{\alpha\beta}$ is inverse to a (nondegenerate) compatible metric g, and we may write $\tilde{\eta}^{\bar{a}c}(t) = \eta^{\bar{a}b} A_b{}^c(t)$. Then the *linear* and *quadratic integrals* $L_t, I_t \colon TM \to \mathbb{R}$ of (the geodesic flow of) g are defined by $L_t(X) := g(\widetilde{K}(t), X) = g_{\alpha\beta}\widetilde{K}^\alpha(t)X^\beta$ and $I_t(X) := \tau_g^{-2}\widetilde{H}(t)(X,X) = \tau_g^{-2}\widetilde{H}_{\alpha\beta}(t)X^\alpha X^\beta$, where $\widetilde{K}(t)$ and $\widetilde{H}(t)$ are the holomorphic Killing fields and c-projective Hermitian symmetric Killing tensors associated to g by Theorem 5.11.

PROPOSITION 5.14. *The integrals I_t, L_t of g (for all $t \in \mathbb{R}$) mutually commute under the Poisson bracket on TM induced by g.*

PROOF. Since, by Theorem 5.11, the canonical Killing fields commute and Lie preserve the canonical Killing tensors, it remains only to show that for any $s, t \in \mathbb{R}$ the quadratic integrals I_s and I_t commute. The Hermitian symmetric tensors of valence $(2,0)$ corresponding to I_t are $Q_{(t)}^{\bar{a}b} := \eta^{\bar{a}d}\eta^{\bar{c}b}H_{d\bar{c}}^{(t)}$, where we write $H_{d\bar{c}}^{(t)}$ instead of $\widetilde{H}_{d\bar{c}}(t)$. It thus suffices to show that for all $s \neq t \in \mathbb{R}$, $Q_{(s)}^{\bar{a}b}$ and $Q_{(t)}^{\bar{a}b}$ have vanishing Schouten–Nijenhuis bracket, which (in barred and unbarred indices) means

$$\left(\nabla_b Q_{(t)}^{d(\bar{a}}\right) Q_{(s)}^{\bar{c})b} - \left(\nabla_b Q_{(s)}^{d(\bar{a}}\right) Q_{(t)}^{\bar{c})b} = 0, \qquad \left(\nabla_{\bar{a}} Q_{(t)}^{\bar{c}(b}\right) Q_{(s)}^{d)\bar{a}} - \left(\nabla_{\bar{a}} Q_{(s)}^{\bar{c}(b}\right) Q_{(t)}^{d)\bar{a}} = 0.$$

We prove the first equation (the second is analogous); taking for ∇ the Levi-Civita connection of g (so $\nabla \eta = 0$), this reduces to:

$$\eta^{\bar{c}f}\left(\nabla_f H_{\bar{a}(b}^{(t)}\right) H_{d)\bar{c}}^{(s)} - \eta^{\bar{c}f}\left(\nabla_f H_{\bar{a}(b}^{(s)}\right) H_{d)\bar{c}}^{(t)} = 0.$$

The key trick is to multiply the left hand side by $s-t$ and observe that $(s-t)\eta^{\bar{c}f} = \tilde{\eta}^{\bar{c}f}(t) - \tilde{\eta}^{\bar{c}f}(s)$. Now using equation (205) for $\tilde{\eta}^{\bar{a}b}(t)$ and $H_{d\bar{c}}^{(t)}$, we obtain

$$\tilde{\eta}^{\bar{c}f}(t)\left(\nabla_f H_{\bar{a}(b}^{(t)}\right) H_{d)\bar{c}}^{(s)} = Y^{\bar{e}} \delta_{\bar{a}}{}^{\bar{c}} H_{\bar{e}(b}^{(t)} H_{d)\bar{c}}^{(s)} - Y^{\bar{c}} H_{\bar{a}(b}^{(t)} H_{d)\bar{c}}^{(s)} = Y^{\bar{c}} H_{\bar{c}(b}^{(t)} H_{d)\bar{a}}^{(s)} - Y^{\bar{c}} H_{\bar{a}(b}^{(t)} H_{d)\bar{c}}^{(s)}$$

for some $Y^{\bar{a}}$. The same reasoning applies to $\tilde{\eta}^{\bar{a}b}(s)$ and $H_{d\bar{c}}^{(s)}$ with the same vector field $Y^{\bar{a}}$ to obtain the same expression with s and t interchanged. These two expressions sum to zero and hence

$$(\tilde{\eta}^{\bar{c}f}(t) - \tilde{\eta}^{\bar{c}f}(s))\left(\left(\nabla_f H_{\bar{a}(b}^{(t)}\right) H_{d)\bar{c}}^{(s)} - \left(\nabla_f H_{\bar{a}(b}^{(s)}\right) H_{d)\bar{c}}^{(t)}\right) = -\tilde{\sigma}(s)\nabla_{(d} H_{b)\bar{a}}^{(s)} - \tilde{\sigma}(t)\nabla_{(d} H_{b)\bar{a}}^{(t)}$$

which vanishes because $\widetilde{H}(s)$ and $\widetilde{H}(t)$ are c-projective Hermitian symmetric Killing tensors. $\square$

We now discuss the question how many of the functions L_t and I_t ($t \in \mathbb{R}$) are *functionally independent* on TM, i.e. have linearly independent differentials. Since TM has dimension $4n$, and the functions L_t and I_t mutually commute (i.e. they span an Abelian subalgebra under the Poisson bracket induced by g), at most $2n$ of these functions can be functionally dependent at each point of TM. If equality holds on the fibres of TM over a dense open subset of M, the geodesic flow of g is said to be *integrable*.

Since $A_a{}^b(t) = A_a{}^b - t\delta_a{}^b$, integrability turns out to be related to the spectral theory of the field $A_a{}^b$ of endomorphisms of $T^{1,0}M$. In particular, using the trivialisation of $\mathcal{E}(1,1)$ determined by g, the determinant $\tilde{\sigma}(t) := \det \tilde{\eta}(t)$ becomes the characteristic polynomial $\chi_A(t) := \det A(t)$ of $A_a{}^b$. Since A is Hermitian, the coefficients of $\chi_A(t)$ are smooth real-valued functions on M. Any complex-valued function μ on an open subset $U \subseteq M$ has an associated *algebraic multiplicity* $m_\mu \colon U \to \mathbb{N}$, where $m_\mu(p)$ is the multiplicity of $\mu(p)$ as a root of $\chi_A(t)$ at $p \in U$, or equivalently the rank of the generalised $\mu(p)$-eigenspace of $A_a{}^b$ in $T_p^{1,0}M$; additionally, its *geometric multiplicity* $d_\mu(p)$ is the dimension of the $\mu(p)$-eigenspace of $A_a{}^b$ in $T_p^{1,0}M$, and its *index* $h_\mu(p)$ is the multiplicity of $\mu(p)$ in the minimal polynomial of $A_a{}^b$ at p.

REMARK 5.5. If $\mu \colon U \to \mathbb{C}$ is smooth with m_μ constant on U, then the restriction of $A_a{}^b - \mu \delta_a{}^b$ to the generalised μ-eigendistribution, defines, using an arbitrary local frame of this distribution, a family of nilpotent $m_\mu \times m_\mu$ matrices N. There are only finitely many conjugacy classes of such matrices, parametrised by partitions of m_μ: we can either use the *Segre characteristics*, which are the sizes of the Jordan blocks of N, or the dual partition by the *Weyr characteristics* $\dim \ker N^k - \dim \ker N^{k-1}$, $k \in \mathbb{Z}^+$. The index h_μ is the first Segre characteristic (i.e. the size of the largest Jordan block), while the geometric multiplicity is the first Weyr characteristic $\dim \ker N$ (so $\max\{d_\mu, h_\mu\} \le m_\mu$ with equality if and only if $d_\mu = 1$ or $h_\mu = 1$). The unique (hence dense) open orbit consists of nilpotent matrices of degree m_μ, whose Jordan normal forms have a single Jordan block, and in general, the orbit closures stratify the nilpotent matrices (with the partial ordering of strata corresponding to the dominance ordering of partitions). Thus N maps a sufficiently small neighbourhood of a point $p \in U$ into a unique minimal stratum, and for generic $p \in U$, this stratum is the orbit closure of $N(p)$. In other words, the type of N may be assumed constant in neighbourhood of any point in a dense open subset of U.

The general theory of families of matrices is considerably simplified here by Proposition 5.12 and the following two lemmas.

LEMMA 5.15. *Suppose U is an open subset of M and $T^{1,0}U = E \oplus F$ where E and F are smooth A-invariant subbundles over U such that the restriction of A to E has a single Jordan block with smooth eigenvalue $\mu \colon U \to \mathbb{C}$. Then the gradient of μ is a section over U of $E \oplus F^\perp \subseteq T^{1,0}U \oplus T^{0,1}U = TU \otimes \mathbb{C}$, where $F^\perp$ denotes the subspace of $T^{0,1}U$ orthogonal to F with respect to g.*

PROOF. In a neighbourhood of any point in U, we may choose a frame $Z^a(1), \ldots, Z^a(m)$ of E such that A is in Jordan normal form on E. We identify E^* with the annihilator of F in $\Omega^{1,0}$ and let $Z_a(1), \ldots, Z_a(m)$ be the dual frame (with $Z_a(i)Z^a(j) = \delta_{ij}$). Then the transpose of A is in Jordan normal form with respect to this dual frame in reverse order: for $k = 1, \ldots m$ we thus have

$$(A_a{}^b - \mu \delta_a{}^b)Z^a(k) = Z^b(k-1), \tag{218}$$
$$(A_a{}^b - \mu \delta_a{}^b)Z_b(k) = Z_b(k+1), \tag{219}$$

where $Z^b(-1) = 0 = Z_b(m+1)$. By (128), the $(0,1)$-derivative of (218) yields

$$(g_{a\bar c}\Lambda^b + \delta_a{}^b \nabla_{\bar c}\mu)Z^a(k) = (A_a{}^b - \mu\delta_a{}^b)\nabla_{\bar c}Z^a(k) - \nabla_{\bar c}Z^b(k-1), \tag{220}$$

which we may contract with $Z_b(k)$, using (219), to obtain

$$-\Lambda^b Z_b(k)g_{a\bar c}Z^a(k) + \nabla_{\bar c}\mu = Z_b(k+1)\nabla_{\bar c}Z^b(k) - Z_b(k)\nabla_{\bar c}Z^b(k-1).$$

Summing from $k = 1$ to m, the right hand side sums to zero, and hence
$$m\nabla^a \mu = \sum_{k=1}^m \Lambda^b Z_b(k) Z^a(k)$$
so the $(1,0)$-gradient of μ is a linear combination of $Z^a(1), \ldots, Z^a(m)$, hence a section of E. Since A is Hermitian, its restriction to $\overline{F}^\perp$ also has a single Jordan block, with eigenvalue $\bar\mu$, and so $\nabla^{\bar a}\mu = \overline{\nabla^a\bar\mu}$ belongs to $F^\perp$ by the same argument. □

It follows that if there is more than one Jordan block with eigenvalue μ, then μ is constant—equivalently, all nonconstant eigenvalues of A have geometric multiplicity one. In fact, a stronger result holds.

LEMMA 5.16. *Let μ be a smooth function on M and let $U \subset M$ be a nonempty open subset on which μ has constant algebraic multiplicity m. If μ is constant and M is connected, then μ has algebraic multiplicity $m_\mu \geq m$ on M. Conversely, if $m \geq 2$ then μ is locally constant on U.*

PROOF. Since A is Hermitian, $\chi_A(t)$ has real coefficients, $\bar\mu$ is an eigenvalue of A with the same algebraic multiplicity as μ, and $\nabla^{\bar a}\mu = \overline{\nabla^a\bar\mu}$. By assumption $\chi_A(t) = (t-\mu)^m q(t)$ on U, where q is smooth with $q(\mu)$ nonvanishing. Since $\chi_A(t)$ is a Killing potential for g, its gradient $\nabla^a \chi_A(t)$ is a holomorphic vector field on M for all t.

Suppose first that μ is constant: we now show by induction on k that if $m \geq k$ then $m_\mu \geq k$ on M, which is trivially true for $k = 0$. So suppose that $m \geq k+1$ and $m_\mu \geq k$, so that $p(t) = \chi_A(t)/(t-\mu)^k$ is a polynomial in t. Since $\nabla^a p(\mu)$ is holomorphic on M and vanishing on U, it vanishes on M, since M is connected. Similarly for $\overline{p(\mu)}$, so that $p(\mu)$ is locally constant on M, hence zero, since M is connected and $p(\mu)$ vanishes on U. Thus $m_\mu \geq k+1$ as required.

For the second part, the $(m-2)$nd derivative in t of $\chi^a(t)$ is also a Killing potential, which may be written
$$\nabla^a \chi_A^{(m-2)}(t) = m!(t-\mu)q(t)\nabla^a\mu + (t-\mu)^2 X^a(t)$$
for some polynomial of vector fields $X^a(t)$. Applying $\nabla^b = g^{b\bar c}\nabla_{\bar c}$ and evaluating at $t = \mu$ yields $\nabla^a\mu\nabla^b\mu = 0$, i.e. $\nabla^a\mu = 0$. Replacing μ by $\bar\mu$, we deduce that μ is locally constant on U. □

In contrast, in the analogous real projective theory of geodesically equivalent pseudo-Riemannian metrics, Jordan blocks with nonconstant eigenvalues can occur: see [15].

In order to apply the above lemmas at a point $p \in M$, we need p to be stable for A in the following sense. First, we need to suppose that the number of distinct eigenvalues of $A_a{}^b$ is constant on some neighbourhood of p. This condition on p is clearly open, and it is also dense: if the number of distinct eigenvalues is not constant near p, then there are points arbitrarily close to p where the number of distinct eigenvalues is larger; repeating this argument, there are points arbitrarily close to p where the number of distinct eigenvalues is locally maximal, hence locally constant. Now, on the dense open set where this condition holds, the eigenvalues of $A_a{}^b$ are smoothly defined, and their algebraic multiplicities are locally constant (since they are all upper semi-continuous). Now a point p in this dense open set is *stable* if in addition the Jordan type (Segre or Weyr characteristics) of each generalised eigenspace of $A_a{}^b$ is constant on a neighbourhood of p. The stable points are open and dense by Remark 5.5.

DEFINITION 5.4. We say $p \in M$ is a *regular point* for the pencil $\tilde{\eta}(t) = \eta^{\bar{a}b}(A_a{}^b - t\delta_a{}^b)$ if it is stable for $A_a{}^b$, and for each smooth eigenvalue μ on an open neighbourhood of p, either $d\mu_p \neq 0$ or μ is constant on an open neighbourhood of p.

Equivalently, the regular points are the open subset of the stable points where the rank of the span of the canonical Killing fields associated to the pencil is maximal, i.e. equal to the order ℓ. Consequently, by Proposition 5.12, the regular points form a dense open subset of M.

COROLLARY 5.17. *Let μ be a smooth eigenvalue of A over the set of stable points. Then*

$$A_a{}^b \nabla_b \mu = \mu \nabla_a \mu \quad \text{and} \quad A_{\bar{a}}{}^{\bar{b}} \nabla_{\bar{b}} \mu = \mu \nabla_{\bar{a}} \mu. \tag{221}$$

If μ is constant, its algebraic multiplicity is constant on the set of regular points.

Indeed, where μ has algebraic multiplicity $m_\mu = 1$, Lemma 5.15 implies that $\nabla_a \mu$ generates the eigenspace of μ, whereas where $m_\mu \geq 2$, Lemma 5.16 implies that μ is locally constant, and hence equations (221) are trivially satisfied. Furthermore, it implies that the algebraic multiplicities of the constant eigenvalues are upper semi-continuous on M, hence constant on the connected set of regular points.

THEOREM 5.18. *Let $(M, J, [\nabla])$ be a c-projective manifold that admits (pseudo-) Kähler metrics $g_{a\bar{b}}$ and $\tilde{g}_{a\bar{b}}$ associated to linearly independent solutions $\eta^{\bar{a}c}$ and $\tilde{\eta}^{\bar{a}c} = \eta^{\bar{a}b} A_b{}^c$ of the metrisability equation (125).*

(1) *The number of functionally independent linear integrals L_s is equal to the number of nonconstant eigenvalues of A at any regular point of M.*
(2) *The number of functionally independent quadratic integrals I_t is equal to the degree of the minimal polynomial of A at any stable point of M.*
(3) *The integrals I_t are functionally independent from the integrals L_s.*

PROOF. Integrals of the form L_s or I_t are functionally independent near $X \in T_pM$ if their derivatives are linearly independent at X. Since L_s is linear along the fibres of $TM \to M$, the restriction of dL_s to $T_X(T_pM) \cong T_pM$ is $g(\widetilde{K}(s), \cdot)$ (at p). Similarly, I_t is quadratic along fibres, and the restriction of dI_t to $T_X(T_pM) \cong T_pM$ is $\tau_g{}^{-2} \widetilde{H}(t)(X, \cdot)$ with $\tau_g = \det \eta$. Hence, for generic $X \in T_pM$, the quadratic integrals I_t are functionally independent from the linear integrals L_s, and the number of functionally independent linear, respectively quadratic, integrals is at least the dimension of the span of $\widetilde{K}(s)$ at p, respectively the dimension of the span of $\widetilde{H}(t)$ at p.

The geodesic flow preserves the integrals and therefore the property of the integrals to be functionally independent. Since any two points of M can be connected by a piecewise geodesic curve, it suffices to compute the dimensions of these spaces at a regular point of p, where the dimensions of the spans of $\widetilde{K}(s)$ and $\widetilde{H}(t)$ are maximal.

At such a point, the number of linearly independent Killing vector fields $\widetilde{K}(s)$ is the number of nonconstant eigenvalues of A, so it remains to compute the number of linearly independent Killing tensors $\widetilde{H}(t)$. For this, recall that $\widetilde{H}_{b\bar{a}}(t) =$

$(\operatorname{adj} A(t))_b{}^c H_{c\bar a}$, with $\operatorname{adj} A(t) = A(t)^{-1} \det A(t)$. Now write $A(t)$ in Jordan canonical form: on an $h \times h$ Jordan block with eigenvalue μ, $(t-\mu)^h A(t)^{-1}$ is a polynomial of degree $h-1$ in t with h linearly independent coefficients. Hence on the generalised μ-eigenspace, $(t-\mu)^{m_\mu} A(t)^{-1}$ is a polynomial in t with h_μ linearly independent coefficients, where m_μ is the geometric multiplicity of μ, and h_μ the index (the multiplicity of μ in the minimal polynomial, i.e. the size of the largest Jordan block). It follows readily that the dimension of the span of $\operatorname{adj} A(t)$ is the degree of the minimal polynomial of A. □

5.6. The local complex torus action

For a c-projective manifold M^{2n} admitting a metrisability pencil with no constant eigenvalues, Theorem 5.18 shows that any metric in the pencil is integrable, i.e. its geodesic flow admits $2n$ functionally independent integrals. Furthermore n of the independent integrals are linear, inducing Hamiltonian Killing vector fields. Hence if M is compact, it is toric (i.e. has an isometric Hamiltonian n-torus action).

When the pencil has constant eigenvalues, there are only ℓ independent linear integrals, where ℓ is the *order* of the pencil (the number of nonconstant eigenvalues), and at most n independent quadratic integrals. In this case the flows of the Hamiltonian Killing vector fields $\widetilde K(t)$ generate a foliation of M whose generic leaves are ℓ-dimensional. If M is compact, one can prove (see [**2**]) that these leaves are the orbits of an isometric Hamiltonian action of an ℓ-torus $U(1)^\ell$, and it is convenient to assume this locally. The complexified action, generated by the commuting holomorphic vector fields $\widetilde K(t)$ and $J\widetilde K(t)$, is then a local holomorphic action of $(\mathbb{C}^\times)^\ell$, and we refer to the leaves of the foliation, which are locally J-invariant submanifolds with generic dimension 2ℓ, as *complex orbits*.

LEMMA 5.19. *The complex orbits through regular points are totally geodesic and their tangent spaces are A-invariant. The c-projectively equivalent metrics g and $\tilde g$ restrict to nondegenerate c-projectively equivalent metrics (with respect to the induced complex structure) on any regular complex orbit $\mathcal{O}^c$. The metrisability pencil $\tilde\eta(t)$ restricts to a metrisability pencil of order ℓ on $\mathcal{O}^c$, using g and its restriction to trivialise $\mathcal{E}_\mathbb{R}(1,1)$, and then $\tilde g$ is a constant multiple of the metric induced by $\tilde\eta$.*

PROOF. Since the complex orbit of any regular point contains only regular points and the tangent space to the orbit is spanned by holomorphic vector fields, it suffices to prove that the $(1,0)$-tangent spaces to regular points are closed under the Levi-Civita connection ∇ of g. These tangent spaces are spanned by eigenvectors Z^a of $A_a{}^b$ with nonconstant eigenvalues μ, and by differentiating the eigenvector equation using (128), as in the proof of Lemma 5.15, we obtain

$$(A_a{}^b - \mu\,\delta_a{}^b)\nabla_c Z^a = (-\Lambda_a \delta_c{}^b + , \delta_a{}^b \nabla_c \mu) Z^a = -(\Lambda_a Z^a)\delta_c{}^b + (\nabla_c \mu) Z^b.$$

Clearly if we contract the right hand side with a $(1,0)$-vector X^c tangent to a complex orbit, we obtain another such vector. Hence $\nabla_X Z$ is a $(1,0)$-vector tangent to the complex orbit as required: the complex orbits are thus totally geodesic.

The tangent spaces to a regular complex orbit $\mathcal{O}^c$ are clearly J-invariant and A-invariant, so that g induces a Kähler metric $\mathcal{O}^c$, with a metrisability pencil spanned by the restrictions of η and $\tilde\eta$, where we use g and its restriction to trivialise $\mathcal{E}_\mathbb{R}(1,1)$. Since $\tilde\eta = \eta \circ A$, and the generalised eigenspaces of A which are not tangent to $\mathcal{O}^c$

have constant eigenvalues, the metric induced by the restriction of $\tilde\eta$ is a constant multiple of the restriction of $\tilde g$. □

Also of interest is the local $(\mathbb{R}^+)^\ell$ action whose local orbits are the leaves of the foliation generated by the vector fields $J\widetilde K(t)$, which we call *real orbits*.

LEMMA 5.20. *The real orbits through regular points are totally geodesic, and their tangent spaces are A-invariant and generated by the gradients of the nonconstant eigenvalues of A. The c-projectively equivalent metrics g and $\tilde g$ restrict to nondegenerate projectively equivalent metrics on any regular real orbit $\mathcal{O}$, and the restriction of A is a constant multiple of the $(1,1)$-tensor $\bigl(\frac{\mathrm{vol}(\tilde g|_\mathcal{O})}{\mathrm{vol}(\tilde g|_\mathcal{O})}\bigr)^{1/(\ell+1)}(\tilde g|_\mathcal{O})^{-1}g|_\mathcal{O}$.*

PROOF. At a regular point p, $X \in T_pM$ is tangent to the real orbit through p if and only if it is tangent to the complex orbit through p and orthogonal to the Killing vector fields $\widetilde K(t)$ at p. Since both properties are preserved along geodesics, the real orbits are totally geodesic with respect to g (hence also $\tilde g$).

Let $\mathcal{O}^c$ be the complex orbit through the regular real orbit $\mathcal{O}$, so that g and $\tilde g$ restrict to c-projectively equivalent Kähler metrics on $\mathcal{O}^c$. Furthermore $(\mathrm{vol}(\tilde g|_{\mathcal{O}^c}))^{1/2(\ell+1)}\tilde g^{-1}|_{\mathcal{O}^c}$ is a constant multiple of $(\mathrm{vol}(g|_{\mathcal{O}^c}))^{1/2(\ell+1)}g^{-1}\circ A|_{\mathcal{O}^c}$. The tangent spaces to $\mathcal{O}$ are generated by the vector fields $\nabla^a\mu$, for nonconstant eigenvalues μ, which are mutually orthogonal and non-null. Hence $T_p\mathcal{O}^c$ is the orthogonal direct sum of $T_p\mathcal{O}$ and $JT_p\mathcal{O}$ (with respect to both g and $\tilde g$). Hence g and $\tilde g$ restrict to nondegenerate metrics on $\mathcal{O}$ and A restricts to a constant multiple of $\bigl(\frac{\mathrm{vol}(\tilde g|_\mathcal{O})}{\mathrm{vol}(\tilde g|_\mathcal{O})}\bigr)^{1/(\ell+1)}(\tilde g|_\mathcal{O})^{-1}g|_\mathcal{O}$. The Levi-Civita connections of g and $\tilde g$ on $\mathcal{O}^c$ are related by (11) for some 1-form Υ_α. If we now restrict to $\mathcal{O}$ (which is totally geodesic in $\mathcal{O}^c$), it follows that the induced Levi-Civita connections ∇ and $\widetilde\nabla$ are related by

$$\widetilde\nabla_\alpha X^\gamma - \nabla_\alpha X^\gamma = \tfrac{1}{2}(\Upsilon_\alpha\delta_\beta{}^\gamma + \delta_\alpha{}^\gamma\Upsilon_\beta),$$

i.e. the metrics on $\mathcal{O}$ are projectively equivalent. □

5.7. Local classification

Let $(M, J, [\nabla])$ be a c-projective $2n$-manifold admitting two compatible non-homothetic (pseudo-)Kähler metrics, and hence a pencil of solutions of the metrisability equation of order $0 \leq \ell \leq n$. Lemma 5.20 shows that the real orbits yield a foliation of the set M^0 of regular points which is transverse and orthogonal to the common level sets of the nonconstant eigenvalues $\xi_1, \ldots \xi_\ell$ of A; these are also the levels of the elementary symmetric functions $\sigma_1, \ldots \sigma_\ell$ of $\xi_1, \ldots \xi_\ell$, which are Hamiltonians for Killing vector fields generating the local isometric Hamiltonian ℓ-torus action on M^0. Indeed, on M^0, $\chi_A(t) = \chi_c(t)\chi_{\mathrm{nc}}(t)$, where $\chi_{\mathrm{nc}}(t) = \prod_{i=1}^\ell(t-\xi_i) = \sum_{r=0}^\ell(-1)^r\sigma_r t^{\ell-r}$, $\chi_c(t)$ has constant coefficients, and $\sigma_0 = 1$. The leaf space S of the foliation of M^0 by the complex orbits may then be identified with the Kähler quotient of M^0 by this local ℓ-torus action.

It is convenient to write $\chi_c(t) = \prod_u \rho_u(t)^{m_u}$, where $\rho_u(t)$ are the distinct irreducible real factors (with $\deg \rho_u = 1$ or 2) and m_u their multiplicities. Then if S is a manifold, its universal cover is a product of complex manifolds S_u of (real) dimension $2m_u\deg\rho_u$.

5.7. LOCAL CLASSIFICATION

These observations lead to a local classification of (pseudo-)Kähler metrics which belong to a metrisability pencil (i.e. admit a c-projectively equivalent metric, or equivalently, a Hamiltonian 2-form), which was obtained in [2] in the Kähler case, and in [16] for general (pseudo-)Kähler metrics. We state the result as follows.

THEOREM 5.21. *Let $(M, J, [\nabla])$ be a c-projective $2n$-manifold, and suppose that g is a (pseudo-)Kähler metric in a metrisability pencil of order ℓ, which we may write as $\tilde{\eta}^{\bar{a}b}(t) = \eta^{\bar{a}b}(A_a{}^b - t\delta_a{}^b)$, where $\eta^{\bar{a}b}$ corresponds to g. Then on any open subset of M^0 for which the leaf space of the complex orbits is a manifold S, we may write:*

$$(222) \quad g = \sum_u g_u(\chi_{\mathrm{nc}}(A_S)\cdot,\cdot) + \sum_{i=1}^\ell \frac{\Delta_j}{\Theta_j(\xi_j)} d\xi_j^2 + \sum_{j=1}^\ell \frac{\Theta_j(\xi_j)}{\Delta_j}\Big(\sum_{r=1}^\ell \sigma_{r-1}(\hat{\xi}_j)\theta_r\Big)^2,$$

$$(223) \quad \omega = \sum_u \omega_u(\chi_{\mathrm{nc}}(A_S)\cdot,\cdot) + \sum_{r=1}^\ell d\sigma_r \wedge \theta_r, \quad \text{with} \quad d\theta_r = \sum_u (-1)^r \omega_u(A_S^{\ell-r}\cdot,\cdot),$$

$$(224) \quad Jd\xi_j = \frac{\Theta_j(\xi_j)}{\Delta_j}\sum_{r=1}^\ell \sigma_{r-1}(\hat{\xi}_j)\,\theta_r, \qquad J\theta_r = (-1)^r \sum_{j=1}^\ell \frac{\xi_j^{\ell-r}}{\Theta_j(\xi_j)}d\xi_j.$$

The ingredients appearing here are as follows, where we lift objects on S to M by identifying the horizontal distribution $\ker(d\xi_1, \ldots d\xi_\ell, \theta_1, \ldots \theta_\ell)$ with the pullback of TS.

- *$\xi_1, \ldots \xi_\ell$ are the nonconstant roots of A, which are smooth complex-valued functions on M^0, functionally independent over $\mathbb{R}$, such that for any $j \in \{1, \ldots \ell\}$, $\bar{\xi}_j = \xi_k$ for some (necessarily unique) k.*
- *$\chi_{\mathrm{nc}}(t) = \prod_{i=1}^\ell (t - \xi_i) = \sum_{r=0}^\ell (-1)^r \sigma_r t^{\ell-r}$, $\sigma_{r-1}(\hat{\xi}_j)$ is the $(r-1)$st elementary symmetric function of $\{\xi_k : k \neq j\}$, and $\Delta_j = \prod_{k \neq j}(\xi_j - \xi_k)$.*
- *For $j \in \{1, \ldots \ell\}$, Θ_j is a smooth nonvanishing complex function on the image of ξ_j such that if $\bar{\xi}_j = \xi_k$ then $\overline{\Theta}_j = \Theta_k$.*
- *For each distinct irreducible real factor ρ_u of χ_c, the metric g_u is induced by a (pseudo-)Kähler metric on the factor S_u of the universal cover of S.*
- *A_S is a parallel Hermitian endomorphism with respect to the local product metric $\sum_u g_u$ on S, preserving the distributions induced by TS_u, on which it has characteristic polynomial $\rho_u(t)^{m_u}$.*

Any such (pseudo-)Kähler metric admits a metrisability pencil of order ℓ, with

$$A = A_S + \sum_{i=1}^\ell \xi_i\Big(d\xi_i \otimes \frac{\partial}{\partial \xi_i} + Jd\xi_i \otimes J\frac{\partial}{\partial \xi_i}\Big).$$

In other words (M, g, J, ω) is locally a bundle over a product S of (pseudo-)Kähler whose fibres (the complex orbits) are totally geodesic toric (pseudo-)Kähler manifolds of a special kind, called "orthotoric". The proof in [2] proceeds by establishing the orthotoric property of the fibres and the special structure of the base S. In contrast, the proof in [16] relies upon the observation (generalising Lemma 5.20) that the local quotient of (M, g) by the real isometric ℓ-torus action admits a projectively equivalent metric: the first two sums in (222) are the general form of such a metric when the nonconstant eigenvalues of the projective pencil have algebraic multiplicity one.

In the Riemannian case, the expression (222) provides a complete local description of the metric: locally, we may assume $S = \prod_u S_u$ is product of open subsets $S_u \subseteq \mathbb{R}^{2m_u}$, and then A_S is a constant multiple of the identity on each factor. In the pseudo-Riemannian case, it remains only to describe explicitly the parallel Hermitian endomorphism A_S on $S = \prod_u S_u$, for which we refer to [18].

REMARK 5.6. In order to understand the compatible metrics corresponding to the general element $\tilde{\eta} - t\eta$ of the metrisability pencil, it is convenient to make a projective change $s = (at+b)/(ct+d)$ of parameter, as in Remark 5.3. The metric corresponding to $c\tilde{\eta} + d\eta$ (assuming this is nondegenerate) must have the same form (222) as g, with respect to the coordinates $\tilde{\xi}_j = (a\xi_j + b)/(c\xi_j + d)$, and with A replaced by $\tilde{A} = (cA+d)^{-1}(aA+b)$. We find in particular that the new functions $\tilde{\Theta}_j$ are related to the old functions by $\tilde{\Theta}_j(s)(ct+d)^{\ell+1} = (ad-bc)^{\ell+1}\Theta_j(t)$—in other words they transform like polynomials of degree $\ell + 1$ (sections of $\mathcal{O}(\ell+1)$ over the projective parameter line).

REMARK 5.7. It is straightforward to show that the restriction of the metric (222) to any complex orbit (a totally geodesic integral submanifold of $\partial_{\xi_j}, J\partial_{\xi_j}$: $j \in \{1, \dots \ell\}$) has constant holomorphic sectional curvature if and only if each $\Theta_j(t)$ is a polynomial independent of j, of degree at most $\ell + 1$: the curvature computations in [**2**] extend readily to the (pseudo-)Kähler case. If we write $\Theta_j(t) = \Theta(t) := \sum_{r=-1}^{\ell} a_r t^{\ell-r}$, then the complex orbits have constant holomorphic sectional curvature $B = \frac{1}{4}a_{-1}$.

Following [**2**], we may introduce holomorphic coordinates $u_r + it_r$ on the complex orbits by writing $\theta_r = dt_r + \alpha_r$ and $Jdu_r = dt_r$ for $r \in \{1, \dots \ell\}$, where α_r are pullbacks of 1-forms on S. Thus

$$Jdu_r = -\alpha_r - (-1)^r \sum_{j=1}^{\ell} \frac{\xi_j^{\ell-r}}{\Theta_j(\xi_j)} Jd\xi_j$$

where $d\alpha_r = \sum_u (-1)^r \omega_u(A_S^{\ell-r} \cdot, \cdot)$, and these formulae extend to any $r \leq \ell$. For $r \geq 1$, $dJdu_r = 0$, whereas $dJdu_0 = -\omega$ and $dJdu_{-1} = \phi + \sigma_1\omega$, where $\phi = g(JA\cdot,\cdot)$.

In particular, if $\Theta_j(t) = \Theta(t)$, then

$$\sum_{r=-1}^{\ell}(-1)^r a_r (Jdu_r + \alpha_r) = -Jd\sigma_1$$

and hence
$$dJd\sigma_1 = a_{-1}(\phi + \sigma_1\omega) + a_0\omega - \sum_u \omega_u(\Theta(A)\cdot,\cdot).$$

However, σ_1 differs from $\mathrm{trace}\, A = A_a{}^a$ by an additive constant, so $d\sigma_1 = -\Lambda$ and hence $dJd\sigma_1 = -2\nabla J\Lambda$, i.e.

(225) $\qquad 2\nabla\Lambda = (a_{-1} + a_0\sigma_1)g + a_{-1}g(A\cdot,\cdot) - \sum_u g_u(\Theta(A)\cdot,\cdot).$

CHAPTER 6

Metric c-projective structures and nullity

Henceforth, we assume that (M, J) is a complex manifold (i.e. with J integrable) of real dimension $2n \geq 4$, equipped with a *metric c-projective structure*, i.e. a c-projective structure $[\nabla]$ containing the Levi-Civita connection of a (pseudo-)Kähler metric g, which we denote by ∇^g, or ∇ if g is understood. We may also consider a metric c-projective structure as an equivalence class $[g]$ of (pseudo-)Kähler metrics on (M, J) having the same J-planar curves.

By Proposition 4.5, the map sending a metric $g \in [g]$ to $\eta = \tau_g^{-1} g^{-1}$ embeds $[g]$ into $\mathfrak{m}_c = \mathfrak{m}_c[\nabla]$ as an open subset of the nondegenerate solutions to the metrisability equation (126). We refer to $\dim \mathfrak{m}_c$ as the mobility of g for any $g \in [g]$, cf. Chapter 4.4, and we are interested in the case that $\dim \mathfrak{m}_c \geq 2$. In Chapter 5, we obtained some consequences of this assumption for the geodesic flow of g on M. We now turn to the relationship between mobility and curvature.

As explained in Chapter 4.5, $\mathfrak{m}_c$ may be identified with the space of parallel sections of the real tractor bundle $\mathcal{V}$ with respect to the *prolongation connection* (157)–(158). However, in [44, Theorem 5], it was shown that if $\dim \mathfrak{m}_c \geq 3$, then $\mathfrak{m}_c$ may also be identified with the space of parallel sections of $\mathcal{V}$ with respect to the connection

$$\nabla_\alpha \begin{pmatrix} A^{\beta\gamma} \\ \Lambda^\beta \\ \rho \end{pmatrix} = \begin{pmatrix} \nabla_\alpha A^{\beta\gamma} + \delta_\alpha{}^{(\beta} \Lambda^{\gamma)} + J_\alpha{}^{(\beta} J_\epsilon{}^{\gamma)} \Lambda^\epsilon \\ \nabla_\alpha \Lambda^\beta + \rho \delta_\alpha{}^\beta - 2B g_{\alpha\gamma} A^{\beta\gamma} \\ \nabla_\alpha \rho - 2B g_{\alpha\beta} \Lambda^\beta \end{pmatrix}$$

for some uniquely determined constant B. In this section we explore this phenomenon, and its implications for the curvature of M. First, as a warm-up, we consider the analogous situation in real projective geometry.

6.1. Metric projective geometry and projective nullity

A *metric projective structure* on a smooth manifold M of dimension $n \geq 2$ is a projective structure $[\nabla]$ containing the Levi-Civita connection of a (pseudo-)Riemannian metric, or (which amounts to the same thing) an equivalence class $[g]$ of (pseudo-)Riemannian metrics with the same geodesic curves. As in the c-projective case (see Chapter 4.3 and Remark 4.4), up to sign, $[g]$ embeds into the space $\mathfrak{m} = \mathfrak{m}[\nabla]$ of solutions to the projective metrisability equation (135) as the open subset of nondegenerate solutions.

A metric projective structure has mobility $\dim \mathfrak{m} \geq 1$, and we are interested in the case that $\dim \mathfrak{m} \geq 2$. However, it is shown in [57] that, on a connected projective manifold $(M, [\nabla])$ with mobility $\dim \mathfrak{m} \geq 3$, there is a constant B such that solutions $\mathcal{A}^{\beta\gamma}$ of the mobility equations may be identified with parallel sections

for the connection
$$(226) \quad \nabla_\alpha \begin{pmatrix} A^{\beta\gamma} \\ \mu^\beta \\ \rho \end{pmatrix} = \begin{pmatrix} \nabla_\alpha A^{\beta\gamma} + 2\delta_\alpha{}^{(\beta}\mu^{\gamma)} \\ \nabla_\alpha \mu^\beta + \rho\delta_\alpha{}^\beta - Bg_{\alpha\gamma}A^{\beta\gamma} \\ \nabla_\alpha \rho - 2Bg_{\alpha\beta}\mu^\beta \end{pmatrix}$$

on the tractor bundle associated to the metrisability equation. This connection is the main tool used in [45] to determine all possible values of the mobility of an n-dimensional simply-connected Lorentzian manifold.

This result is an example of a general phenomenon: in metric projective geometry, solutions to first BGG equations are often in bijection with parallel sections of tractor bundles for a much simpler (albeit somewhat mysterious) connection than the prolongation connection. We illustrate this with a toy example. The operator
$$\Gamma(TM(-1)) \ni \eta^\beta \mapsto (\nabla_\alpha \eta^\beta)_\circ = \nabla_\alpha \eta^\beta - \tfrac{1}{n}\delta_\alpha{}^\beta \nabla_\gamma \eta^\gamma$$
is projectively invariant, where $TM(-1)$ denotes the bundle of vector fields of projective weight -1; its kernel consists of solutions to the *concircularity equation*
$$(227) \quad (\nabla_\alpha \eta^\beta)_\circ = 0,$$
called *concircular vector fields*. This equation is especially congenial in that its prolongation connection yields a connection on a vector bundle that is equivalent to the Cartan connection. Indeed, following [7], for any solution η^α of (227) there is a unique function ρ (of projective weight -1) such that $\nabla_\alpha \eta^\beta = -\delta_\alpha{}^\beta \rho$, namely $\rho = -\tfrac{1}{n}\nabla_\gamma \eta^\gamma$. We then have
$$(228) \quad R_{\alpha\beta}{}^\gamma{}_\delta \eta^\delta = (\nabla_\alpha \nabla_\beta - \nabla_\beta \nabla_\alpha)\eta^\gamma = \delta_\alpha{}^\gamma \nabla_\beta \rho - \delta_\beta{}^\gamma \nabla_\alpha \rho,$$
and tracing over $\alpha\gamma$ yields $\mathrm{Ric}_{\beta\delta}\eta^\delta = (n-1)\nabla_\beta \rho$. We conclude that η^α lifts uniquely to parallel section of the standard tractor bundle for the connection
$$(229) \quad \nabla_\alpha \begin{pmatrix} \eta^\beta \\ \rho \end{pmatrix} = \begin{pmatrix} \nabla_\alpha \eta^\beta + \delta_\alpha{}^\beta \rho \\ \nabla_\alpha \rho - \mathsf{P}_{\alpha\beta}\eta^\beta \end{pmatrix}$$
induced by the (normal) Cartan connection, where $\mathsf{P}_{\alpha\beta} \equiv \tfrac{1}{n-1}\mathrm{Ric}_{\alpha\beta}$.

The simpler connection arising in the metric projective case is described as follows.

THEOREM 6.1. *Let $(M,[\nabla])$ be a metric projective manifold, and for any $p \in M$, let N_p be the dimension of the span at p of the local solutions of (227). Then for any metric g with Levi-Civita connection $\nabla^g \in [\nabla]$, there is a function B on M, which is uniquely determined and smooth where $N_p \geq 1$, such that every concircular vector field lifts uniquely to a parallel section of the standard tractor bundle for the connection*
$$(230) \quad \nabla_\alpha \begin{pmatrix} \eta^\beta \\ \rho \end{pmatrix} = \begin{pmatrix} \nabla^g_\alpha \eta^\beta + \delta_\alpha{}^\beta \rho \\ \nabla^g_\alpha \rho - Bg_{\alpha\beta}\eta^\beta \end{pmatrix}.$$

Moreover B is locally constant on the open set where $N_p \geq 2$, which is empty or dense in each connected component of M. If M is connected and B is locally constant on a dense open set, it is actually constant on M.

PROOF. We take $\nabla = \nabla^g$ and use g to raise and lower indices. Suppose that
$$\nabla_\alpha \eta^\beta + \delta_\alpha{}^\beta \rho = 0 \quad \text{and} \quad \nabla_\alpha \widetilde{\eta}^\beta + \delta_\alpha{}^\beta \widetilde{\rho} = 0$$

6.1. METRIC PROJECTIVE GEOMETRY AND PROJECTIVE NULLITY

for solutions $\eta^\alpha, \widetilde{\eta}^\alpha$ of (227). Then (228) implies that

(231) $R^g_{\alpha\beta\gamma\delta}\eta^\delta = g_{\alpha\gamma}\nabla_\beta\rho - g_{\beta\gamma}\nabla_\alpha\rho$ and $R^g_{\alpha\beta\gamma\delta}\widetilde{\eta}^\delta = g_{\alpha\gamma}\nabla_\beta\widetilde{\rho} - g_{\beta\gamma}\nabla_\alpha\widetilde{\rho},$
and so $2\widetilde{\eta}_{[\alpha}\nabla_{\beta]}\rho = R^g_{\alpha\beta\gamma\delta}\widetilde{\eta}^\gamma\eta^\delta = -R^g_{\alpha\beta\gamma\delta}\eta^\gamma\widetilde{\eta}^\delta = 2\eta_{[\beta}\nabla_{\alpha]}\widetilde{\rho}.$

In particular, $\eta_{[\alpha}\nabla_{\beta]}\rho = 0$ and so there is a unique smooth function B on the open set where $\eta^\alpha \neq 0$ such that

(232) $$\nabla_\alpha\rho - Bg_{\alpha\beta}\eta^\beta = 0$$

on M for any extension of B over the zero-set of η^α (since $\nabla_\alpha\rho$ also vanishes there). Equation (231) now implies that any two concircular vector fields have the same function B where both functions are determined. Thus B is uniquely determined and smooth where $N_p \geq 1$. Differentiating (232) on the open set where B is smooth gives

$$\nabla_\beta\nabla_\alpha\rho - \eta_\alpha\nabla_\beta B + Bg_{\alpha\beta}\rho = 0,$$

and so $\eta_{[\alpha}\nabla_{\beta]}B = 0$. Hence $\nabla_\alpha B = 0$ on the open set where $N_p \geq 2$. This subset is empty or dense in each component of M, since two solutions of (227) that are pointwise linearly dependent on an open set are linearly dependent on that open set.

It remains to show that if M is connected and B is locally constant on a dense open subset U (which could be disconnected), then it is actually constant. To see this, we use only that

$$\mathsf{P}_{\beta\gamma}\eta^\gamma = B\eta_\beta \quad \text{and} \quad \nabla_\alpha B = 0$$

on U, for then we may differentiate once more to conclude that

$$(\nabla_\alpha\mathsf{P}_{\beta\gamma})\eta^\gamma + \mathsf{P}_{\beta\gamma}\nabla_\alpha\eta^\gamma = B\nabla_\alpha\eta_\beta$$

and hence that

$$(\nabla_\alpha\mathsf{P}_{\beta\gamma})\eta^\gamma - \mathsf{P}_{\alpha\beta}\rho = -Bg_{\alpha\beta}\rho$$

on U. Tracing over $\alpha\beta$ yields

$$(\nabla_\alpha\mathsf{P}^\alpha{}_\gamma)\eta^\gamma - \mathsf{P}_\alpha{}^\alpha\rho = -nB\rho$$

and hence that

(233) $$\begin{pmatrix} \mathsf{P}^\alpha{}_\beta & 0 \\ -\frac{1}{n}\nabla_\alpha\mathsf{P}^\alpha{}_\beta & \frac{1}{n}\mathsf{P}_\alpha{}^\alpha \end{pmatrix}\begin{pmatrix} \eta^\beta \\ \rho \end{pmatrix} = B\begin{pmatrix} \eta^\alpha \\ \rho \end{pmatrix}$$

on U. Although this equation was derived on U, it is a valid stipulation everywhere on M. Moreover, the tractor

$$\begin{pmatrix} \eta^\beta \\ \rho \end{pmatrix}$$

is nowhere vanishing on M (else in (229), the vector field η^β would vanish identically). From this point of view, we see that B extends as a smooth function on M. Finally, since B is locally constant on U, it is locally constant and hence constant on M. □

The connection (230) of Theorem 6.1 differs from the tractor connection (229) by the endomorphism-valued 1-form

$$\begin{pmatrix} 0 & 0 \\ \mathsf{P}_{\alpha\beta} - Bg_{\alpha\beta} & 0 \end{pmatrix} : X^\alpha \otimes \begin{pmatrix} \eta^\beta \\ \rho \end{pmatrix} \mapsto \begin{pmatrix} 0 \\ (\mathsf{P}_{\alpha\beta} - Bg_{\alpha\beta})X^\alpha\eta^\beta \end{pmatrix}$$

The connections agree on the flat model. Specifically, on the unit sphere we have

$$R_{\alpha\beta\gamma\delta} = g_{\alpha\gamma}g_{\beta\delta} - g_{\beta\gamma}g_{\alpha\delta} \quad \text{whence} \quad \mathsf{P}_{\alpha\beta} = g_{\alpha\beta},$$

so that the connections coincide with $B = 1$.

The proof of Theorem 6.1 may be broken down into two steps. First, one shows that the connection (230) has the required lifting property for some function B, which may only be uniquely determined and smooth on an open set. Secondly, one establishes sufficient regularity to determine the connection globally on M (in this case, with B constant). In the literature, the second step has often been carried out by probing M with geodesics. In the above proof we advocate an alternative line of argument that we believe to be simpler and more generally applicable.

REMARK 6.1. For example, we may apply the same technique to the mobility equations (226), where the replacement for (233) has the form

$$\begin{pmatrix} R & 0 & 0 \\ \nabla R & R & 0 \\ \nabla\nabla R + R \bowtie R & \nabla R & R \end{pmatrix} \begin{pmatrix} A^{\beta\gamma} \\ \mu^{\beta} \\ \rho \end{pmatrix} = B \left[\begin{pmatrix} A^{\beta\gamma} \\ \mu^{\beta} \\ \rho \end{pmatrix} - \frac{A_\delta{}^\delta}{n} \begin{pmatrix} g^{\beta\gamma} \\ 0 \\ \frac{1}{n}\mathsf{P}_\gamma{}^\gamma \end{pmatrix} \right].$$

As above, this is sufficient to show that B is constant if it is locally constant on a dense open set. One striking difference between this case and Theorem 6.1, however, is that the connection (230) actually has the same covariant constant sections as does the standard Cartan or prolongation connection (229). For the mobility equations, however, not only is the resulting connection (226) different from the prolongation connection [43] but also their covariant constant sections are generally different. Nevertheless, all solutions of the mobility equations lift uniquely as covariant constant sections with respect to either of these connections (and this is their crucial property).

We next seek to elucidate the first step in the proof of Theorem 6.1. Here we observe that the key equations (231) used to establish the uniqueness of B may be viewed as a characterisation of B in terms of the curvature R^g of g, namely that

$$R^g_{\alpha\beta\gamma\delta}\eta^\delta = B(g_{\alpha\gamma}\eta_\beta - g_{\beta\gamma}\eta_\alpha).$$

This motivates the introduction of some terminology, following Gray [50].

DEFINITION 6.1. Let (M, g) be a (pseudo-)Riemannian manifold and suppose that the tensor $R_{\alpha\beta\gamma\delta}$ has the symmetries of the Riemannian curvature of g. Then a *nullity vector* of R at $p \in M$ is a tangent vector $v^\alpha \in T_pM$ with $R_{\alpha\beta\gamma\delta}v^\delta = 0$, and the *nullity space* of R at p is the set of such nullity vectors. We say R has *nullity* at p if the nullity space is nonzero, i.e. the nullity index is positive.

In particular, if R^g is the Riemannian curvature of g, then at each $p \in M$, there is at most one scalar $B \in \mathbb{R}$ such that $R^B_{\alpha\beta\gamma\delta} := R^g_{\alpha\beta\gamma\delta} - B(g_{\alpha\gamma}g_{\beta\delta} - g_{\beta\gamma}g_{\alpha\delta})$ has nullity at p. Indeed if v^α and $\widetilde{v}^\alpha$ are nullity vectors for R^B and $R^{\widetilde{B}}$ respectively then

$$0 = (B - \widetilde{B})(g_{\alpha\gamma}g_{\beta\delta} - g_{\beta\gamma}g_{\alpha\delta})\widetilde{v}^\beta v^\delta = (B - \widetilde{B})(v_\beta\widetilde{v}^\beta g_{\alpha\gamma} - v_\alpha\widetilde{v}_\gamma),$$

which implies that $B = \widetilde{B}$ unless v^α or $\widetilde{v}^\alpha$ are zero.

DEFINITION 6.2. Let (M, g) be a (pseudo-)Riemannian manifold. Then the (*projective*) *nullity distribution* of g is the union of the nullity spaces of $R^B_{\alpha\beta\gamma\delta}$ over

6.1. METRIC PROJECTIVE GEOMETRY AND PROJECTIVE NULLITY

$B \in \mathbb{R}$ and $p \in M$. We say that g has *(projective) nullity* at $p \in M$ if there is a nonzero $v^\alpha \in T_p M$ in the nullity distribution of g, i.e.

$$(234) \qquad \left(R^g_{\alpha\beta\gamma\delta} - B(g_{\alpha\gamma}g_{\beta\delta} - g_{\beta\gamma}g_{\alpha\delta})\right)v^\delta = 0,$$

for some $B \in \mathbb{R}$, uniquely determined by p.

The definition of B is reminiscent of an eigenvalue; indeed, the $\alpha\gamma$ trace of (234) is

$$\mathsf{P}_\alpha{}^\beta v^\alpha = B v^\beta,$$

so B is an eigenvalue of the endomorphism $\mathsf{P}_\alpha{}^\beta$. On the other hand the trace-free part of (234) provides a projectively invariant characterisation, using the projective Weyl tensor $P_{\alpha\beta}{}^\gamma{}_\delta := R_{\alpha\beta}{}^\gamma{}_\delta - \delta_\alpha{}^\gamma \mathsf{P}_{\beta\delta} + \delta_\beta{}^\gamma \mathsf{P}_{\alpha\delta}$, as follows (cf. [**49**]).

PROPOSITION 6.2. *Let (M, g) be a (pseudo-)Riemannian manifold of dimension $n \geq 2$, and let $v^\delta \in T_p M$ be nonzero. Then the following statements are equivalent:*
  (1) *v^δ is a projective nullity vector at p*
  (2) *there exists $B \in \mathbb{R}$ such that $P_{\alpha\beta}{}^\gamma{}_\delta v^\beta = (\mathsf{P}_{\alpha\delta} - B g_{\alpha\delta})v^\gamma$*
  (3) *$P_{\alpha\beta}{}^\gamma{}_\delta v^\delta = 0$.*

PROOF. (1)$\Rightarrow$(2). Since $\mathsf{P}_{\alpha\beta}v^\beta = B g_{\alpha\beta}v^\beta$, $R_{\alpha\beta\gamma\delta} = R_{\gamma\delta\alpha\beta}$ and $\mathsf{P}_{\alpha\beta} = \mathsf{P}_{\beta\alpha}$, we have

$$P_{\alpha\beta\gamma\delta}v^\beta = R_{\gamma\delta\alpha\beta}v^\beta - g_{\alpha\gamma}\mathsf{P}_{\beta\delta}v^\beta + g_{\beta\gamma}\mathsf{P}_{\alpha\delta}v^\beta$$
$$= B(g_{\alpha\gamma}g_{\beta\delta} - g_{\beta\gamma}g_{\alpha\delta})v^\beta - Bg_{\alpha\gamma}g_{\beta\delta}v^\beta + g_{\beta\gamma}\mathsf{P}_{\alpha\delta}v^\beta = (\mathsf{P}_{\alpha\delta} - Bg_{\alpha\delta})g_{\beta\gamma}v^\beta,$$

and (2) follows by raising the index γ.

(2)$\Rightarrow$(3). Since $P_{[\alpha\beta}{}^\gamma{}_{\delta]} = 0$, which follows easily from $R_{[\alpha\beta}{}^\gamma{}_{\delta]} = 0$,

$$P_{\alpha\beta}{}^\gamma{}_\delta v^\delta = (P_{\alpha\delta}{}^\gamma{}_\beta - P_{\beta\delta}{}^\gamma{}_\alpha)v^\delta,$$

which vanishes by (2), since $\mathsf{P}_{\alpha\beta} - Bg_{\alpha\beta}$ is symmetric in $\alpha\beta$.

(3)$\Rightarrow$(1). Observe that $0 = R_{\alpha\beta\gamma\delta}v^\gamma v^\delta = (g_{\alpha\gamma}\mathsf{P}_{\beta\delta} - g_{\beta\gamma}\mathsf{P}_{\alpha\delta})v^\gamma v^\delta = v_{[\alpha}\mathsf{P}_{\beta]\delta}v^\delta$. Hence there exists $B \in \mathbb{R}$ such that $\mathsf{P}_{\beta\delta}v^\delta = Bg_{\beta\delta}v^\delta$, and hence

$$R_{\alpha\beta\gamma\delta}v^\delta = (g_{\alpha\gamma}\mathsf{P}_{\beta\delta} - g_{\beta\gamma}\mathsf{P}_{\alpha\delta})v^\delta = B(g_{\alpha\gamma}g_{\beta\delta} - g_{\beta\gamma}g_{\alpha\delta})v^\delta,$$

i.e. v^δ is in the projective nullity at p. □

In particular, this shows that the projective nullity distribution is a metric projective invariant, as is the expression $\mathsf{P}_{\alpha\beta} - Bg_{\alpha\beta}$ wherever there is projective nullity, hence so is the special tractor connection (230). The above argument for this fact is given in [**49**], where the special connection on the standard tractor bundle is also discussed.

REMARK 6.2. In the 2-dimensional case, all metrics have nullity at all points and B is the Gaußian curvature. On the unit n-sphere the nullity distribution is the tangent bundle and $B \equiv 1$. Condition (234) may be written as

$$C_{\alpha\beta\gamma\delta}v^\delta = \left(\tfrac{1}{(n-1)(n-2)}\mathrm{Scal} - \tfrac{1}{n-2}B\right)(g_{\alpha\gamma}v_\beta - g_{\beta\gamma}v_\alpha) - \tfrac{1}{n-2}(\mathrm{Ric}_{\alpha\gamma}v_\beta - \mathrm{Ric}_{\beta\gamma}v_\alpha)$$

where $C_{\alpha\beta\gamma\delta}$ is conformal Weyl curvature tensor. For a Riemannian metric, we may orthogonally diagonalise the Ricci tensor to see that if $C_{\alpha\beta\gamma\delta} = 0$ (as it is in three dimensions or in the conformally flat case in higher dimensions) then $R_{\alpha\beta\gamma\delta}$ has nullity if and only if all but possibly one of the eigenvalues of $\mathsf{P}_\alpha{}^\beta$ coalesce with B being the possible exception. So in the three-dimensional Riemannian case $R_{\alpha\beta\gamma\delta}$

has nullity if and only if the discriminant of the characteristic polynomial of $\mathsf{P}_\alpha{}^\beta$ vanishes:

$$(\mathsf{P}_\alpha{}^\alpha)^6 - 9(\mathsf{P}_\alpha{}^\alpha)^4(\mathsf{P}_\beta{}^\gamma \mathsf{P}_\gamma{}^\beta) + 21(\mathsf{P}_\alpha{}^\alpha)^2(\mathsf{P}_\beta{}^\gamma \mathsf{P}_\gamma{}^\beta)^2 - 3(\mathsf{P}_\beta{}^\gamma \mathsf{P}_\gamma{}^\beta)^3$$
$$+ 8(\mathsf{P}_\alpha{}^\alpha)^3(\mathsf{P}_\delta{}^\epsilon \mathsf{P}_\epsilon{}^\gamma \mathsf{P}_\gamma{}^\delta) - 36(\mathsf{P}_\alpha{}^\alpha)(\mathsf{P}_\beta{}^\gamma \mathsf{P}_\gamma{}^\beta)(\mathsf{P}_\delta{}^\epsilon \mathsf{P}_\epsilon{}^\zeta \mathsf{P}_\zeta{}^\delta) + 18(\mathsf{P}_\delta{}^\epsilon \mathsf{P}_\epsilon{}^\gamma \mathsf{P}_\gamma{}^\delta)^2 = 0.$$

Indeed, in three dimensions (where $R_{\alpha\beta\gamma\delta}$ is determined by $\mathsf{P}_{\alpha\beta}$) it is also the case in Lorentzian signature that $R_{\alpha\beta\gamma\delta}$ has nullity if and only if $\mathsf{P}_\alpha{}^\beta$ is diagonalisable with eigenvalues distributed in this manner. In any case, in three dimensions it follows that B is a continuous function and is smooth except perhaps at points where $\mathsf{P}_{\alpha\beta}$ is a multiple of $g_{\alpha\beta}$. In the four-dimensional Riemannian case, one can check that if $R_{\alpha\beta\gamma\delta}$ has nullity and the eigenvalues of $\mathsf{P}_\alpha{}^\beta$ are $B, \lambda_2, \lambda_3, \lambda_4$, then

$$I \equiv C_{\alpha\beta\gamma\delta} C^{\alpha\beta\gamma\delta} = 6\big((\lambda_2 - \lambda_3)^2 + (\lambda_3 - \lambda_4)^2 + (\lambda_4 - \lambda_2)^2\big)$$

and if this expression is nonzero, then

$$B = \tfrac{1}{4}\mathsf{P}_\alpha{}^\alpha + \tfrac{1}{4I}\big(C_{\alpha\beta}{}^{\gamma\delta} C_{\gamma\delta}{}^{\epsilon\zeta} C_{\epsilon\zeta}{}^{\alpha\beta} - 18 C_{\alpha\beta}{}^{\gamma\delta} \mathsf{P}_\gamma{}^\alpha \mathsf{P}_\delta{}^\beta\big).$$

It follows that B is smooth on $\{I \neq 0\}$ whilst on $\{I = 0\}$ three of the four eigenvalues of $\mathsf{P}_\alpha{}^\beta$ merge as above and B is the odd one out unless $\mathsf{P}_{\alpha\beta} \propto g_{\alpha\beta}$. Therefore, as in three dimensions, it follows that B extends as a continuous function that is smooth except where $\mathsf{P}_{\alpha\beta}$ is a multiple of $g_{\alpha\beta}$. We anticipate similar behaviour in general but, for the moment, the regularity of B remains unknown.

6.2. C-projective nullity

We return now to metric c-projective geometry, where we seek to develop analogous interconnections between curvature and special tractor connections to those in the metric projective case. In order to do this, we first develop a notion of c-projective nullity for (pseudo-)Kähler metrics, modelled on the curvature of complex projective space (48) in the same way that projective nullity for (pseudo-)Riemannian metrics is modelled on the curvature of the unit sphere.

We suppose therefore that (M, J, g) is a (pseudo-)Kähler manifold with ∇ the Levi-Civita connection of $g_{\alpha\beta}$ and $\Omega_{\alpha\beta} = J_\alpha{}^\gamma g_{\gamma\beta}$ the Kähler form. Further let us write

$$(235) \qquad S_{\alpha\beta\gamma\delta} \equiv g_{\alpha\gamma} g_{\beta\delta} - g_{\beta\gamma} g_{\alpha\delta} + \Omega_{\alpha\gamma} \Omega_{\beta\delta} - \Omega_{\beta\gamma} \Omega_{\alpha\delta} + 2\Omega_{\alpha\beta} \Omega_{\gamma\delta}$$

for the Kähler curvature tensor of constant sectional holomorphic curvature 4. As in the (pseudo-)Riemannian case, at each $p \in M$, there is at most one scalar $B \in \mathbb{R}$ such that $G^B_{\alpha\beta\gamma\delta} := R_{\alpha\beta\gamma\delta} - B S_{\alpha\beta\gamma\delta}$ has nullity at p. Indeed, if v^α and $\widetilde{v}^\alpha$ are nullity vectors for $G^B_{\alpha\beta\gamma\delta}$ and $G^{\widetilde{B}}_{\alpha\beta\gamma\delta}$ respectively then

$$0 = (B - \widetilde{B}) S_{\alpha\beta\gamma\delta} \widetilde{v}^\beta v^\delta$$
$$= (B - \widetilde{B})(v_\beta \widetilde{v}^\beta g_{\alpha\gamma} - v_\alpha \widetilde{v}_\gamma + J_\beta{}^\delta v_\delta \widetilde{v}^\beta \Omega_{\alpha\gamma} + J_\alpha{}^\delta v_\delta J_\gamma{}^\beta \widetilde{v}_\beta + 2 J_\gamma{}^\delta v_\delta J_\alpha{}^\beta \widetilde{v}_\beta)$$

which implies that $B = \widetilde{B}$ unless v^α or $\widetilde{v}^\alpha$ are zero. By analogy with Definition 6.2, and again following Gray [50] (who used the term "holomorphic constancy"), we therefore define c-projective nullity as follows.

DEFINITION 6.3. The *(c-projective) nullity distribution* $\mathcal{N}$ of a (pseudo-)Kähler manifold (M, J, g) is the union of the nullity spaces of $G^B_{\alpha\beta\gamma\delta}$ over $B \in \mathbb{R}$ and $p \in M$,

6.2. C-PROJECTIVE NULLITY

and for each $p \in M$, we write $\mathcal{N}_p$ for the (*c-projective*) *nullity space* $\mathcal{N} \cap T_p M$. We say that (J, g) has (*c-projective*) *nullity* at $p \in M$ if $\mathcal{N}_p$ is nonzero, i.e.

$$(236) \qquad \left(R_{\alpha\beta\gamma\delta} - B S_{\alpha\beta\gamma\delta} \right) v^\delta = 0,$$

for some $B \in \mathbb{R}$, uniquely determined by p, and some nonzero $v^\alpha \in T_p M$.

Thus $\mathcal{N}_p$ is the kernel of the linear map

$$v^\delta \mapsto G^B_{\alpha\beta\gamma\delta} v^\delta,$$

for some $B \in \mathbb{R}$ depending on p. Let us remark that, since $G = G^B$ has the symmetries of the curvature tensor of a Kähler metric, $\mathcal{N}_p$ is a J-invariant subspace of $T_p M$ (i.e. $v^\delta \in \mathcal{N}_p$ implies $J_\alpha{}^\delta v^\alpha \in \mathcal{N}_p$), hence is even dimensional.

Bearing in mind the discussion of Chapter 4.1, we may write (236) in barred and unbarred indices. We find that

$$(237) \qquad \begin{aligned} \left(R_{a\bar{b}c\bar{d}} + 2B(g_{a\bar{b}} g_{c\bar{d}} + g_{c\bar{b}} g_{a\bar{d}}) \right) v^{\bar{d}} &= 0 \\ \left(R_{a\bar{b}\bar{c}d} - 2B(g_{a\bar{c}} g_{d\bar{b}} + g_{a\bar{b}} g_{d\bar{c}}) \right) v^d &= 0. \end{aligned}$$

As in the projective case, tracing (236) over $\alpha\gamma$ yields an eigenvalue equation

$$(238) \qquad \mathrm{Ric}^\beta{}_\delta v^\delta = 2(n+1) B v^\beta, \quad \text{equivalently } \mathsf{P}^\beta{}_\delta v^\delta = 2B v^\beta,$$

since $\mathsf{P}_{\alpha\beta} = \frac{1}{n+1} \mathrm{Ric}_{\alpha\beta}$ by (25) and (109). This can equivalently be expressed in barred and unbarred indices as

$$(239) \qquad \mathsf{P}^b{}_d v^d = 2B v^b, \qquad \text{or as} \qquad \mathsf{P}^{\bar{b}}{}_{\bar{d}} v^{\bar{d}} = 2B v^{\bar{b}}.$$

Of course, we may derive (239) also directly by tracing the second equation of (237), respectively its conjugate, with respect to $a\bar{c}$, respectively $\bar{a}c$. Further, note that the symmetries of the Ricci tensor of a (pseudo-)Kähler metric show that (239) can be also equivalently written as $\mathsf{P}_d{}^b v^d = 2B v^b$, respectively $\mathsf{P}_{\bar{d}}{}^{\bar{b}} v^{\bar{d}} = 2B v^{\bar{b}}$.

Now assume that (237) is satisfied and decompose $R_{a\bar{b}}{}^c{}_d$ according to (28) as

$$R_{a\bar{b}}{}^c{}_d = H_{a\bar{b}}{}^c{}_d + \delta_a{}^c \mathsf{P}_{\bar{b}d} + \delta_d{}^c \mathsf{P}_{\bar{b}a}.$$

Then equation (239) implies

$$H_{a\bar{b}}{}^c{}_d v^{\bar{b}} = (R_{a\bar{b}}{}^c{}_d - \delta_a{}^c \mathsf{P}_{\bar{b}d} - \delta_d{}^c \mathsf{P}_{\bar{b}a}) v^{\bar{b}} = 2B(\delta_a{}^c v_d + \delta_d{}^c v_a) - (\delta_a{}^c \mathsf{P}_{\bar{b}d} + \delta_d{}^c \mathsf{P}_{\bar{b}a}) v^{\bar{b}} = 0.$$

Furthermore,

$$\begin{aligned} H_{a\bar{b}}{}^c{}_d v^d &= (R_{a\bar{b}}{}^c{}_d - \delta_a{}^c \mathsf{P}_{\bar{b}d} - \delta_d{}^c \mathsf{P}_{\bar{b}a}) v^d \\ &= (2B(g_{a\bar{b}} v^c + \delta_a{}^c v_{\bar{b}}) - 2B \delta_a{}^c v_{\bar{b}} - \mathsf{P}_{\bar{b}a} v^c) \\ &= (2B g_{a\bar{b}} - \mathsf{P}_{\bar{b}a}) v^c, \end{aligned}$$

which implies $H_{a\bar{b}}{}^c{}_d v^a v^d = 0$. It fact these two conditions are also sufficient for nullity.

PROPOSITION 6.3. *Let (M, J, g) be a (pseudo-)Kähler manifold of dimension $2n \geq 4$, and let $v^d \in T_p^{1,0} M \cong T_p M$ be a nonzero tangent vector. Then the following statements are equivalent*:
  (1) $v^d \in \mathcal{N}_p$
  (2) *there exists $B \in \mathbb{R}$ such that* $H_{a\bar{b}}{}^c{}_d v^d = (2B g_{a\bar{b}} - \mathsf{P}_{a\bar{b}}) v^c$
  (3) $H_{a\bar{b}}{}^c{}_d v^a v^d = 0 \quad and \quad H_{a\bar{b}}{}^c{}_d v^{\bar{b}} = 0,$

where, as in (28), $H_{a\bar{b}}{}^c{}_d$ is the trace-free part of $R_{a\bar{b}}{}^c{}_d \equiv -g^{\bar{e}c} R_{a\bar{b}d\bar{e}}$.

PROOF. We have just observed that (1) implies (2) and (3). Note, moreover that taking the trace with respect to a and c in (2) gives (239), which shows immediately that (2) implies (1). Hence, it remains to show that (3) implies (1). If (3) holds, then

$$R_{a\bar{b}}{}^c{}_d v^a v^d = (\delta_a{}^c \mathsf{P}_{\bar{b}d} + \delta_d{}^c \mathsf{P}_{\bar{b}a}) v^a v^d \;\Rightarrow\; R_{a\bar{b}\bar{c}d} v^a v^d = 2 v_{\bar{c}} \mathsf{P}_{\bar{b}d} v^d$$

so $0 = v_{[\bar{c}} \mathsf{P}_{\bar{b}]d} v^d$ and we conclude that $\mathsf{P}_{\bar{c}d} v^d = 2\bar{B} v_{\bar{c}}$, for some constant $\bar{B}$. Substituting the conjugate conclusion $\mathsf{P}_{d\bar{c}} v^{\bar{d}} = 2B v_c$ into $R_{a\bar{b}}{}^c{}_d v^{\bar{b}}$ gives

$$R_{a\bar{b}}{}^c{}_d v^{\bar{b}} = (H_{a\bar{b}}{}^c{}_d + \delta_a{}^c \mathsf{P}_{\bar{b}d} + \delta_d{}^c \mathsf{P}_{\bar{b}a}) v^{\bar{b}} = 2B \delta_a{}^c v_d + 2B \delta_d{}^c v_a$$

which, after lowering the index c is equivalent to (237), as required. (Note that B is necessarily real, since $R_{a\bar{b}}{}^c{}_d$ and $S_{a\bar{b}}{}^c{}_d$ are real tensors.) $\square$

COROLLARY 6.4. *At any $p \in M$, the nullity distribution of $\mathcal{N}_p$ is a metric c-projective invariant, i.e. the same for c-projectively equivalent (pseudo-)Kähler metrics $g_{a\bar{b}}$ and $\tilde{g}_{a\bar{b}}$. Furthermore, if $\mathcal{N}_p$ is nonzero, and $B, \widetilde{B} \in \mathbb{R}$ are the corresponding scalars in the definition of $\mathcal{N}_p$ with respect to $g, \tilde{g}$ respectively, then $\widetilde{\mathsf{P}}_{a\bar{b}} - 2\widetilde{B} \tilde{g}_{a\bar{b}} = \mathsf{P}_{a\bar{b}} - 2B g_{a\bar{b}}$.*

PROOF. By Proposition 2.13, criterion (3) of Proposition 6.3 is c-projectively invariant. In fact, by Proposition 4.4, $H_{a\bar{b}}{}^c{}_d$ is precisely the harmonic curvature of the underlying c-projective structure. The last part follows immediately from criterion (2). $\square$

REMARK 6.3. For later use, we apply the projectors of Chapter 1 to reformulate the equivalent conditions of Proposition 6.3 directly in terms of $v^\delta \in T_p M$ as follows:
(1) $v^\delta \in \mathcal{N}_p$
(2) there exists a constant $B \in \mathbb{R}$ such that $H_{\alpha\beta}{}^\gamma{}_\delta v^\delta = (J_\alpha{}^\epsilon \mathsf{P}_{\epsilon\beta} - 2B \Omega_{\alpha\beta}) J_\delta{}^\gamma v^\delta$
(3) $H_{\alpha\beta}{}^\gamma{}_\delta v^\alpha v^\delta = 0$ and $(H_{\alpha\beta}{}^\gamma{}_\delta + J_\beta{}^\epsilon J_\zeta{}^\gamma H_{\alpha\epsilon}{}^\zeta{}_\delta) v^\beta = 0$,

where $H_{\alpha\beta}{}^\gamma{}_\delta \equiv R_{\alpha\beta}{}^\gamma{}_\delta - \delta_{[\alpha}{}^\gamma \mathsf{P}_{\beta]\delta} + J_{[\alpha}{}^\gamma \mathsf{P}_{\beta]\zeta} J_\delta{}^\zeta + J_\alpha{}^\zeta \mathsf{P}_{\beta\zeta} J_\delta{}^\gamma$ and $\mathsf{P}_{\beta\delta} \equiv \frac{1}{n+1} R_{\alpha\beta}{}^\alpha{}_\delta$.

PROPOSITION 6.5. *Let (M, J, g) be a (pseudo-)Kähler manifold of dimension $2n \geq 4$, and B a smooth function on an open subset U. Then for any (real) vector field v in the nullity of $G = G^B$ on U, if v is non-null at $p \in U$, then $dB = 0$ there.*

PROOF. The differential Bianchi identity $\nabla_{[a} R_{b]\bar{c}d}{}^e = 0$ on U implies that

$$\nabla_{[a} G_{b]\bar{c}d}{}^e = 2(\nabla_{[a} B) g_{b]\bar{c}} \delta_d{}^e + 2(\nabla_{[a} B) \delta_{b]}{}^e g_{d\bar{c}}.$$

Since v^a and $v^{\bar{a}}$ belong to the nullity of G, we may contract with $v^{\bar{c}} v^d$ to obtain

$$0 = 2(\nabla_{[a} B) g_{b]\bar{c}} v^{\bar{c}} v^e + 2(\nabla_{[a} B) \delta_{b]}{}^e g_{d\bar{c}} v^d v^{\bar{c}}.$$

A further contraction with $v_e = g_{e\bar{f}} v^{\bar{f}}$ yields $(\nabla_{[a} B) v_{b]} g_{d\bar{c}} v^d v^{\bar{c}} = 0$, so if v is non-null at p, $(\nabla_{[a} B) v_{b]} = 0$ there; hence $(\nabla_{[a} B) \delta_{b]}{}^e = 0$, which implies $\nabla_a B = 0$, i.e. $dB = 0$. $\square$

6.3. Mobility, nullity, and the special tractor connection

Our aim in this section is to show that, under certain conditions, the solutions of the mobility equation (129) on a (pseudo-)Kähler manifold (M, J, g) lift uniquely to parallel sections of $\mathcal{V} \subseteq \mathcal{V}_\mathbb{C}$ for the special tractor connection:

$$
\begin{aligned}
(240) \quad & \nabla_a^{\mathcal{V}_\mathbb{C}} \begin{pmatrix} A^{\bar{b}c} \\ \Lambda^b \mid \Lambda^{\bar{b}} \\ \mu \end{pmatrix} = \begin{pmatrix} \nabla_a A^{\bar{b}c} + \delta_a{}^c \Lambda^{\bar{b}} \\ \nabla_a \Lambda^b + \mu \delta_a{}^b - 2B A_a{}^b \mid \nabla_a \Lambda^{\bar{b}} \\ \nabla_a \mu - 2B \Lambda_a \end{pmatrix} \\
& \nabla_{\bar{a}}^{\mathcal{V}_\mathbb{C}} \begin{pmatrix} A^{\bar{b}c} \\ \Lambda^b \mid \Lambda^{\bar{b}} \\ \mu \end{pmatrix} = \begin{pmatrix} \nabla_{\bar{a}} A^{\bar{b}c} + \delta_{\bar{a}}{}^{\bar{b}} \Lambda^c \\ \nabla_{\bar{a}} \Lambda^b \mid \nabla_{\bar{a}} \Lambda^{\bar{b}} + \mu \delta_{\bar{a}}{}^{\bar{b}} - 2B A^{\bar{b}}{}_{\bar{a}} \\ \nabla_{\bar{a}} \mu - 2B \Lambda_{\bar{a}} \end{pmatrix},
\end{aligned}
$$

where ∇ is the Levi-Civita connection for $g_{a\bar{b}}$ and B is a smooth function on M. Here $\mathcal{V}_\mathbb{C}$ is identified via $g_{a\bar{b}}$ with a direct sum of unweighted tensor bundles, and we write the connection in barred and unbarred indices, so that for sections of $\mathcal{V} \subseteq \mathcal{V}_\mathbb{C}$, the two lines of (240) are conjugate.

REMARK 6.4. By Theorem 4.6, we know already that any solution $A^{\bar{a}b}$ of the mobility equation (129) lifts uniquely to a parallel section of $\mathcal{V}$ for the more complicated prolongation connection (157)–(158). If it also lifts to a parallel section for (240), then (cf. Remark 6.1) the two lifts may differ, albeit only in the last component. More precisely the last component μ of the parallel lift for the special tractor connection is given by $\mu = \mu' - \frac{1}{n}(\mathsf{P}_{a\bar{b}} - 2B g_{a\bar{b}}) A^{\bar{a}b}$, where μ' is the last component of the parallel lift for the prolongation connection. Note that if the metric $g^{\bar{a}b}$ itself lifts to a parallel section for (240), then B must be locally constant.

In [**44**, Theorem 5], it is shown that if the mobility of (M, g, J) is at least three, then there is a constant B such that all solutions of the mobility equation lift uniquely to parallel sections of $\mathcal{V}$ for (240). Before developing this, and related results, it is useful to establish some basic properties of special tractor connections (240) and their parallel sections. Throughout this section we set, for a given function B,

$$(241) \qquad G_{a\bar{b}c\bar{d}} := R_{a\bar{b}c\bar{d}} + 2B(g_{a\bar{b}} g_{c\bar{d}} + g_{c\bar{b}} g_{a\bar{d}}).$$

The equations satisfied by parallel sections of (240) are

$$(242\text{a}) \qquad \nabla_a A_b{}^c = -\delta_a{}^c \Lambda_b$$
$$(242\text{b}) \qquad \nabla_a \Lambda_{\bar{c}} = -\mu g_{a\bar{c}} + 2B A_{a\bar{c}} \quad \text{and} \quad \nabla_a \Lambda_b = 0,$$
$$(242\text{c}) \qquad \nabla_a \mu = 2B \Lambda_a$$

and their complex conjugates. Of course, the first line is simply the mobility equation. In particular, from (145)–(146) (and the symmetry of $\mathsf{P}_{b\bar{c}}$) we have

$$
(243) \quad \begin{aligned}
g_{d\bar{b}} \nabla_a \Lambda_{\bar{c}} - g_{a\bar{c}} \nabla_{\bar{b}} \Lambda_d &= R_{a\bar{b}e\bar{c}} A_d{}^e - R_{a\bar{b}d\bar{e}} A^{\bar{e}}{}_{\bar{c}} \\
&= W_{a\bar{b}e\bar{c}} A_d{}^e - W_{a\bar{b}d\bar{e}} A^{\bar{e}}{}_{\bar{c}} - g_{a\bar{c}} \mathsf{P}_{e\bar{b}} A_d{}^e + g_{d\bar{b}} \mathsf{P}_{a\bar{e}} A^{\bar{e}}{}_{\bar{c}},
\end{aligned}
$$

where $W_{a\bar{b}e\bar{c}} = H_{a\bar{b}e\bar{c}}$, since J is integrable.

100   6. METRIC C-PROJECTIVE STRUCTURES AND NULLITY

LEMMA 6.6. *If $A_b{}^c$ and Λ_b satisfy (242a)–(242b) for smooth functions B, μ, then*

(244) $$G_{a\bar{b}c}{}^e A_e{}^d = G_{a\bar{b}e}{}^d A_c{}^e$$

(245) $$G_{a\bar{b}c}{}^d \Lambda_d = -g_{c\bar{b}}(\nabla_a \mu - 2B\Lambda_a) + 2(\nabla_a B)A_{c\bar{b}}.$$

PROOF. We substitute (242b) into (243) to obtain
$$R_{a\bar{b}e\bar{c}} A_d{}^e - R_{a\bar{b}d\bar{e}} A^{\bar{e}}{}_{\bar{c}} = g_{a\bar{c}}(\mu g_{d\bar{b}} - 2BA_{d\bar{b}}) - g_{d\bar{b}}(\mu g_{a\bar{c}} - 2BA_{a\bar{c}})$$
$$= 2B(g_{d\bar{b}} g_{a\bar{e}} A^{\bar{e}}{}_{\bar{c}} - g_{a\bar{c}} g_{e\bar{b}} A_d{}^e),$$

and (244) follows from (241). To obtain (245), we apply (241) instead to the identity
$$R_{a\bar{b}c}{}^d \Lambda_d = (\nabla_a \nabla_{\bar{b}} - \nabla_{\bar{b}} \nabla_a) \Lambda_c = \nabla_a(-\mu g_{c\bar{b}} + 2BA_{c\bar{b}})$$
$$= 2(\nabla_a B)A_{c\bar{b}} - 2Bg_{a\bar{b}} \Lambda_c - g_{c\bar{b}}(\nabla_a \mu). \qquad \square$$

In Theorem 6.11, we show that if (M, g, J) of mobility ≥ 2 has c-projective nullity, then all solutions of the mobility equation lift uniquely to parallel sections of $\mathcal{V}$ for (240), where B is characterised by nullity. First we establish the following converse and regularity result.

THEOREM 6.7. *Let (M, J, g) be a connected (pseudo-)Kähler manifold with a non-parallel solution $A^{\bar{a}b}$ of the mobility equation, which lifts, over a dense open subset U of M, to a real parallel section $(A^{\bar{a}b}, \Lambda^a, \mu)$ for (240) with B locally constant. Then:*

(1) *B is constant and $(A^{\bar{a}b}, \Lambda^a, \mu)$ extends to a parallel section over M;*
(2) *$G_{a\bar{b}c}{}^d \Lambda_d = 0$, and hence (J, g) has c-projective nullity on M.*

PROOF. As noted in Remark 6.4, Theorem 4.6 provides a real section $(A^{\bar{a}b}, \Lambda^a, \mu')$ of $\mathcal{V}$ (defined on all of M) which is parallel for the connection given by (157) and (158). On U we compute, using (242b) and (243), that

(246) $$R_{a\bar{b}}{}^d{}_f A^{\bar{c}f} + R_{a\bar{b}}{}^{\bar{c}}{}_{\bar{e}} A^{\bar{e}d} = 2B(\delta_a{}^d A^{\bar{c}}{}_{\bar{b}} - \delta_{\bar{b}}{}^{\bar{c}} A_a{}^d),$$

which implies

(247) $$\tfrac{1}{n}(\mathrm{Ric}_{\bar{b}f} A^{\bar{c}f} + R_{a\bar{b}}{}^{\bar{c}}{}_{\bar{e}} A^{\bar{e}a}) = 2B(A^{\bar{c}}{}_{\bar{b}} - \tfrac{1}{n}\delta_{\bar{b}}{}^{\bar{c}} A_a{}^a).$$

Applying $\nabla_{\bar{d}}$ to (247) and taking the trace with respect to $\bar{d}$ and $\bar{c}$ shows that

(248) $$\tfrac{1}{(1-n)(n+1)}\big((\nabla_{\bar{d}} \mathrm{Ric}_{\bar{b}f}) A^{\bar{d}f} + (\nabla_{\bar{d}} R_{a\bar{b}}{}^{\bar{d}}{}_{\bar{e}}) A^{\bar{e}a} + (1-n)\mathrm{Ric}_{\bar{b}f} \Lambda^f\big) = 2B\Lambda_{\bar{b}}.$$

Recall that
$$-\mu \delta_a{}^b + 2BA_a{}^b = \nabla_a \Lambda^b = -\mu' \delta_a{}^b + \mathsf{P}_{a\bar{c}} A^{\bar{c}b} - \tfrac{1}{n} H_{a\bar{c}}{}^b{}_d A^{\bar{c}d}$$

and that $\mathsf{P}_{a\bar{b}} = \tfrac{1}{n+1} \mathrm{Ric}_{a\bar{b}}$. Hence, applying ∇_h to (248) and taking trace shows, together with the identities (247) and (248), that we have an identity of the form

(249) $$\begin{pmatrix} R & 0 & 0 \\ \nabla R & R & 0 \\ \nabla\nabla R + R \bowtie R & \nabla R & R \end{pmatrix} \begin{pmatrix} A^{\bar{a}b} \\ \Lambda^a \\ \mu' \end{pmatrix} = 2B\left[\begin{pmatrix} A^{\bar{a}b} \\ \Lambda^a \\ \mu' \end{pmatrix} - \frac{A_c{}^c}{n} \begin{pmatrix} g^{\bar{a}b} \\ 0 \\ \tfrac{1}{n} \mathsf{P}_d{}^d \end{pmatrix}\right].$$

Note that the section of $\mathcal{V}$ given by the expression inside the square brackets on the right-hand side is defined and nowhere vanishing on all of M: suppose it were vanishing at a point $x \in M$ and denote by c the trace of $A_a{}^b$ at x. Then Theorem 4.6 implies that $(A^{\bar{a}b}, \Lambda^a, \mu')$ and $\tfrac{c}{n}(g^{\bar{a}b}, 0, \tfrac{1}{n}\mathsf{P}_{\bar{a}b} g^{\bar{a}b})$ are two parallel sections of the connection given by (157) and (158) that coincide at x and hence everywhere, which contradicts our assumption that $A^{\bar{a}b}$ is not a constant multiple of $g^{\bar{a}b}$. Moreover, since the left-hand side of (249) is defined on all of M, the identity (249) can be used to extend B smoothly as a function to all of M. Since B is locally constant on U and M is connected, B is actually a constant and $(A^{\bar{a}b}, \Lambda^a, \mu)$ extends smoothly to a parallel section of the connection (240) on all of M.

The formula in the second part is immediate from (245) with $\nabla_a B = 0$ and $\nabla_a \mu = 2 B \Lambda_a$. It follows that (J, g) has c-projective nullity on the dense open set where Λ^a is nonzero. However, since B is constant (hence continuous), the set where (J, g) has c-projective nullity is closed. $\square$

REMARK 6.5. One may give an alternative proof of (1) of Theorem 6.7 without using the last line of (249). Specifically, the second line of (249) can be already used to extend B to a smooth function on the set $\{\Lambda^\alpha \neq 0\}$. Since $J_\alpha{}^\beta \Lambda^\alpha$ is a holomorphic Killing field, one can show that its zero locus consists of submanifolds of even codimension and hence $\{\Lambda^\alpha \neq 0\}$ is connected.

REMARK 6.6. When (J, g) has c-projective nullity, $\mathsf{P}_{a\bar{b}} - 2B g_{a\bar{b}}$ (with B given by (236)) is a metric c-projective invariant by Corollary 6.4, and hence the connection (240) is metric c-projectively invariant. In particular, by Theorem 6.7, the connection is metric c-projective invariant if B is constant and it admits a parallel section with Λ_a nonzero.

On the other hand, if the connection (240) admits a parallel section with $\Lambda_a = 0$, then (242b) shows that $B = 0$ unless the corresponding solution $A_{a\bar{b}}$ of the mobility equation is a (necessarily locally constant) multiple of $g_{a\bar{b}}$. Thus a parallel solution of the mobility equation which is not a multiple of g lifts to a parallel section for (240) if and only if $B = 0$.

THEOREM 6.8. *Let $(M, J, [\nabla])$ be a connected metric c-projective manifold of dimension $2n \geq 4$ arising from a (pseudo-)Kähler metric g with mobility ≥ 3. Then either (J, g) has c-projective nullity on M with B constant in (236), or $2n \geq 6$ and all metrics c-projectively equivalent to g are affinely equivalent to g (i.e. have the same Levi-Civita connection).*

To prove this theorem, we use a couple of lemmas, the first of which is a purely algebraic (pointwise) result.

LEMMA 6.9. *Suppose that $R_{a\bar{b}c\bar{d}}$ is a tensor which has Kähler symmetries (108) with respect to $g_{a\bar{b}}$. Let $A_{a\bar{b}}$, $\widetilde{A}_{a\bar{b}}$, $\Lambda_{a\bar{b}}$ and $\widetilde{\Lambda}_{a\bar{b}}$ be (real) tensors that satisfy*

$$(250) \qquad R_{a\bar{b}\bar{c}e} A_d{}^e + R_{a\bar{b}d\bar{f}} A^{\bar{f}}{}_{\bar{c}} = g_{a\bar{c}} \Lambda_{d\bar{b}} - g_{d\bar{b}} \Lambda_{a\bar{c}}$$

$$(251) \qquad R_{a\bar{b}\bar{c}e} \widetilde{A}_d{}^e + R_{a\bar{b}d\bar{f}} \widetilde{A}^{\bar{f}}{}_{\bar{c}} = g_{a\bar{c}} \widetilde{\Lambda}_{d\bar{b}} - g_{d\bar{b}} \widetilde{\Lambda}_{a\bar{c}}.$$

If $A_{a\bar{b}}$, $\widetilde{A}_{a\bar{b}}$ and $g_{a\bar{b}}$ are linearly independent, then $\Lambda_{a\bar{b}}$, respectively $\widetilde{\Lambda}_{a\bar{b}}$, are linear combinations of $g_{a\bar{b}}$ and $A_{a\bar{b}}$, respectively $g_{a\bar{b}}$ and $\widetilde{A}_{a\bar{b}}$, with the same second coefficient.

PROOF. Note first that these equations remain unchanged if we add scalar multiplies of $g_{a\bar{b}}$ to the tensors $A_{a\bar{b}}$, $\widetilde{A}_{a\bar{b}}$, $\Lambda_{a\bar{b}}$ and $\widetilde{\Lambda}_{a\bar{b}}$. Hence, we can assume without loss of generality that the trace of these tensors vanishes. We then have to show that $\Lambda_{a\bar{b}}$ and $\widetilde{\Lambda}_{a\bar{b}}$ are a common scalar multiple of $A_{a\bar{b}}$ and $\widetilde{A}_{a\bar{b}}$ respectively.

102 6. METRIC C-PROJECTIVE STRUCTURES AND NULLITY

From equation (250) it follows immediately that

(252) $\widetilde{A}_a{}^h(R_{h\bar{b}\bar{c}e}A_d{}^e + R_{h\bar{b}d\bar{f}}A^{\bar{f}}{}_{\bar{c}}) - \widetilde{A}^{\bar{i}}{}_{\bar{b}}(R_{a\bar{i}\bar{c}e}A_d{}^e + R_{a\bar{i}d\bar{f}}A^{\bar{f}}{}_{\bar{c}})$
$$= \widetilde{A}_{a\bar{c}}\Lambda_{d\bar{b}} + \widetilde{A}_{d\bar{b}}\Lambda_{a\bar{c}} - g_{a\bar{c}}\widetilde{A}^{\bar{i}}{}_{\bar{b}}\Lambda_{d\bar{i}} - g_{d\bar{b}}\widetilde{A}_a{}^h\Lambda_{h\bar{c}}.$$

By the symmetries (108) of $R_{a\bar{b}c\bar{d}}$, the left-hand side of identity (252) equals

$$(\widetilde{A}_a{}^h R_{h\bar{b}\bar{c}e} - \widetilde{A}^{\bar{i}}{}_{\bar{b}}R_{a\bar{i}\bar{c}e})A_d{}^e + (\widetilde{A}_a{}^h R_{h\bar{b}d\bar{f}} - \widetilde{A}^{\bar{i}}{}_{\bar{b}}R_{a\bar{i}d\bar{f}})A^{\bar{f}}{}_{\bar{c}}$$
$$= (\widetilde{A}_a{}^h R_{h\bar{b}\bar{c}e} + \widetilde{A}^{\bar{i}}{}_{\bar{b}}R_{e\bar{c}a\bar{i}})A_d{}^e - (\widetilde{A}_a{}^h R_{d\bar{f}bh} + \widetilde{A}^{\bar{i}}{}_{\bar{b}}R_{d\bar{f}a\bar{i}})A^{\bar{f}}{}_{\bar{c}}$$
(253) $\qquad = A_{a\bar{c}}\widetilde{\Lambda}_{d\bar{b}} + A_{d\bar{b}}\widetilde{\Lambda}_{a\bar{c}} - g_{a\bar{c}}A^{\bar{i}}{}_{\bar{b}}\widetilde{\Lambda}_{d\bar{i}} - g_{d\bar{b}}A_a{}^h\widetilde{\Lambda}_{h\bar{c}}$

where the last equality follows from (252). From (252) and (253) we therefore obtain

(254) $\qquad g_{a\bar{c}}\tau_{d\bar{b}} + g_{d\bar{b}}\tau_{a\bar{c}} = A_{d\bar{b}}\widetilde{\Lambda}_{a\bar{c}} + A_{a\bar{c}}\widetilde{\Lambda}_{d\bar{b}} - \widetilde{A}_{d\bar{b}}\Lambda_{a\bar{c}} - \widetilde{A}_{a\bar{c}}\Lambda_{d\bar{b}},$

where $\tau_{a\bar{b}} = A_a{}^e\widetilde{\Lambda}_{e\bar{b}} - \widetilde{A}^{\bar{f}}{}_{\bar{b}}\Lambda_{a\bar{f}}$. Taking the trace with respect to a and c yields

(255) $\qquad n\tau_{d\bar{b}} + g_{d\bar{b}}\tau_a{}^a = 0,$

which shows that $\tau_{a\bar{b}} = 0$. Therefore, we conclude from (254) that

$$A_{d\bar{b}}\widetilde{\Lambda}_{a\bar{c}} + A_{a\bar{c}}\widetilde{\Lambda}_{d\bar{b}} = \widetilde{A}_{d\bar{b}}\Lambda_{a\bar{c}} + \widetilde{A}_{a\bar{c}}\Lambda_{d\bar{b}}.$$

Since any nonzero tensor of this form determines its factors up to scale, and $A_{a\bar{b}}$ and $\widetilde{A}_{a\bar{b}}$ are linearly independent, we conclude that $\Lambda_{a\bar{b}}$ and $\widetilde{\Lambda}_{a\bar{b}}$ are the same multiple of $A_{a\bar{b}}$ and $\widetilde{A}_{a\bar{b}}$ respectively. □

We next relate linear dependence to pointwise linear dependence.

LEMMA 6.10. *Let (M, J, g) be a connected (pseudo-)Kähler $2n$-manifold ($n \geq 2$) and let $A_a{}^b$ be a solution of (128) such that $\widetilde{A}_a{}^b := p\delta_a{}^b + qA_a{}^b$ is also a solution for real functions p and q. Then p and q are constant or $A_a{}^b = \xi\delta_a{}^b$ for constant ξ.*

PROOF. By assumption, we have $\nabla_a A_b{}^c = -\delta_a{}^c\Lambda_b$ and $\nabla_a \widetilde{A}_b{}^c = -\delta_a{}^c\widetilde{\Lambda}_b$, hence

$$\nabla_a p\,\delta_b{}^c + \nabla_a q\,A_b{}^c = -(\widetilde{\Lambda}_b - q\Lambda_b)\delta_a{}^c$$

If $\nabla_a q = 0$, it follows easily that p and q are locally constant hence constant. Otherwise, contracting this expression with a nonzero tangent vector X^a in the kernel of $\nabla_a q$, we deduce that $\widetilde{\Lambda}_b = q\Lambda_b$ and $\nabla_a p = -\xi\nabla_a q$ for some function ξ. Thus $\nabla_a q\,(A_b{}^c - \xi\delta_b{}^c) = 0$. If $A_b{}^c = \xi\delta_b{}^c$, it follows from what we have already proven that ξ is constant. Otherwise, we deduce that p and q are constant. □

PROOF OF THEOREM 6.8. Suppose that $A_{a\bar{b}}$ and $\widetilde{A}_{a\bar{b}}$ are nondegenerate solutions of the mobility equation such that $g_{a\bar{b}}, A_{a\bar{b}}$ and $\widetilde{A}_{a\bar{b}}$ are linearly independent. At each point of M, (243) implies that $A_{a\bar{b}}$ and $\widetilde{A}_{a\bar{b}}$ satisfy (250)–(251), with $\Lambda_{a\bar{c}} = \nabla_a\Lambda_{\bar{c}}$ and $\widetilde{\Lambda}_{a\bar{c}} = \nabla_a\widetilde{\Lambda}_{\bar{c}}$. By Lemma 6.10, $g_{a\bar{b}}, A_{a\bar{b}}$ and $\widetilde{A}_{a\bar{b}}$ are pointwise linearly independent on a dense open set U', and hence, on U', Lemma 6.9 implies that $A_{a\bar{b}}$ and $\widetilde{A}_{a\bar{b}}$ lift to smooth solutions $(A_{a\bar{b}}, \Lambda_a, \mu)$ and $(\widetilde{A}_{a\bar{b}}, \widetilde{\Lambda}_a, \widetilde{\mu})$ of (242a)–(242b) for the same smooth function B. Thus we may apply Lemma 6.6.

The trace-free parts of $A_a{}^b$ and $\widetilde{A}_a{}^b$ are pointwise linearly independent on U', hence if $n = 2$, their common centraliser at each $p \in U'$ consists only of multiples of the identity. By (244), $G_{a\bar{b}c}{}^d$ is a multiple $\alpha_{a\bar{b}}$ of $\delta_a{}^d$, hence zero, since

6.3. MOBILITY, NULLITY, AND THE SPECIAL TRACTOR CONNECTION

$G_{a\bar{b}c}{}^d = G_{c\bar{b}a}{}^d$. Thus g has constant holomorphic sectional curvature, which proves the theorem for $2n = 4$.

To prove the theorem for $2n \geq 6$, we substitute (245) into $G_{a\bar{b}c}{}^d = G_{c\bar{b}a}{}^d$ to obtain

$$g_{c\bar{b}}(\nabla_a \mu - 2B\Lambda_a) - 2(\nabla_a B)A_{c\bar{b}} = g_{a\bar{b}}(\nabla_c \mu - 2B\Lambda_c) - 2(\nabla_c B)A_{a\bar{b}}.$$

If we contract this equation with a vector Y^c in the kernel of $\nabla_c B$, then since $n \geq 3$, we obtain a degenerate Hermitian form on the left hand side, equal to a multiple of $g_{a\bar{b}}$. Hence both sides vanish, i.e. Y^c is in the kernel of $\nabla_c \mu - 2B\Lambda_c$ and we have

$$A_{c\bar{b}}Y^c = \xi g_{c\bar{b}} Y^c \quad \text{and} \quad \nabla_a \mu - 2B\Lambda_a = 2\xi(\nabla_a B)$$

for some function ξ on U'. Hence (245) now reads

$$G_{a\bar{b}c}{}^d \Lambda_d = 2(\nabla_a B)(A_{c\bar{b}} - \xi g_{c\bar{b}}) = 2(\nabla_c B)(A_{a\bar{b}} - \xi g_{a\bar{b}}).$$

If $\nabla_c B$ is nonzero on an open subset of U', it follows that $A_a{}^b - \xi \delta_a{}^b$ has (complex) rank at most one there, with image spanned by $\nabla^a B$ and kernel containing the kernel of $\nabla_a B$. Since the same holds for $\widetilde{A}_a{}^b - \widetilde{\xi}\delta_a{}^b$ for some function $\widetilde{\xi}$, we have that $\delta_a{}^b, A_a{}^b$ and $\widetilde{A}_a{}^b$ are linearly dependent, a contradiction. Hence $\nabla_c B$ is identically zero on U', i.e. B is locally constant. The result now follows from Theorem 6.7. □

REMARK 6.7. The above proof shows (for mobility ≥ 3) that any solution of the mobility equation (129) lifts to a parallel section for (240), where B is given by (236), unless all solutions are parallel (i.e. affine equivalent to g). However, in the latter case, any solution of (129) lifts to a parallel section for (240) with $B = 0$ (cf. Remark 6.6). This establishes [**44**, Theorem 5]; the next result may be seen as a strengthening of this theorem in which c-projective nullity is brought to the fore, cf. also [**26**, Theorem 2].

THEOREM 6.11. *Let (M, J, g) be a connected (pseudo-)Kähler manifold admitting a solution of the mobility equation that is not a constant multiple of g. Assume that there is a dense open subset $U \subseteq M$ on which (J, g) has c-projective nullity and denote by B the function in (236). Then the following hold:*

- *B is constant and (J, g) has c-projective nullity on M;*
- *any solution $A^{\bar{a}b}$ of the mobility equation lifts uniquely to a section of $\mathcal{V}$ which is parallel for the special tractor connection (240).*

We divide the proof of Theorem 6.11 into several propositions.

PROPOSITION 6.12. *Under the assumptions of Theorem 6.11, there is a dense open subset $U' \subseteq U$ on which B is smooth, and for any solution $A_{b\bar{c}}$ of the mobility equation:*

(1) *there is a smooth real-valued function μ on U' such that (242b) holds, and if B is locally constant then (242c) also holds on U';*
(2) *for any vector v^a in the nullity distribution of (J, g),*

(256) $$2(\nabla_a B)A_{c\bar{b}}v^c + v_{\bar{b}}(2B\Lambda_a - \nabla_a \mu) = 0.$$

In particular, if v^a is not in any eigenspace of $A_a{}^b$ then B is locally constant.

PROOF. To see that (1) holds for $A_{b\bar c}$, first recall that Λ^a, given by (242a), is holomorphic. Next, by assumption, at any $p \in U$ there is a nonzero tangent vector $v^{\bar b}$ such that $v^{\bar b} G_{a\bar b c \bar d} = 0$. Hence, by equation (243), on U we have

$$
\begin{aligned}
-v_c \nabla_a \Lambda_{\bar d} + g_{a\bar d} v^{\bar b} \nabla_{\bar b} \Lambda_c &= v^{\bar b} R_{a\bar b c\bar e} A^{\bar e}{}_{\bar d} - v^{\bar b} R_{a\bar b e \bar d} A_c{}^e \\
&= -2B(g_{a\bar e} v_c + g_{c\bar e} v_a) A^{\bar e}{}_{\bar d} + 2B(g_{a\bar d} v_e + g_{e\bar d} v_a) A_c{}^e \\
&= 2B(g_{a\bar d} v^{\bar b} A_{c\bar b} - v_c A_{a\bar d})
\end{aligned}
\tag{257}
$$

and so

$$v_c V_{a\bar d} - g_{a\bar d} v^{\bar b} \overline{V}_{\bar b c} = 0,$$

where $V_{a\bar b} \equiv \nabla_a \Lambda_{\bar b} - 2B A_{a\bar b}$. As $v^{\bar b} \neq 0$ on U, it follows that $V_{a\bar b}$ is pure trace, i.e. the second equation of (242b) holds pointwise. By assumption, there is a dense open subset $U' \subseteq U$ on which $g_{a\bar b}$ and $A_{a\bar b}$ are pointwise linearly independent for some solution $A_{a\bar b}$, from which it follows that B is a smooth real-valued function on U'. Hence (242a)–(242b) hold on U' for any solution, with μ smooth on U'.

By Lemma 6.6, any solution satisfies (245), which implies (256). Now if $\nabla_a B = 0$ it follows immediately from the existence of nullity that (242c) holds on U'. □

Proposition 6.12 and Theorem 6.7 have the following immediate consequence.

COROLLARY 6.13. *The conclusions of Theorem 6.11 hold unless the nullity distribution is contained in an eigendistribution of every solution of the mobility equations.*

It remains to show that Theorem 6.11 also holds when the nullity distribution is contained in an eigendistribution of every solution of the mobility equations, and for this it suffices to show that $\nabla_a B = 0$ on a dense open set. Suppose then that v^b is a nonzero nullity vector such that $A_a{}^b v^a = \xi v^b$ for some smooth function ξ, so that $\nabla_a \mu - 2B \Lambda_a = 2\xi \nabla_a B$ by (256) and hence (245) reads

$$G_{a\bar b c}{}^d \Lambda_d = 2(\nabla_a B)(A_{c\bar b} - \xi g_{c\bar b}) = 2(\nabla_c B)(A_{a\bar b} - \xi g_{a\bar b}) \tag{258}$$

as in the proof of Theorem 6.8. Since v^a is an eigenvector of $A_a{}^b$ with eigenvalue ξ, $v^{\bar b}$ is an eigenvector of $A^{\bar a}{}_{\bar b}$ with eigenvalue $\bar\xi$. However $v^{\bar b}$ is in the nullity of G, so the contraction of (258) with $v^{\bar b}$ yields $(\nabla_c B)(A_{a\bar b} - \xi g_{a\bar b}) v^{\bar b} = 0$. If we now combine these observations with Proposition 6.5, we obtain that either $\nabla_a B = 0$ on a dense open set (and we are done) or there is an open set on which $\widetilde{A}_a{}^b := A_a{}^b - \xi \delta_a{}^b$ has (complex) rank at most one, v^a is a null vector in its kernel, and ξ is real. Hence the generalised ξ-eigenspace of $A_a{}^b$ is nondegenerate, and so has (complex) dimension at least two, which implies that ξ is locally constant by Lemma 5.16. Now $\widetilde{A}_a{}^b$ is a rank one solution of the mobility equation with a nonzero (but null) nullity vector in its kernel, and so Theorem 6.11 is a consequence of the following proposition.

PROPOSITION 6.14. *Suppose (M, J, g) is a connected (pseudo-)Kähler manifold of dimension $2n \geq 4$ admitting a non-parallel solution of the mobility equation $A_{a\bar b}$ such that $A_a{}^b$ is a complex endomorphism of rank 1. Assume that g has nullity on some dense open set $U \subseteq M$ and that there is a nonzero vector in the nullity distribution that is in the kernel of $A_a{}^b$. Then the function B defined as in (236) is a constant and the conclusions of Theorem 6.11 hold.*

Before we give a proof of Proposition 6.14 we collect some crucial information about solutions of the mobility equation of rank 1:

LEMMA 6.15. *Suppose (M, J, g) is a connected (pseudo-)Kähler manifold of dimension $2n \geq 4$ admitting a non-parallel solution of the mobility equation $A_{a\bar{b}}$ such that $A_a{}^b$ is a complex endomorphism of rank 1. Assume that g has nullity on some dense open set $U \subseteq M$ and that there is a nonzero vector in the nullity distribution that is in the kernel of $A_a{}^b$. Denote by B the function defined as in (236) and let $\Lambda_a = \nabla_a \lambda$ with $\lambda = -A_a{}^a$. Then the following holds on a dense open subset $U' \subseteq U$:*

(1) *the triple $(A_a{}^b, \Lambda^a, \mu)$ satisfies system (242) (and its conjugate) for some smooth nonvanishing real-valued function μ;*
(2) $A_{a\bar{b}} = \mu^{-1} \Lambda_a \Lambda_{\bar{b}}$ *and* $-\mu\lambda = \Lambda_a \Lambda^a$;
(3) $\nabla_a B$ *is proportional to Λ_a, and at any $x \in U'$ either $\nabla_a B = 0$ or the nullity space of g at x lies in the kernel of $A_a{}^b$.*

PROOF. Statement (1) follows immediately from (132), Proposition 6.12 and the existence of a nullity vector in the kernel of $A_a{}^b$. Since $A_a{}^b$ has rank 1, its nonzero eigenvalue is $-\lambda$, and Λ^a is a nonzero section of the corresponding eigenspace by Corollary 5.17. Thus on the dense open subset $U' \subseteq U$ where $\Lambda^a \neq 0$, $A_a{}^b = \xi \Lambda_a \Lambda^b$, with $\xi = -\lambda/(\Lambda_a \Lambda^a)$, and differentiating this identity using (242) yields

$$(\nabla_a \xi + 2B\xi^2 \Lambda_a)\Lambda_{\bar{c}} = (\xi\mu - 1)g_{a\bar{c}}.$$

Since the left hand side is simple and $g_{a\bar{b}}$ nondegenerate both sides must vanish, which shows that $\xi = \mu^{-1}$, and hence (2) holds. The identity (245) may now be written

$$G_{a\bar{b}c}{}^d \Lambda_d = 2(\nabla_a B) A_{c\bar{b}} = 2\mu^{-1}(\nabla_a B) \Lambda_c \Lambda_{\bar{b}}$$

This immediately implies the second statement of (3), while the first statement follows from the symmetry of $G_{a\bar{b}c}{}^d$ in a and c. □

PROOF OF PROPOSITION 6.14. We have already noted that to prove Proposition 6.14 it suffices to show that B is locally constant. By Lemma 6.15, $A_\alpha{}^\beta$ is of the form

(259) $$A_\alpha{}^\beta = \tfrac{1}{2\mu}(\Lambda_\alpha \Lambda^\beta + \Omega_{\gamma\alpha} \Lambda^\gamma J_\delta{}^\beta \Lambda^\delta).$$

Let us write $D \subset TM$ for the distribution defined by the kernel of $A_\alpha{}^\beta$ and

$$P_\alpha{}^\beta = \delta_\alpha{}^\beta - \tfrac{1}{\Lambda_\gamma \Lambda^\gamma}(\Lambda_\alpha \Lambda^\beta + \Omega_{\delta\alpha} \Lambda^\delta J_\zeta{}^\beta \Lambda^\zeta) = \delta_\alpha{}^\beta + \tfrac{1}{2\mu\lambda}(\Lambda_\alpha \Lambda^\beta + \Omega_{\delta\alpha} \Lambda^\delta J_\zeta{}^\beta \Lambda^\zeta)$$

for the orthogonal projection $P: TM \to D$, where we use Lemma 6.15(2) to rewrite $\Lambda_\gamma \Lambda^\gamma = 2\Lambda_c \Lambda^c = -2\lambda\mu$. Note that (J, g) induces by restriction a complex structure J^D and a J^D-invariant metric g_D on D. The projection P also determines a linear connection on D by

$$\nabla_\alpha^D X^\beta = P_\gamma{}^\beta \nabla_\alpha X^\gamma, \quad \text{for} \quad X \in \Gamma(D)$$

which preserves this Hermitian structure on D. Since Λ and $J\Lambda$ commute and preserve D, $\mathcal{L}_\Lambda P = 0$ and $\Gamma(D)$ is generated by sections commuting with Λ and $J\Lambda$, which are called *basic*. For any basic $X \in \Gamma(D)$, Lemma 6.15(1) implies

(260) $$\nabla_\Lambda X = \nabla_X \Lambda = -\mu X$$

and hence for any other basic element $Y \in \Gamma(D)$ we compute

$$(\mathcal{L}_\Lambda g_D)(X, Y) = \mathcal{L}_\Lambda(g_D(X, Y)) = \nabla_\Lambda(g_D(X, Y)) = -2\mu g_D(X, Y).$$

Since $\mathcal{L}_\Lambda \lambda = \nabla_\Lambda \lambda = g(\Lambda, \Lambda) = -2\lambda\mu$, it follows that $\mathcal{L}_\Lambda(\lambda^{-1} g_D) = 0$. Let us now regard ∇^D as a *partial connection* on D, i.e. an operator $\nabla^D : \Gamma(D) \to \Gamma(D^* \otimes D)$. Since $\nabla^D_X \lambda = \nabla_X \lambda = 0$ for any $X \in \Gamma(D)$, the partial connection ∇^D preserves $\lambda^{-1} g_D$. Furthermore, its *partial torsion*, given by

$$\nabla^D_X Y - \nabla^D_Y X - P([X,Y]) \quad \text{for} \quad X, Y \in \Gamma(D),$$

vanishes. It follows that $\mathcal{L}_\Lambda \nabla^D$ is a section of $D^* \otimes D^* \otimes D \cong D^* \otimes D^* \otimes D^*$, which is symmetric in the first two entries and skew in the last two entries, which implies it vanishes identically. We conclude that $\mathcal{L}_\Lambda R^D = 0$, where the *horizontal curvature* R^D of ∇^D is defined, for $X, Y, Z \in \Gamma(D)$, by

$$R^D(X,Y)(Z) = \nabla^D_X \nabla^D_Y Z - \nabla^D_Y \nabla^D_X Z - \nabla^D_{P([X,Y])} Z.$$

For basic $X, Y, Z \in \Gamma(D)$ we compute via (260) that

$$\nabla^D_X \nabla^D_Y Z = P(\nabla_X \nabla_Y Z) - \tfrac{1}{2\lambda} g(\nabla_Y Z, \Lambda) X - \tfrac{1}{2\lambda} g(\nabla_Y Z, J\Lambda) JX$$
$$\nabla^D_{\nabla^D_X Y} Z = P(\nabla_{\nabla_X Y} Z) - \tfrac{1}{2\lambda} g(\nabla_X Y, \Lambda) Z - \tfrac{1}{2\lambda} g(\nabla_X Y, J\Lambda) JZ.$$

Using $g(Y, \Lambda) = 0 = g(Y, J\Lambda)$ for $Y \in \Gamma(D)$, we also obtain, for $X \in \Gamma(D)$, that

$$g(\nabla_X Y, \Lambda) = \mu g(X, Y) \quad \text{and} \quad g(\nabla_X Y, J\Lambda) = \mu g(Y, JX),$$

from which we deduce, for (basic) $X, Y, Z \in \Gamma(D)$, that

(261) $$R^D(X,Y)Z = P(R(X,Y)Z) - \tfrac{\mu}{2\lambda} S(X,Y) Z,$$

where S is the constant holomorphic sectional curvature tensor defined as in (235).

Let us write $\mathrm{Ric}^D(Y,Z) = \mathrm{trace}(X \mapsto R^D(X,Y)Z)$ and $\mathrm{Ric}^P(Y,Z) = \mathrm{trace}(X \mapsto P(R(X,Y)Z))$ for the Ricci-type contractions of R^D and $P(R(X,Y)Z)$. Via the inverse g_D^{-1} of g_D, we view Ric^D and Ric^P as endomorphism of D, from which viewpoint equation (261) implies that they are related as follows:

(262) $$\mathrm{Ric}^D = \mathrm{Ric}^P - \tfrac{n\mu}{\lambda} \mathrm{Id}_D.$$

By assumption, at each point of a dense open subset, there is a vector V in D that lies in the nullity distribution of g. Inserting V into equation (261) yields

(263) $$R^D(X,V)Z = \bigl(B - \tfrac{\mu}{2\lambda}\bigr) S(X,V) Z,$$

which implies that

(264) $$\lambda \mathrm{Ric}^D(V) = 2n\bigl(B\lambda - \tfrac{\mu}{2}\bigr) V.$$

Set $C := B\lambda - \tfrac{\mu}{2}$. By (1) and (3) of Lemma 6.15 we see that $\nabla_X C = 0$ for all $X \in \Gamma(D)$ and that $\nabla_{J\Lambda} C = 0$. Equation (264) shows that V is an eigenvector of $\lambda \mathrm{Ric}^D$ with eigenvalue $2C$. Since $\mathcal{L}_\Lambda R^D = 0$ and $\mathcal{L}_\Lambda(\lambda g_D^{-1}) = 0$, it follows that $\mathcal{L}_\Lambda(\lambda \mathrm{Ric}^D) = 0$, and hence $\nabla_\Lambda C = \mathcal{L}_\Lambda C = 0$ as well. Thus C is locally constant, which implies that B is locally constant by Lemma 6.15(1), and this completes the proof. $\square$

REMARK 6.8. Proposition 6.14 may alternatively be proved as follows. Starting with the usual equations

(265) $$\begin{aligned} \nabla_\alpha \Lambda_\beta &= -\mu g_{\alpha\beta} + 2B A_{\alpha\beta} \\ \nabla_\alpha \mu &= 2B \Lambda_\alpha \\ \nabla_\alpha \lambda &= \Lambda_\alpha \end{aligned} \quad \text{where} \quad \begin{aligned} A_{\alpha\beta} &= \frac{\Lambda_\alpha \Lambda_\beta + K_\alpha K_\beta}{2\mu} \\ K^\alpha &= J_\beta{}^\alpha \Lambda^\beta \\ \Lambda_\alpha \Lambda^\alpha &+ 2\lambda\mu = 0 \end{aligned}$$

we may consider the new metric

(266) $$\tilde{g}^{\alpha\beta} \equiv \lambda g^{\alpha\beta} + (1+2\mu)A^{\alpha\beta}$$

and verify from (265) that

- $\mathcal{L}_\Lambda \tilde{g}^{\alpha\beta} = 0$,
- for any v^β such that $A_{\alpha\beta}v^\beta = 0$, we have

(267) $$\widetilde{\mathrm{Ric}}_{\alpha\beta}v^\beta = \mathrm{Ric}_{\alpha\beta}v^\beta - \left(2B + \frac{n\mu}{\lambda} + \frac{1}{2\lambda}\right)v_\alpha,$$

where $\widetilde{\mathrm{Ric}}_{\alpha\beta}$ is the Ricci tensor of $\tilde{g}^{\alpha\beta}$.

Hence, if v^β is a nullity vector for $g_{\alpha\beta}$ so that in addition $\mathrm{Ric}_{\alpha\beta}v^\beta = 2(n+1)Bv_\alpha$, then

(268) $$\tilde{g}^{\alpha\gamma}\widetilde{\mathrm{Ric}}_{\gamma\beta}v^\beta = \left(n(2B\lambda - \mu) - \tfrac{1}{2}\right)v^\alpha.$$

Now, since $\mathcal{L}_\Lambda(\tilde{g}^{\alpha\gamma}\widetilde{\mathrm{Ric}}_{\gamma\beta}) = 0$, it follows that any eigenvalue of this endomorphism is preserved by the flow of Λ^α. Therefore,

$$0 = \mathcal{L}_\Lambda\left(n(2B\lambda - \mu) - \tfrac{1}{2}\right) = n\left(2\lambda\mathcal{L}_\Lambda B + \Lambda^\alpha(2B\nabla_\alpha\lambda - \nabla_\alpha\mu)\right) = 2n\lambda\mathcal{L}_\Lambda B.$$

But from (265) we see that $0 = \nabla_{[\alpha}\nabla_{\beta]}\mu = 2(\nabla_{[\alpha}B)\Lambda_{\beta]}$ whence $\nabla_\alpha B = 0$, as required.

The only drawback with this proof is that verifying (267), though straightforward, is computationally severe, whereas the corresponding identity (264) in the previous proof is more easily established. The previous proof may be seen as a limiting case of the reasoning just given. Specifically, for any constant $c \neq 0$, consider the metric

(269) $$\tilde{g}_{\alpha\beta} \equiv \frac{1}{\lambda}g_{\alpha\beta} + \frac{1}{\lambda^2}\left(1 + \frac{c}{\mu}\right)A_{\alpha\beta} \quad \text{with inverse} \quad \tilde{g}^{\alpha\beta} \equiv \lambda g^{\alpha\beta} + \left(1 + \frac{\mu}{c}\right)A^{\alpha\beta}$$

to arrive at

(270) $$\widetilde{\mathrm{Ric}}_{\alpha\beta}v^\beta = \mathrm{Ric}_{\alpha\beta}v^\beta - \left(2B + \frac{n\mu}{\lambda} + \frac{c}{\lambda}\right)v_\alpha$$

instead of (267), an equation in which one can take a sensible limit as $c \to 0$ essentially to arrive at (264) instead of (268). The metrics (269) and their invariance $\mathcal{L}_\Lambda \tilde{g}_{\alpha\beta} = 0$ can also be recognised in the previous proof. More precisely, the first equation from (265) can be expressed as $\mathcal{L}_\Lambda g_{\alpha\beta} = -2\mu g_{\alpha\beta} + 4BA_{\alpha\beta}$ or, more compactly, as

$$\mathcal{L}_\Lambda(\lambda^{-1}g_{\alpha\beta}) = 4\lambda^{-1}BA_{\alpha\beta},$$

which implies, using our earlier terminology, that the metric $\lambda^{-1}g_{\alpha\beta}$ restricted to D is invariant under the flow of Λ^α. We also observed in the previous proof that orthogonal projection $P_\alpha{}^\beta = \delta_\alpha{}^\beta + \lambda^{-1}A_\alpha{}^\beta$ onto D is invariant under this flow. We are therefore led to invariance of the covariant quadratic form

$$P_\alpha{}^\gamma P_\beta{}^\epsilon \lambda^{-1}g_{\gamma\epsilon} = \lambda^{-1}(g_{\alpha\beta} + \lambda^{-1}A_{\alpha\beta}),$$

which is the limit of (269) as $c \to 0$ whilst the nondegenerate metric $\tilde{g}_{\alpha\beta}$ is obtained by decreeing that the remaining vectors Λ^α and K^α at each point be orthogonal to D and each other and satisfy $\tilde{g}_{\alpha\beta}\Lambda^\alpha\Lambda^\beta = \tilde{g}_{\alpha\beta}K^\alpha K^\beta = 2c$. The metric (266) is the case that Λ^α and K^α are taken to be orthonormal. In any case, it follows that $\mathcal{L}_\Lambda \tilde{g}_{\alpha\beta} = 0$. □

6.4. The standard tractor bundle for metric c-projective structures

The metric theory of the standard tractor bundle $\mathcal{T}$ turns out to be rather degenerate. For a metric c-projective structure $(M, J, [\nabla])$ induced by the Levi-Civita connection ∇ of a Kähler metric g, we have $\mathsf{P}_{ab} = 0$ and so the standard tractor connection (52) is given by

$$(271) \qquad \nabla_a^\mathcal{T} \begin{pmatrix} X^b \\ \rho \end{pmatrix} = \begin{pmatrix} \nabla_a X^b + \rho \delta_a{}^b \\ \nabla_a \rho \end{pmatrix} \qquad \nabla_{\bar{a}}^\mathcal{T} \begin{pmatrix} X^b \\ \rho \end{pmatrix} = \begin{pmatrix} \nabla_{\bar{a}} X^b \\ \nabla_{\bar{a}} \rho - \mathsf{P}_{\bar{a}b} X^b \end{pmatrix}.$$

The kernel $\ker D^\mathcal{T}$ of the first BGG operator (59) consists of vector fields X^b with c-projective weight $(-1, 0)$ which satisfy

$$(272) \qquad \nabla_a X^b + \rho \delta_a{}^b = 0 \qquad \nabla_{\bar{a}} X^b = 0$$

for some section ρ of $\mathcal{E}(-1, 0)$; then $\rho = -\frac{1}{n} \nabla_a X^a$, and, setting the torsion to zero in Proposition 3.3, (X^a, ρ) defines a parallel section for the tractor connection (271). This is similar to the projective case, with the following distinction: although the tensor η^α in Theorem 6.1 is projectively weighted, the bundle $\mathcal{E}(1)$ is canonically trivialised by a choice of metric; here, in contrast, it is the real line bundle $\mathcal{E}(1,1)$ that enjoys such a trivialisation, and not the complex line bundle $\mathcal{E}(1,0)$.

However, taking care to use (45) (see also Proposition 2.13), it follows that any solution of (272) satisfies

$$(273) \qquad \begin{aligned} -\delta_b{}^c \nabla_{\bar{a}} \rho &= (\nabla_{\bar{a}} \nabla_b - \nabla_b \nabla_{\bar{a}}) X^c = R_{\bar{a}b}{}^c{}_d X^d + \mathsf{P}_{\bar{a}b} X^c \\ &= H_{\bar{a}b}{}^c{}_d X^d - \delta_b{}^c \mathsf{P}_{\bar{a}d} X^d, \end{aligned}$$

where $H_{\bar{a}b}{}^c{}_d = H_{\bar{a}d}{}^c{}_b$ and $H_{\bar{a}b}{}^b{}_d = 0$. We may rearrange this as

$$(274) \qquad \delta_b{}^c \nabla_{\bar{a}} \rho = \delta_b{}^c \mathsf{P}_{\bar{a}d} X^d - H_{\bar{a}b}{}^c{}_d X^d;$$

then the trace over b and c shows that $\nabla_{\bar{a}} \rho = \mathsf{P}_{\bar{a}b} X^b$ (as in Proposition 3.3) and hence that $H_{\bar{a}b}{}^c{}_d X^d = 0$. Following the projective case (Theorem 6.1), we lower an index in (273) and (274) to obtain

$$(275) \qquad R_{\bar{a}b\bar{c}d} X^d = -g_{b\bar{c}} \nabla_{\bar{a}} \rho - \mathsf{P}_{\bar{a}b} X_{\bar{c}}.$$

It follows that for any solutions (X^a, ρ) and $(\widetilde{X}^a, \tilde{\rho})$ of (272),

$$R_{\bar{a}b\bar{c}d} \widetilde{X}^b X^d = -\widetilde{X}_{\bar{c}} \nabla_{\bar{a}} \rho - \mathsf{P}_{\bar{a}b} \widetilde{X}^b X_{\bar{c}} = -\widetilde{X}_{\bar{c}} \nabla_{\bar{a}} \rho - X_{\bar{c}} \nabla_{\bar{a}} \tilde{\rho}$$

and hence, by symmetry,

$$X_{[\bar{a}} \nabla_{\bar{c}]} \tilde{\rho} = \widetilde{X}_{[\bar{c}} \nabla_{\bar{a}]} \rho.$$

As in Theorem 6.1, by first taking $\widetilde{X} = X$, we conclude that there is a real function B, uniquely determined and smooth on the union of the open sets where some solution X^a of (272) is nonzero, such that for any solution (X^a, ρ) of (272),

$$(276) \qquad \nabla_{\bar{a}} \rho = \mathsf{P}_{\bar{a}b} X^b = 2 B X_{\bar{a}}.$$

THEOREM 6.16. *Let $(M, J, [\nabla])$ be a connected c-projective manifold, where ∇ preserves a (pseudo-)Kähler metric $g_{a\bar{b}}$. Suppose that $\dim \ker D^\mathcal{T} \geq 2$. Then there is a unique constant B such that any element of the kernel of $D^\mathcal{T}$ lifts to a parallel section of $\mathcal{T}$ for the connection*

$$(277) \qquad \nabla_a \begin{pmatrix} X^b \\ \rho \end{pmatrix} = \begin{pmatrix} \nabla_a X^b + \rho \delta_a{}^b \\ \nabla_a \rho \end{pmatrix} \qquad \nabla_{\bar{a}} \begin{pmatrix} X^b \\ \rho \end{pmatrix} = \begin{pmatrix} \nabla_{\bar{a}} X^b \\ \nabla_{\bar{a}} \rho - 2 B g_{\bar{a}b} X^b \end{pmatrix}.$$

PROOF. By Proposition 3.3 and (276), it remains only to show that the smooth function B is actually a constant. Differentiating the equation $2Bg_{\bar{c}b}X^b = \nabla_{\bar{c}}\rho$ and using (272) gives

(278) $\quad 2(\nabla_a B)X_{\bar{c}} - 2Bg_{\bar{c}a}\rho = \nabla_a \nabla_{\bar{c}}\rho \quad$ and $\quad 2(\nabla_{\bar{a}}B)X_{\bar{c}} = \nabla_{\bar{a}}\nabla_{\bar{c}}\rho.$

With (46), the second equation of (278) implies that $X_{[\bar{c}}\nabla_{\bar{a}]}B = 0$. Where there are two nonzero solutions X^a and $\widetilde{X}^a$, it follows from (272) that the sections $X^a, \widetilde{X}^b \in \Gamma(M, T^{1,0}(-1,0))$ and therefore $X^{[a}\widetilde{X}^{b]} \in \Gamma(M, T^{2,0}(-2,0))$ are holomorphic. Consequently, $U = \{X^{[a}\widetilde{X}^{b]} \neq 0\}$ is the complement of an analytic subvariety and is thus connected. On U we also have $\widetilde{X}_{[\bar{c}}\nabla_{\bar{a}]}B = 0$ whence $\nabla_{\bar{a}}B = 0$, i.e. B is locally constant on U, hence constant. □

By analogy with the projective case, one might now expect c-projective nullity to appear. However, when we combine (275) and (276), we obtain

$$\left(R_{\bar{a}b\bar{c}d} + \mathsf{P}_{\bar{a}b}g_{\bar{c}d} + 2Bg_{\bar{c}b}g_{\bar{a}d}\right)X^d = 0,$$

which is a halfway house on the way to (237). Underlying this degeneracy is the fact that $\mathcal{T}$ is associated to a holomorphic representation of $\mathrm{SL}(n+1,\mathbb{C})$.

Nevertheless the constant B in Theorem 6.16 is generically characterised by c-projective nullity in the following degenerate sense.

THEOREM 6.17. *Let (M, J, g) be a connected (pseudo-)Kähler manifold admitting a non-parallel solution X^a of (272). For any function B, the following are equivalent*:

(1) B *is characterised by c-projective nullity (237) on a dense open subset*;
(2) B *is constant and X^b lifts to a section of $\mathcal{T}$ parallel for (277)*;
(3) $\mathsf{P}_{ab} = 2Bg_{\bar{a}b}$.

In particular, g is an Einstein metric, and the connections (271) and (277) coincide.

PROOF. (1)⇒(2). This follows from Theorem 6.11 because $X^{\bar{a}} \otimes X^b$ is a solution of the mobility equation which is not a constant multiple of $g^{\bar{a}b}$, and by contracting (273) by a nullity vector v^b.

(2)⇒(3). The identity (45) implies

$$\mathsf{P}_{\bar{a}b}\rho = (\nabla_{\bar{a}}\nabla_b - \nabla_b\nabla_{\bar{a}})\rho = -2Bg_{\bar{a}c}\nabla_b X^c = 2Bg_{\bar{a}b}\rho,$$

which establishes (3) on the dense open subset $\{\rho \neq 0\}$, hence everywhere.

(3)⇒(1). Since $\nabla_{\bar{a}}\rho = \mathsf{P}_{\bar{a}d}X^d = 2Bg_{\bar{a}d}X^d$, equation (273) implies

(279) $\quad -2Bg_{\bar{a}d}X^d\delta_b{}^c = R_{\bar{a}b}{}^c{}_d X^d + \mathsf{P}_{\bar{a}b}X^c = R_{\bar{a}b}{}^c{}_d X^d + 2Bg_{\bar{a}b}X^c,$

and we deduce that $G_{\bar{a}b}{}^c{}_d X^d = 0$. Hence, X^d/ρ is a nullity vector for g on the dense open subset $\{\rho \neq 0\}$. □

6.5. Special tractor connections and the complex cone

Let $(M, J, [\nabla])$ be a metric c-projective structure. Then for any compatible metric g and any function B, there is a special tractor connection on $\mathcal{T}$ defined by (277). We first observe that the induced connection on $\mathcal{V}_{\mathbb{C}} = \overline{\mathcal{T}} \otimes \mathcal{T}$ is the special tractor connection (240) (for the given g and B). This can be seen easily by taking $A^{\bar{b}c} = \overline{X^b}X^c$, $\Lambda^a = \bar{\rho}X^a$ and $\mu = \bar{\rho}\rho$ in (240). Consequently, parallel sections for the special tractor connection on $\mathcal{V}$ define parallel Hermitian forms on $\mathcal{T}^*$. This was used in [26] to characterise, for Kähler manifolds (M, J, g), the presence of nontrivial parallel sections for (240) in terms of the local classification of [2] (see Section 5.7). Using the extension of this classification (pseudo-)Kähler manifolds [16], together with Remark 5.7, and Theorems 6.7 and 6.11, we have the following more general characterisation.

THEOREM 6.18. *Let (M, J, g) be a connected (pseudo-)Kähler manifold admitting a solution A of the mobility equation which is not parallel (i.e. $\Lambda \neq 0$), and let B be a function on M. Then the following are equivalent.*

(1) *For the given function B, (J,g) has c-projective nullity on a dense open set.*
(2) *Any solution to the mobility equation lifts to a global parallel section for (240) with B constant.*
(3) *On a dense open subset of M, A lifts to a parallel section for (240) with B locally constant.*
(4) *Λ lies in the c-projective nullity of (J,g) with constant B.*
(5) *B is constant and its c-projective nullity distribution contains the complex span (tangent to the complex orbits) of the Killing vector fields of the pencil $A + t\,\mathrm{Id}$.*
(6) *Near any regular point, (J,g) is given by (222), where for all $j \in \{1, \ldots \ell\}$, $\Theta_j(t) = \Theta(t)$, a polynomial of degree $\ell + 1$, with constant coefficients, leading coefficient $4B$, and divisible by any constant coefficient factor of the minimal polynomial of A.*

PROOF. (1) $\Rightarrow$ (2) by Theorem 6.11, (2) $\Rightarrow$ (3) trivially, and (3) $\Rightarrow$ (4) by Theorem 6.7, and (4) $\Rightarrow$ (1) because Λ is nonzero on a dense open subset of M. Clearly (5) $\Rightarrow$ (4), and conversely, (1)–(4) imply (244), i.e. $G_{a\bar{b}c}{}^d = R_{a\bar{b}c}{}^d + 2B(g_{a\bar{b}}\delta_c{}^d + g_{c\bar{b}}\delta_a{}^d)$ commutes with $A_a{}^b$: therefore, since $G_{a\bar{b}c}{}^d \Lambda_d = 0$ and Λ^α is the sum of the gradients of the nonconstant eigenvalues $\xi_1, \ldots \xi_\ell$ of $A_a{}^b$, which are sections of the corresponding eigenspaces, it follows that $\mathrm{grad}_g \xi_j$ and $J\mathrm{grad}_g \xi_j$ are in the nullity for $j \in \{1, \ldots \ell\}$; this is the tangent distribution to the complex orbits.

Now (5) implies that the restriction of the metric to the complex orbits has constant holomorphic sectional curvature, and hence the functions $\Theta_j(t)$ are equal to a common polynomial $\Theta(t)$ of degree $\ell + 1$. Now comparing (242b) with (225), we conclude that $\Theta(t)$ has leading coefficient $a_{-1} = 4B$, and that $g_u(\Theta(A)\cdot,\cdot) = 0$ for all irreducible constant coefficient factors ρ_u of χ_A; thus all constant coefficient factors of the minimal polynomial of A are also factors of Θ.

Conversely, given (6), (225) implies (242b) with B constant, and hence, by equation (245) of Lemma 6.6, and the Bianchi symmetry $G_{a\bar{b}c}{}^d = G_{c\bar{b}a}{}^d$, we have

$$Y_{\bar{b}}(\nabla_a \mu - 2B\Lambda_a) = -G_{a\bar{b}c}{}^d \Lambda_d Y^c = g_{a\bar{b}}(\nabla_c \mu - 2B\Lambda_c)Y^c$$

for any $(1,0)$-vector Y^c. Since the left hand side is degenerate, whereas $g_{a\bar{b}}$ is nondegenerate, we conclude that $\nabla_a \mu = 2B\Lambda_a$, thus establishing (3). □

Secondly, we observe that by Lemma 3.4, the special tractor connection on $\mathcal{T}$ induces a complex affine connection on the complex affine cone $\pi_\mathcal{C}\colon \mathcal{C} \to M$ described in Chapter 3.2. Combining these observations, as in Remark 4.6, any solution A^{bc} of the mobility equation which lifts to a parallel section of $\mathcal{V}$ for the special tractor connection on $\mathcal{V}$, and is nondegenerate as a Hermitian form on $\mathcal{T}^*$, induces a Hermitian metric on $\mathcal{C}$ which is parallel for the induced complex affine connection on $\mathcal{C}$.

In particular, if B is constant, then the metric g itself induces a parallel section of $\mathcal{V}$ with $A^{b\bar{c}} = g^{b\bar{c}}$, $\Lambda^a = 0$ and $\mu = 2B$, which is clearly nondegenerate on $\mathcal{T}$ if and only if $B \neq 0$.

PROPOSITION 6.19. *Let $(M, J, [\nabla])$ be a metric c-projective structure. Then any compatible metric g and any real constant $B \neq 0$ induce a Hermitian metric on the complex affine cone $\mathcal{C} \to M$ defined by the c-projective structure.*

For $B > 0$, this Hermitian metric is, up to scale, a metric cone $dr^2 + r^2 \hat{g}$ over a (pseudo-)Sasakian metric $\hat{g} = g + (d\psi + \alpha)^2$ on a (local) circle bundle over (M, J, g) whose curvature $d\alpha$ is a multiple of the Kähler form ω of g (see e.g. [**19**]). In the present context, as observed by [**79**, **82**], any solution of the mobility equation which lifts to parallel section for the special tractor connection (240) on $\mathcal{V}$ (with $B \neq 0$ constant) induces a Hermitian $(0, 2)$ tensor on M which is parallel for the cone metric of g, B or $-g, -B$. Combining this with Theorem 6.18, we have the following extension of one of the key results of [**79**] to the (pseudo-)Kähler case.

THEOREM 6.20. *Let (M, J, g) be a connected (pseudo-)Kähler manifold admitting a non-parallel solution of the mobility equation, and assume that there is a dense open subset $U \subseteq M$ on which (J, g) has c-projective nullity. Then, perhaps after replacing g by a c-projectively equivalent metric, there is a nonzero constant B such that solutions of the mobility equation on M are in bijection with Hermitian $(0, 2)$ tensors on the cone $\mathcal{C}$ that are parallel for the Hermitian metric determined by g and B as in Proposition 6.19.*

PROOF. Under these assumptions, the equivalent conditions of Theorem 6.18 apply. Hence the constant $4B$ is the leading coefficient of the common polynomial $\Theta_j(t) = \Theta(t)$ of degree $\ell + 1$ appearing in (222), which therefore vanishes if and only if Θ has a root at infinity. However, by Remark 5.6, Θ transforms as a polynomial of degree $\ell + 1$ over projective changes of pencil parameter, and since Θ is not identically zero, we may change the pencil parameter so that ∞ is not a root, while keeping the metrisability solution nondegenerate. We thus obtain a c-projectively equivalent metric with nullity constant $B \neq 0$, and the rest follows from Theorem 6.18(2), Proposition 6.19 and the subsequent observations (above) applied to this metric. □

By Theorem 6.8, the hypotheses of this theorem are satisfied when (M, J, g) has mobility at least 3 and the compatible metrics are not all affinely equivalent. In the case that g is positive definite, this allowed V. Matveev and S. Rosemann [**79**] to obtain the following classification of the possible mobilities of Kähler metrics, using the de Rham (or de Rham–Wu) decomposition of the cone.

THEOREM 6.21. *Let (M, J, g) be a simply connected Kähler manifold of dimension $2n \geq 4$ admitting a non-parallel solution of the mobility equation. Then the mobility of (g, J) has the form $k^2 + \ell$ for $k, \ell \in \mathbb{N}$ with $0 \leq k \leq n - 1$, $1 \leq \ell \leq (n + 1 - k)/2$ and $(k, \ell) \neq (0, 1)$, unless (g, J) has constant holomorphic sectional curvature. Furthermore, any such value arises in this way.*

6.6. The c-projective Hessian, nullity, and the Tanno equation

Let (M, J, g) be a (pseudo-)Kähler manifold of dimension $2n \geq 4$ and denote by ∇ the Levi-Civita connection of g. Recall that for any solution $A_{a\bar{b}}$ of the mobility equation of g the gradient, respectively the skew gradient, of the function $A_a{}^a = -\lambda$ is a holomorphic vector field, respectively a holomorphic Killing field, which is, by Corollary 5.5, equivalent to the real section $\sigma = \lambda \mathrm{vol}(g)^{-\frac{1}{n+1}} \in \mathcal{E}(1,1)$ being in the kernel of the c-projective Hessian. With respect to the Levi-Civita connection and the trivialisation $\mathrm{vol}(g)^{-\frac{1}{n+1}}$ of $\mathcal{E}(1,1)$, the c-projective Hessian equation reads as

$$\nabla_a \nabla_b \lambda = 0, \quad \text{respectively} \quad \nabla_{\bar{a}} \nabla_{\bar{b}} \lambda = 0. \tag{280}$$

In Section 4.7 we prolonged the c-projective Hessian equation and have seen that any (real) solution of this equation lifts to a unique parallel section of the connection (178)–(179), which shows that (280) implies that the function $\lambda = \bar\lambda$ satisfies also the following system of equations

$$\begin{aligned}
\nabla_a \nabla_b \nabla_c \lambda &= 0 & \nabla_a \nabla_{\bar b} \nabla_c \lambda &= -\mathsf{P}_{c\bar b} \nabla_a \lambda - \mathsf{P}_{a\bar b} \nabla_c \lambda - H_{a\bar b}{}^d{}_c \nabla_d \lambda \\
\nabla_{\bar a} \nabla_b \nabla_c \lambda &= 0 & \nabla_{\bar a} \nabla_{\bar b} \nabla_c \lambda &= -\mathsf{P}_{c\bar b} \nabla_{\bar a} \lambda - \mathsf{P}_{c\bar a} \nabla_{\bar b} \lambda - H_{\bar a c}{}^{\bar d}{}_{\bar b} \nabla_{\bar d} \lambda.
\end{aligned} \tag{281}$$

Suppose now that g has mobility ≥ 2 and that g has nullity on a dense open subset of M. Then Theorem 6.11 shows that the function B defined as in (236), is actually constant and any solution $A_{a\bar b}$ lifts to a (real) parallel section of $\mathcal{V}_\mathbb{C}$ for the connection (240). The connection (240) induces a connection on the dual vector bundle $\mathcal{W}_\mathbb{C} = \mathcal{V}_\mathbb{C}^*$, which is given by

$$\nabla_a^{\mathcal{W}_\mathbb{C}} \begin{pmatrix} \lambda \\ \mu_b \mid \nu_{\bar b} \\ \zeta_{b\bar c} \end{pmatrix} = \begin{pmatrix} \nabla_a \lambda - \mu_a \\ \nabla_a \mu_b \mid \nabla_a \nu_{\bar b} + 2B g_{a\bar b} \lambda - \zeta_{a\bar b} \\ \nabla_a \zeta_{b\bar c} + 2B g_{a\bar c} \mu_b \end{pmatrix} \tag{282}$$

$$\nabla_{\bar a}^{\mathcal{W}_\mathbb{C}} \begin{pmatrix} \lambda \\ \mu_b \mid \nu_{\bar b} \\ \zeta_{b\bar c} \end{pmatrix} = \begin{pmatrix} \nabla_{\bar a} \lambda - \nu_{\bar a} \\ \nabla_{\bar a} \mu_b + 2B g_{\bar a b} \lambda - \zeta_{b\bar a} \mid \nabla_{\bar a} \nu_{\bar b} \\ \nabla_{\bar a} \zeta_{b\bar c} + 2B g_{\bar a b} \nu_{\bar c} \end{pmatrix}, \tag{283}$$

where $\mathcal{W}_\mathbb{C}$ is identified via g with a direct sum of unweighted tensor bundles. By Proposition 2.13 the two equations on the right-hand side of (281) can be also written as

$$\nabla_a \nabla_{\bar b} \nabla_c \lambda = -R_{a\bar b}{}^d{}_c \nabla_d \lambda \quad \text{and} \quad \nabla_{\bar a} \nabla_{\bar b} \nabla_c \lambda = -R_{\bar a c}{}^{\bar d}{}_{\bar b} \nabla_{\bar d} \lambda.$$

Comparison of these equations with (282) and (283), shows immediately that any function λ satisfying (280) lifts to a parallel section for the special connection (282)–(283) if and only if the gradient of λ lies in the nullity distribution of g. Note that, by Theorem 6.7, for any solution of the mobility equation $A_{a\bar b}$ the function $A_a{}^a = -\lambda$ has this property. Moreover, in fact, the following Proposition holds.

PROPOSITION 6.22. *Fix $0 \neq B \in \mathbb{R}$ and set $G_{a\bar b c\bar d} \equiv R_{a\bar b c\bar d} + 2B(g_{a\bar b} g_{c\bar d} + g_{c\bar b} g_{a\bar d})$. Let λ be a smooth real-valued function and write $\Lambda_a = \nabla_a \lambda$ for its derivative. Then the following statements are equivalent:*

(1)
$$G_{a\bar b c}{}^d \Lambda_d = 0 \quad \text{and} \quad \nabla_a \Lambda_b = 0,$$

(2) *λ lifts uniquely to a parallel section of the connection (282) and (283) (and B is characterised by (236)),*

(3)
$$\begin{aligned}
\nabla_a \nabla_b \Lambda_c &= 0 & \text{and} & & \nabla_a \nabla_{\bar b} \Lambda_c &= -2B(\Lambda_a g_{c\bar b} + \Lambda_c g_{a\bar b}) \\
\nabla_{\bar a} \nabla_b \Lambda_c &= 0 & \text{and} & & \nabla_{\bar a} \nabla_{\bar b} \Lambda_c &= -2B(\Lambda_{\bar a} g_{c\bar b} + \Lambda_{\bar b} g_{c\bar a}).
\end{aligned} \tag{284}$$

6.6. THE C-PROJECTIVE HESSIAN, NULLITY, AND THE TANNO EQUATION

PROOF. In the discussion above we have already observed that (1) is equivalent to (2). Let us now show that (1) is equivalent to (3), as shown also by Tanno [**98**, Prop. 10.3]. Suppose first that (1) holds. Then, obviously also the first two equations on the left-hand side of (284) hold. Moreover, we immediately deduce from (1) that

$$(285) \qquad -2B(\Lambda_a g_{c\bar{b}} + \Lambda_c g_{a\bar{b}}) = R_{a\bar{b}c}{}^d \Lambda_d = \nabla_a \nabla_{\bar{b}} \Lambda_c,$$

which shows that the first equation of the right-hand side of (284) holds. The conjugate of (285) implies the second equation of the right-hand side of (284), since $\nabla_{\bar{a}} \nabla_{\bar{b}} \nabla_c \lambda = \nabla_{\bar{a}} \nabla_c \nabla_{\bar{b}} \lambda$. Conversely, suppose now that (3) holds. Then, obviously the identity (285) is satisfied, which shows that $G_{a\bar{b}c}{}^d \Lambda_d = 0$. Hence, it remains to show that Λ^a is a holomorphic vector field. From (284) we deduce that

$$\nabla_a \nabla_b \nabla_{\bar{c}} \Lambda_d = -2B((\nabla_a \Lambda_b) g_{c\bar{d}} + (\nabla_a \Lambda_d) g_{b\bar{c}}).$$

Since $R_{ab}{}^c{}_d = 0$ and $R_{ab}{}^{\bar{c}}{}_{\bar{d}} = 0$, skewing in a and b yields

$$0 = -B((\nabla_a \Lambda_d) g_{b\bar{c}} - (\nabla_b \Lambda_d) g_{a\bar{c}}),$$

which implies $0 = -B(n-1)\nabla_a \Lambda_d$. □

REMARK 6.9. If a function λ satisfies (1) for $B = 0$, then this is still equivalent to (2), and the equivalent statements (1) and (2) imply (3), but the implication from (3) to (1) is not necessarily true.

The system of equations (284) can also be written as

$$(286) \quad \nabla_\alpha \nabla_\beta \nabla_\gamma \lambda = -B(2\Lambda_\alpha g_{\beta\gamma} + g_{\alpha\beta}\Lambda_\gamma + g_{\alpha\gamma}\Lambda_\beta - \Omega_{\alpha\beta} J_\gamma{}^\delta \Lambda_\delta - \Omega_{\alpha\gamma} J_\beta{}^\delta \Lambda_\delta),$$

where $\Lambda_\alpha = \nabla_\alpha \lambda$. Since on Kähler manifolds equation (286) (respectively (284)) was intensively studied by Tanno in [**98**], we refer to this equation as the *Tanno equation*.

CHAPTER 7

Global results

We now turn to the global theory of (pseudo-)Kähler manifolds of mobility at least two. In Section 5.7 we presented a local classification [2, 16] which shows that such a (pseudo-)Kähler manifold (M, J, g) is locally a bundle of toric (in fact, "orthotoric") (pseudo-)Kähler manifolds over a local product S of (pseudo-)Kähler manifolds. When M is compact, and g is positive definite, this is not far from being true globally.

Indeed, under these assumptions, several simplifications occur. Firstly, any compact smooth orthotoric Kähler 2ℓ-manifold is biholomorphic (though not necessarily isometric) to $\mathbb{CP}^\ell$ [3]. Secondly, when g is positive definite, the Hermitian endomorphism $A_a{}^b$ is diagonalisable, and the local classification (222) simplifies to give

$$g = \sum_u \chi_{\mathrm{nc}}(\eta_u) g_u + \sum_{i=1}^\ell \frac{\Delta_j}{\Theta_j(\xi_j)} d\xi_j^2 + \sum_{j=1}^\ell \frac{\Theta_j(\xi_j)}{\Delta_j} \Big(\sum_{r=1}^\ell \sigma_{r-1}(\hat{\xi}_j) \theta_r \Big)^2,$$

$$\omega = \sum_u \chi_{\mathrm{nc}}(\eta_u) \omega_u + \sum_{r=1}^\ell d\sigma_r \wedge \theta_r, \quad \text{with} \quad d\theta_r = \sum_u (-1)^r \eta_u^{\ell-r} \omega_u$$

where η_u are the (real) constant eigenvalues of A, while the nonconstant eigenvalues ξ_j, and functions Θ_j, are all real valued. Thirdly, the eigenvalues are globally ordered and do not cross. However, if η_u is a root of Θ_j (for some u, j) and g_u is a Fubini–Study metric on $\mathbb{CP}^{m_u}$, it is possible to have $\xi_j = \eta_u$ along a critical submanifold of ξ_j, in which case $\chi_{\mathrm{nc}}(\eta_u) = 0$ along that submanifold, and the corresponding factor of the base manifold S collapses. We thus have the following global description [3].

THEOREM 7.1. *Let (M, J, g) be a compact connected Kähler manifold admitting a c-projectively equivalent Kähler metric that generates a metrisability pencil $\eta \circ (A - t\,\mathrm{Id})$ of order ℓ. Then the blow-up of M, along the subvarieties where nonconstant and constant eigenvalues of A coincide, is a toric $\mathbb{CP}^\ell$-bundle over a complex manifold S covered by a product, over the distinct constant eigenvalues, of complete Kähler manifolds with integral Kähler classes.*

Conversely, such complex manifolds do admit Kähler metrics of mobility at least two, but we refer to [3] for a more precise description. In particular, examples are plentiful, and have been used to construct explicit extremal Kähler metrics [4], including in particular, weakly Bochner-flat Kähler metrics [5].

As soon as we impose nullity—for instance, by requiring metrics of mobility at least three—this plenitude disappears, even in the (pseudo-)Kähler case: a compact connected (pseudo-)Kähler $2n$-manifold satisfying the equivalent conditions of

Theorem 6.18 is isometric to $\mathbb{CP}^n$, equipped with a constant multiple of the Fubini–Study metric [**44**], and this rigidity result extends to compact orbifolds (see [**26**]).

In the remainder of this section, we focus on positive definite complete Kähler metrics, and begin by showing, in Chapter 7.1, that the rigidity result for Kähler metrics with nullity also holds in this case. In Section 7.2, we then discuss the group of c-projective transformations and the Yano–Obata Conjecture. This was established for compact (pseudo-)Kähler metrics in [**78**], and here we show that it also holds for complete Kähler metrics.

7.1. Complete Kähler metrics with nullity

In this section we prove the following.

THEOREM 7.2. *Let g be a complete Kähler metric on a connected complex manifold (M, J) of real dimension $2n \geq 4$ that has nullity on a dense open subset of M. Then any complete Kähler metric on (M, J) that is c-projectively equivalent to g, is affinely equivalent to g, unless there is a positive constant $c \in \mathbb{R}$ such that the Kähler manifold (M, J, cg) is isometric to $(\mathbb{CP}^n, J_{\mathrm{can}}, g_{FS})$.*

Since, by Theorem 6.8, a connected Kähler manifold (M, J, g) of degree of mobility at least 3 has nullity on a dense open set unless all c-projectively equivalent metrics are affinely equivalent to g, we obtain the following immediate corollary, which in the case of closed Kähler manifolds was proved in [**44**, Theorem 2].

COROLLARY 7.3. *Let g be a complete Kähler metric on a connected complex manifold (M, J) of real dimension $2n \geq 4$ with mobility at least 3. Then any complete Kähler metric on (M, J) that is c-projectively equivalent to g, is affinely equivalent to g, unless there is a positive constant $c \in \mathbb{R}$ such that the Kähler manifold (M, J, cg) is isometric to $(\mathbb{CP}^n, J_{\mathrm{can}}, g_{FS})$.*

REMARK 7.1. It is easy to construct a complete Kähler manifold (M, J, g) of nonconstant sectional holomorphic curvature with mobility ≥ 3 such that all complete Kähler metrics on (M, J) are affinely equivalent to g. Indeed, take the direct product
$$(M_1, g_1, J_1) \times (M_2, g_2, J_2) \times (M_3, g_3, J_3)$$
of three Kähler manifolds. It is again a Kähler manifold (M, J, g) with complex structure $J := J_1 + J_2 + J_3$ and Kähler metric $g := g_1 + g_2 + g_3$. Obviously, (M, J, g) has mobility ≥ 3, since $c_1 g_1 + c_2 g_2 + c_3 g_3$ is again a Kähler metric on (M, J) for any constants $c_1, c_2, c_3 > 0$. Note also that all these Kähler metrics are affinely equivalent to g and that, if (M_i, g_i, J_i) is complete for $i = 1, 2, 3$ the metric $c_1 g_1 + c_2 g_2 + c_3 g_3$ is also complete for any constants $c_1, c_2, c_3 > 0$.

REMARK 7.2. Suppose that $\tilde{g}$ is (pseudo-)Kähler metric that is compatible with $[\nabla^g]$, then we may write
$$(287) \qquad \mathrm{vol}(\tilde{g}) = e^{(n+1)f} \mathrm{vol}(g) \qquad \text{and} \qquad \tau_{\tilde{g}} = e^{-f} \tau_g,$$
where $f = \frac{1}{n+1} \log \left| \frac{\mathrm{vol}(\tilde{g})}{\mathrm{vol}(g)} \right|$. We have seen in Chapter 2.1 that the Levi-Civita connections $\widetilde{\nabla}$ and ∇ of $\tilde{g}$ and g are related by (11) with $\Upsilon_\alpha = \nabla_\alpha f$. For later use, note that that $\widetilde{\nabla}_\alpha \tilde{g}_{\beta\gamma} = 0$ implies that the derivative $\nabla_\alpha f$ of f satisfies the equation
$$(288) \qquad \nabla_\alpha \tilde{g}_{\beta\gamma} = (\nabla_\alpha f) \tilde{g}_{\beta\gamma} + \tilde{g}_{\alpha(\beta} \nabla_{\gamma)} f - J_\alpha{}^\delta \tilde{g}_{\delta(\beta} J_{\gamma)}{}^\epsilon \nabla_\epsilon f.$$
Thus $\nabla = \widetilde{\nabla}$, i.e. g and $\tilde{g}$ are affinely equivalent, if and only if $\nabla_\alpha f = 0$.

7.1. COMPLETE KÄHLER METRICS WITH NULLITY

In order to prove Theorem 7.2 suppose that g is a complete Kähler metric on a complex connected manifold (M, J) of dimension $2n \geq 4$ with nullity on a dense open set. Further we may assume that g has mobility at least 2, since otherwise Theorem 7.2 is trivially satisfied. Then Theorem 6.11 implies that the function B defined as in (237) is actually a constant. We shall see that for $B > 0$, the theory of the Tanno equation implies that (M, J, g) has positive constant holomorphic sectional curvature. For $B \leq 0$, we shall show that any complete Kähler metric that is c-projectively equivalent to g, is necessarily affinely equivalent to g. The proof makes essential use of the following two Lemmas. Recall for this purpose that a geodesic which is orthogonal to a Killing vector field at one point is orthogonal to it at all points.

LEMMA 7.4. *Suppose (M, J, g) is a Kähler manifold of dimension $2n \geq 4$ with mobility ≥ 2 and nullity on a dense open subset of M. Let $\tilde{g}$ be a Kähler metric that is c-projectively equivalent to g and f the function defined as in (287), whose derivative $\Upsilon_\alpha \equiv \nabla_\alpha f$ relates the Levi-Civita connections of $\tilde{g}$ and g as in (11). Consider a geodesic $c = c(t)$ that is orthogonal to the canonical Killing fields of the pair $(g, \tilde{g})$, defined as in Theorem 5.11. Then, for the constant $B \in \mathbb{R}$ defined as in Theorem 6.11, the function $f(t) = f(c(t))$ satisfies the following ordinary differential equation:*

$$\dddot{f}(t) = -4B g(\dot{c}, \dot{c})\dot{f}(t) + 3\dot{f}(t)\ddot{f}(t) - (\dot{f}(t))^3. \tag{289}$$

PROOF. By Corollary 6.4 and identity (26) we have

$$\nabla_\alpha \Upsilon_\beta = -2\widetilde{B}\tilde{g}_{\alpha\beta} + 2B g_{\alpha\beta} + \tfrac{1}{2}(\Upsilon_\alpha \Upsilon_\beta - J_\alpha{}^\gamma J_\beta{}^\delta \Upsilon_\gamma \Upsilon_\delta). \tag{290}$$

Differentiating (290) and inserting (288) yields

$$\nabla_\alpha \nabla_\beta \Upsilon_\gamma = \\ - 2\widetilde{B}(\Upsilon_\alpha \tilde{g}_{\beta\gamma} + \tilde{g}_{\alpha(\beta} \Upsilon_{\gamma)} - J_\alpha{}^\delta \tilde{g}_{\delta(\beta} J_{\gamma)}{}^\epsilon \Upsilon_\epsilon) + \tfrac{1}{2}(\nabla_\alpha (\Upsilon_\beta \Upsilon_\gamma) \\ - \nabla_\alpha(J_\beta{}^\delta J_\gamma{}^\epsilon \Upsilon_\delta \Upsilon_\epsilon)).$$

Substituting for $-2\widetilde{B}\tilde{g}$ the expression (290) we therefore obtain

$$\nabla_\alpha \nabla_\beta \Upsilon_\gamma = (\nabla_\beta \Upsilon_\gamma)\Upsilon_\alpha - 2B g_{\beta\gamma}\Upsilon_\alpha - \tfrac{1}{2}(\Upsilon_\beta \Upsilon_\gamma - J_\beta{}^\delta J_\gamma{}^\epsilon \Upsilon_\delta \Upsilon_\epsilon)\Upsilon_\alpha \\ + (\nabla_\alpha \Upsilon_{(\beta})\Upsilon_{\gamma)} - 2B g_{\alpha(\beta}\Upsilon_{\gamma)} - \tfrac{1}{2}(\Upsilon_\alpha \Upsilon_\beta \Upsilon_\gamma - J_\alpha{}^\delta \Upsilon_\delta \Upsilon_\epsilon J_{(\beta}{}^\epsilon \Upsilon_{\gamma)}) \\ - J_\alpha{}^\delta (\nabla_\delta \Upsilon_{(\beta}) J_{\gamma)}{}^\zeta \Upsilon_\zeta - 2B g_{\delta(\beta} J_{\gamma)}{}^\zeta \Upsilon_\zeta \\ - \tfrac{1}{2}(\Upsilon_\delta \Upsilon_{(\beta} J_{\gamma)}{}^\zeta \Upsilon_\zeta - J_\delta{}^\eta \Upsilon_\eta \Upsilon_\epsilon J_{(\beta}{}^\epsilon J_{\gamma)}{}^\zeta \Upsilon_\zeta)) \\ + \tfrac{1}{2}(\nabla_\alpha(\Upsilon_\beta \Upsilon_\gamma) - \nabla_\alpha(J_\beta{}^\delta J_\gamma{}^\epsilon \Upsilon_\delta \Upsilon_\epsilon)).$$

Note that the determinant of the complex endomorphism $A_a{}^b$ relating g and $\tilde{g}$ as in (133) is given by e^{-f}. Hence the canonical Killing field $\widetilde{K}^\beta(0) = J^{\alpha\beta}\nabla_\alpha \det A = J^{\alpha\beta}\nabla_\alpha e^{-f}$, defined as in (209), is proportional to $J^{\alpha\beta}\Upsilon_\alpha = J^{\alpha\beta}(\nabla_\alpha f)$ and we have $\Upsilon_\beta J_\alpha{}^\beta \dot{c}^\alpha = 0$. Thus, contracting the above equation with $\dot{c}^\alpha \dot{c}^\beta \dot{c}^\gamma$ all terms involving the complex structure J disappear and we derive that $f(t)$ satisfies the desired ODE. □

LEMMA 7.5. *Let g and $\tilde{g}$ be c-projectively equivalent metrics on a complex manifold (M, J) of dimension $2n \geq 4$. Let c be a geodesic of g that is orthogonal to all canonical Killing fields of $(g, \tilde{g})$ (as defined in Theorem 5.11). Then there is* a reparametrisation ϕ such that $\tilde{c}(t) = c(\phi(t))$ is a geodesic of $\tilde{g}$. The inverse τ of the reparametrisation ϕ satisfies the formula

$$\frac{d}{dt} f(t) = \frac{d}{dt} \log\left|\frac{d\tau}{dt}\right|, \tag{291}$$

where $f(t) = f(c(t))$ with f defined as in (287).

PROOF. Let c be a geodesic of g that is orthogonal to the canonical Killing fields associated to $(g, \tilde{g})$. Then formula (11) for the difference of the Levi-Civita connections $\widetilde{\nabla}$ and ∇ implies

$$\widetilde{\nabla}_{\dot{c}} \dot{c} = \Upsilon(\dot{c})\dot{c} - \Upsilon(J\dot{c})J\dot{c} = \Upsilon(\dot{c})\dot{c},$$

where the last identity follows from the fact that $\Upsilon_\beta J^{\alpha\beta}$ is proportional to a canonical Killing field, as explained in proof of Lemma 7.4. Hence, $\widetilde{\nabla}_{\dot{c}}\dot{c}$ is a multiple of $\dot{c}$ and therefore there exists a reparametrisation ϕ of c such that $\tilde{c}(t) = c(\phi(t))$ is geodesic of $\tilde{g}$. Differentiating $c(t) = \tilde{c}(\tau(t))$, where τ denotes the inverse of ϕ, gives $\dot{c}(t) = \dot{\tau}(t)\dot{\tilde{c}}(\tau(t))$ and hence

$$0 = \nabla_{\dot{c}(t)}\dot{c}(t) = \nabla_{\dot{c}(t)}(\dot{\tau}(t)\dot{\tilde{c}}(\tau(t))) = \dot{\tau}(t)^2 \nabla_{\dot{\tilde{c}}(\tau(t))}\dot{\tilde{c}}(\tau(t)) + \ddot{\tau}(t)\dot{\tilde{c}}(\tau(t))$$
$$= \left(\ddot{\tau}(t) - \dot{\tau}(t)^2 \Upsilon(\dot{\tilde{c}}(\tau(t)))\right)\dot{\tilde{c}}(\tau(t)),$$

since $\tilde{c}$ is a geodesic of $\tilde{g}$. This implies that $\ddot{\tau}(t) = \dot{\tau}(t)^2 \Upsilon(\dot{\tilde{c}}(\tau(t))) = \dot{\tau}(t)\Upsilon(\dot{c}(t))$, which is equivalent to (291), since $\Upsilon_\alpha = \nabla_\alpha f$. $\square$

PROOF OF THEOREM 7.2. Let (M, J, g) be a connected complete Kähler manifold with nullity on a dense open set. We may assume that (M, J, g) has mobility at least 2. Hence, by Theorem 6.11 any solution of the mobility equation lifts uniquely to a parallel section of the connection (240) with B constant. This in turn, by Theorem 6.7, shows that for any solution $A^{\bar{b}c}$ of the mobility equation of g the function $\lambda = -A_a{}^a$ satisfies (1) of Proposition 6.22. Hence, by Proposition 6.22 and Remark 6.9 λ satisfies the Tanno equation (284). Tanno showed in [98] that on a complete connected Kähler manifold and for a constant $B > 0$, the existence of a nonconstant solution λ of the Tanno equation (284) implies that (M, J, g) has positive constant holomorphic sectional curvature, which in turn implies that (M, J, g) is actually closed and isometric to $(\mathbb{CP}^n, J_{\text{can}}, cg_{FS})$ for some positive constant c. Since any metric that is c-projectively but not affinely equivalent to g gives rise to a non-parallel solution of the mobility equation of g and hence to a nonconstant solution of the Tanno equation, Theorem 7.2 holds provided B is positive.

It remains to consider the case that the constant B defined as in Theorem 6.11 is nonpositive. Let $\tilde{g}$ be another complete Kähler metric on (M, J), which is c-projectively equivalent to g. Denote by f again the function defined as in (287), which has the property that $\Upsilon_\alpha = \nabla_\alpha f$ relates the Levi-Civita connections of g and $\tilde{g}$ as in (11). We now show that $B \leq 0$ implies $\Upsilon_\alpha \equiv 0$, that is, g and $\tilde{g}$ are necessarily affinely equivalent.

Note first that, since the canonical Killing fields associated to $(g, \tilde{g})$ are Killing for both metrics by Theorem 5.11 and f is constructed in a natural way only from the pair $(g, \tilde{g})$, the local flows of the canonical Killing fields preserve f. Hence, the canonical Killing fields lie in the kernel of Υ. To show that Υ is identically zero, it therefore remains to show that Υ vanishes when inserting vector fields orthogonal to the canonical Killing fields.

Consider a parametrised geodesic c of g, which at one (and hence at all points) is orthogonal to the canonical Killing fields. Since g and $\tilde{g}$ are complete, c is defined for all times and τ from Lemma 7.5 is a diffeomorphism of $\mathbb{R}$. Without loss of generality we assume that $\dot{\tau}$ is positive, otherwise replace t by $-t$. By Lemma 7.5, the function $\tau\colon \mathbb{R} \to \mathbb{R}$ satisfies (291), which we rewrite as

$$\tag{292} f(t) = \log(\dot{\tau}(t)) + \mathrm{const}_0.$$

Now let us consider equation (289) and set $\dot{\tau}(t) = (p(t))^{-1}$. Substituting (292) into (289) yields

$$\tag{293} \dddot{p} = -4Bg(\dot{c},\dot{c})\dot{p}.$$

If $B = 0$, the equation simplifies to $\dddot{p} = 0$ and its general solution is of the form

$$p(t) = C_2 t^2 + C_1 t + C_0,$$

where C_i is a real constant for $i = 0, 1, 2$. Hence, we get

$$\tag{294} \tau(t) = \int_{t_0}^{t} \frac{d\xi}{C_2 \xi^2 + C_1 \xi + C_0} + \mathrm{const}.$$

If the polynomial $p(t) = C_2 t^2 + C_1 t + C_0$ has real roots (which is always the case if $C_2 = 0$, $C_1 \neq 0$), then the integral starts to be infinite in finite time. If the polynomial has no real roots, but $C_2 \neq 0$, the function τ is bounded. Thus, the only possibility for τ to be a diffeomorphism is $C_2 = C_1 = 0$ implying $\dot{\tau} = \frac{1}{C_0}$, which shows that f is constant along the geodesic c.

If $B < 0$, the general solution of equation (293) is

$$\tag{295} C + C_+ e^{2\sqrt{-Bg(\dot{c},\dot{c})}\cdot t} + C_- e^{-2\sqrt{-Bg(\dot{c},\dot{c})}\cdot t},$$

for real constants C, C_+ and C_-. Hence, τ is of the form

$$\tag{296} \tau(t) = \int_{t_0}^{t} \frac{d\xi}{C + C_+ e^{2\sqrt{-Bg(\dot{c},\dot{c})}\xi} + C_- e^{-2\sqrt{-Bg(\dot{c},\dot{c})}\xi}} + \mathrm{const}.$$

If one of the constants C_+, C_- is not zero, the integral (296) is bounded from one side, or starts to be infinite in finite time. In both cases, τ is not a diffeomorphism of $\mathbb{R}$. The only possibility for $\tau\colon \mathbb{R} \to \mathbb{R}$ to be a diffeomorphism is when $C_+ = C_- = 0$, in which case $\dot{\tau}$ is constant implying f is constant along the geodesic c. Hence, in both cases ($B = 0$ and $B < 0$) the one form $\Upsilon_\alpha = \nabla_\alpha f$ vanishes when inserting vector fields orthogonal to the canonical Killing fields. $\square$

7.2. The Yano–Obata Conjecture for complete Kähler manifolds

For a Kähler manifold (M, J, g) let us write $\mathrm{Isom}(J, g)$, $\mathrm{Aff}(J, g)$ and $\mathrm{CProj}(J, g)$ for the group of complex isometries, the group of complex affine transformations (i.e. of complex diffeomorphisms preserving the Levi-Civita connection) and the group of c-projective transformations of (M, J, g) respectively. By definition of these groups we obtain the following inclusions

$$\mathrm{Isom}(J,g) \subseteq \mathrm{Aff}(J,g) \subseteq \mathrm{CProj}(J,g)$$

and consequently we also have

$$\mathrm{Isom}_0(J,g) \subseteq \mathrm{Aff}_0(J,g) \subseteq \mathrm{CProj}_0(J,g),$$

where subscript 0 denotes the connected component of the identity.

Recall that Lie groups of affine transformations of complete Riemannian manifolds are well understood; see for example [**68**, Chapter IV]. As explained there, if a connected Lie group G acts on a simply-connected complete Riemannian manifold (M^n, g) by affine transformations, then there exists a Riemannian decomposition

$$(M^n, g) = (M_1^{n_1}, g_1) \times (\mathbb{R}^{n_2}, g_{\text{euc}})$$

of (M^n, g) into a direct product of a Riemannian manifold $(M_1^{n_1}, g_1)$ and a Euclidean space $(\mathbb{R}^{n_2}, g_{\text{euc}})$ such that G acts by isometries on the first factor and by affine transformations (i.e. compositions of linear isomorphisms and parallel translations) on the second. Note that this implies that for closed simply-connected Riemannian manifolds one always has $\text{Isom}_0(J, g) = \text{Aff}_0(J, g)$, which holds in fact for any closed (not necessarily simply-connected) Riemannian manifold; see [**103**, Theorem 4]. If in addition (M^n, g) is Kähler for a complex structure J, and G is a connected Lie group of complex affine transformations, then $(M_1^{n_1}, J_1, g_1)$ and $(\mathbb{R}^{n_2}, J_{\text{can}}, g_{\text{euc}})$ are also Kähler; furthermore, G acts on $(M_1^{n_1}, J_1, g_1)$ by complex isometries and on $(\mathbb{R}^{n_2}, J_{\text{can}}, g_{\text{euc}})$ by complex affine transformations. For the complex projective space $\mathbb{CP}^n$ equipped with its natural complex structure J_{can} and the Fubini–Study metric g_{FS} we have $\text{Aff}(J_{\text{can}}, g_{FS}) \neq \text{CProj}(J_{\text{can}}, g_{FS})$. To see this recall (see Proposition 2.10) that group of c-projective transformations of $(\mathbb{CP}^n, J_{\text{can}}, g_{FS})$ coincides with the connected Lie group $\text{PGL}(n+1, \mathbb{C}) \cong \text{PSL}(n+1, \mathbb{C})$ acting on $\mathbb{CP}^n$ in its usual way. Note that an element in $\text{GL}(n+1, \mathbb{C})$ induces a complex isometry on $(\mathbb{CP}^n, J_{\text{can}}, g_{FS})$ if and only if it is proportional to a unitary automorphism of $\mathbb{C}^{n+1}$, which shows that $\text{Isom}(J_{\text{can}}, g_{FS})$ can be identified with the connected Lie group $\text{PU}(n+1) = \text{U}(n+1)/\text{U}(1)$ of projective unitary transformations. For $(\mathbb{CP}^n, J_{\text{can}}, g_{FS})$ we also clearly have $\text{Isom}(J_{\text{can}}, g_{FS}) = \text{Aff}(J_{\text{can}}, g_{FS})$.

In [**78**] in conjunction with [**44**] it was shown that any closed connected Kähler manifold (M, J, g) of dimension $2n \geq 4$ with the property that $\text{CProj}_0(J, g)$ contains $\text{Isom}_0(J, g) = \text{Aff}_0(J, g)$ as a proper subgroup is actually isometric to $(\mathbb{CP}^n, J, cg_{FS})$ for some positive constant c. This rigidity result answers affirmatively in the case of closed Kähler manifolds the so-called *Yano–Obata Conjecture*, which is a c-projective analogue of the *Projective* and *Conformal Lichnerowicz–Obata Conjectures*; see the introductions of the papers [**74**, **78**] for a historical overview. Most recently the conjecture has been proved for closed (pseudo-)Kähler manifolds of all signatures in [**16**]. We now show that the Yano–Obata Conjecture also holds for complete connected Kähler manifolds.

THEOREM 7.6 (Yano–Obata Conjecture). *Let (M, g, J) be a complete connected Kähler manifold of real dimension $2n \geq 4$. Then $\text{Aff}_0(g, J) = \text{CProj}_0(g, J)$, unless (M, g, J) is actually compact and isometric to $(\mathbb{CP}^n, J_{\text{can}}, cg_{FS})$ for some positive constant $c \in \mathbb{R}$.*

REMARK 7.3. While this article was under review, the third and fourth author have established in [**77**, Theorem 1.2] a stronger version of the Yano–Obata conjecture as formulated in Theorem 7.6: on a complete Kähler manifold of dimension $2n \geq 4$ the index of $\text{Aff}(g, J)$ in $\text{CProj}(g, J)$ is at most two unless the Kähler manifold is isometric to $\mathbb{CP}^n$ equipped with a positive constant multiple of the Fubini–Study metric. The proof is based on the present article, in particular on Theorem 7.2 and the circle of ideas used in Section 7.3, and on ideas from [**76**, **106**], which study the index of the group of affine transformations in the group of projective transformations on complete Riemannian manifolds.

7.3. The proof of the Yano–Obata Conjecture

Note that if the mobility of (M, J, g) is 1, then any metric $\tilde{g}$ that is c-projectively equivalent to g is homothetic to g. In particular, any c-projective transformation has to preserve the Levi-Civita connection of g and hence $\mathrm{Aff}_0(g, J) = \mathrm{CProj}_0(g, J)$ in this case. On the other hand, since the pullback of a complete Kähler metric by a c-projective transformation is again a complete Kähler metric, Theorem 7.6 follows from Corollary 7.3 or Theorem 7.2 in the case that g has mobility ≥ 3 or has mobility ≥ 2 and nullity on a dense open set.

For the rest of this section we therefore assume that (M, J, g) is a connected complete Kähler manifold of dimension $2n \geq 4$ with mobility 2. To show that Theorem 7.6 holds in this case (which for closed Kähler manifolds was proved in [78]), let us write $\mathrm{Sol}(g)$ for the 2-dimensional solution space of the mobility equation of g, which we view as a linear subspace of the space of J-invariant sections in $S^2 TM(-1, -1)$.

Suppose now that $\mathrm{Aff}_0(g, J)$ does not coincide with $\mathrm{CProj}_0(g, J)$. Then there exists a complete c-projective vector field V that is not affine. Since the flow Φ_t of V acts on (M, J, g) by c-projective transformations, for any $t \in \mathbb{R}$ and for any $\eta \in \mathrm{Sol}(g)$ the pullback $\Phi_t^* \eta$ is an element of the vector space $\mathrm{Sol}(g)$. Hence, the Lie derivative $\mathcal{L}_V \eta = \frac{d}{dt}|_{t=0} \Phi_t^* \eta$ can also be identified with an element of $\mathrm{Sol}(g)$, which implies that $\mathcal{L}_V$ induces a linear endomorphism of $\mathrm{Sol}(g)$. By the Jordan normal form, in a certain basis $\eta, \tilde{\eta} \in \mathrm{Sol}(g)$, the linear endomorphism $\mathcal{L}_V : \mathrm{Sol}(g) \to \mathrm{Sol}(g)$ corresponds to a matrix of one of the following three forms:

$$(297) \qquad \begin{pmatrix} a & 0 \\ 0 & b \end{pmatrix} \quad \begin{pmatrix} a & b \\ -b & a \end{pmatrix} \quad \begin{pmatrix} a & 1 \\ 0 & a \end{pmatrix},$$

where $a, b \in \mathbb{R}$. We deal with these three cases separately. The last two cases are easy and will be considered in Chapter 7.3.1. The challenging case is the first one, which will be treated in Chapter 7.3.2.

7.3.1. $\mathcal{L}_V$ has complex-conjugate eigenvalues or a nontrivial Jordan block.

Suppose first that the endomorphism $\mathcal{L}_V : \mathrm{Sol}(g) \to \mathrm{Sol}(g)$ has two complex-conjugated eigenvalues. Hence, in some basis $\eta, \tilde{\eta} \in \mathrm{Sol}(g)$ the endomorphism $\mathcal{L}_V$ corresponds to a matrix of the second type in (297). If $b = 0$, then V acts by homotheties on any element in $\mathrm{Sol}(g)$ and hence preserves in particular the Levi-Civita connection of g, which contradicts our assumption that V is not affine. Therefore, we can assume that $b \neq 0$. Then the evolution of the solutions $\eta, \tilde{\eta}$ along the flow Φ_t of V is given by

$$\begin{aligned} \Phi_t^* \eta &= e^{at} \cos(bt) \eta + e^{at} \sin(bt) \tilde{\eta} \\ \Phi_t^* \tilde{\eta} &= -e^{at} \sin(bt) \eta + e^{at} \cos(bt) \tilde{\eta} \end{aligned}.$$

Write $g^{-1} \mathrm{vol}(g)^{\frac{1}{n+1}} \in \mathrm{Sol}(g)$ as $c\eta + d\tilde{\eta}$ for some real constants c and d. Then one has

$$\begin{aligned} \Phi_t^*(c\eta + d\tilde{\eta}) &= c(e^{at} \cos(bt)\eta + e^{at} \sin(bt)\tilde{\eta}) \\ &\quad + d(-e^{at} \sin(bt)\eta + e^{at} \cos(bt)\tilde{\eta}) \\ &= e^{at}\sqrt{c^2 + d^2}(\cos(bt + \alpha)\eta + \sin(bt + \alpha)\tilde{\eta}), \end{aligned}$$

where $\alpha = \arccos(\frac{c}{\sqrt{c^2+d^2}})$. Since g is a Riemannian metric, for any point $x \in M$ there exists a basis of $T_x M$ in which η and $\tilde{\eta}$ are diagonal matrices. Hence, in this basis the i-th entry of $\Phi_t^*(c\eta + d\tilde{\eta})$ is given by $e^{at}\sqrt{c^2+d^2}(\cos(bt+\alpha)e_i + \sin(bt+\alpha)\tilde{e}_i)$, where e_i and $\tilde{e}_i$ are the i-th diagonal entries of η and $\tilde{\eta}$. Therefore, we see that $\Phi_t^*(c\eta + d\tilde{\eta})$ is degenerate for some t, which contradicts the fact that $c\eta + d\tilde{\eta} = g^{-1}\mathrm{vol}(g)^{\frac{1}{n+1}}$ is nondegenerate. The obtained contradiction shows that $\mathcal{L}_V$ can not have two complex-conjugate eigenvalues.

Suppose now that $\mathcal{L}_V \colon \mathrm{Sol}(g) \to \mathrm{Sol}(g)$ is with respect to some basis $\eta, \tilde{\eta} \in \mathrm{Sol}(g)$ a matrix of the third type in (297). Then the evolution of η and $\tilde{\eta}$ along Φ_t is given by

$$\Phi_t^*\eta = e^{at}\eta + te^{at}\tilde{\eta}$$
$$\Phi_t^*\tilde{\eta} = e^{at}\tilde{\eta}.$$

We assume again that $g^{-1}\mathrm{vol}(g)^{\frac{1}{n+1}} = c\eta + d\tilde{\eta}$. Then we obtain

$$\begin{aligned}\Phi_t^*(c\eta + d\tilde{\eta}) &= c(e^{at}\eta + e^{at}t\tilde{\eta}) + d(e^{at}\tilde{\eta})\\ &= e^{at}(c\eta + (d + ct)\tilde{\eta}).\end{aligned}$$

Hence, we see again that for $c \neq 0$ there exists t such that $\Phi_t^*(c\eta + d\tilde{\eta})$ is degenerate which contradicts the fact that g is nondegenerate. Now, if $c = 0$, then Φ_t acts by homotheties on g, which contradicts our assumption that V is not affine. Thus $\mathcal{L}_V$ can also not be of the third type in (297).

7.3.2. $\mathcal{L}_V$ has two real eigenvalues. Now we consider the remaining case, namely the one where $\mathcal{L}_V$ has two different real eigenvalues a and b (the case where the two eigenvalues are equal was already excluded in the previous section). Without loss of generality, we can assume that at least one of the eigenvalues is positive, since we can otherwise just replace V by $-V$. Hence, we can assume without loss of generality that $a > b$ and that $a > 0$. Since $\phi_t^*\eta = e^{at}\eta$ and $\phi_t^*\tilde{\eta} = e^{bt}\tilde{\eta}$, we see that neither η nor $\tilde{\eta}$ can equal $g^{-1}\mathrm{vol}(g)^{\frac{1}{n+1}}$, since otherwise ϕ_t acts by homotheties on g, which contradicts our assumption that V is not affine. Hence, $g^{-1}\mathrm{vol}(g)^{\frac{1}{n+1}} = c\eta + d\tilde{\eta}$ for constant $c, d \neq 0$. By rescaling η and $\tilde{\eta}$, we can therefore assume without loss of generality that

$$g^{\alpha\beta}\mathrm{vol}(g)^{\frac{1}{n+1}} = \eta^{\alpha\beta} + \tilde{\eta}^{\alpha\beta}.$$

Let us write $D^{\alpha\beta} = \eta^{\alpha\beta}\mathrm{vol}(g)^{-\frac{1}{n+1}}$ and $\tilde{D}^{\alpha\beta} = \tilde{\eta}^{\alpha\beta}\mathrm{vol}(g)^{-\frac{1}{n+1}}$ such that

$$g^{\alpha\beta} = D^{\alpha\beta} + \tilde{D}^{\alpha\beta}.$$

Note that for a Kähler manifold (M, J, g) of mobility 2, the dense open subset of regular points (Definition 5.4) does not depend on the choice of non-proportional c-projectively equivalent metrics from the c-projective class of g. In particular, the set of regular points is invariant under c-projective transformations and hence under the action of the flow of a c-projective vector field. Thus if we fix a regular point $x_0 \in M$ and consider the integral curve $\phi_t(x_0)$ of our complete c-projective vector field V through x_0, then there is an open neighbourhood U of the curve $\phi_t(x_0)$ in the set of regular points, and, since g is positive definite, a frame of TU, in which g corresponds to the identity matrix and D and $\tilde{D}$ to diagonal matrices:

$$(298) \quad D = \begin{pmatrix} d_1 & & & & \\ & d_1 & & & \\ & & \ddots & & \\ & & & d_n & \\ & & & & d_n \end{pmatrix}, \quad \tilde{D} = \begin{pmatrix} \tilde{d}_1 & & & & \\ & \tilde{d}_1 & & & \\ & & \ddots & & \\ & & & \tilde{d}_n & \\ & & & & \tilde{d}_n \end{pmatrix},$$

where d_i and $\tilde{d}_i$ are smooth real-valued functions on U such that $d_i + \tilde{d}_i = 1$ for $i = 1, \ldots, n$. Then in the local frame the tensor $A_t^{\alpha\beta} = \phi_t^*(\eta^{\alpha\beta} + \tilde{\eta}^{\alpha\beta})\mathrm{vol}(g)^{-\frac{1}{n+1}}$ corresponds to the following diagonal matrix:

(299)
$$A_t = \begin{pmatrix} e^{at}d_1 + e^{bt}\tilde{d}_1 & & & & \\ & e^{at}d_1 + e^{bt}\tilde{d}_1 & & & \\ & & \ddots & & \\ & & & e^{at}d_n + e^{bt}\tilde{d}_n & \\ & & & & e^{at}d_n + e^{bt}\tilde{d}_n \end{pmatrix}.$$

Since g and $\phi_t^* g$ are positive definite, all diagonal entries of (299) are positive for all $t \in \mathbb{R}$. Hence, $d_i + e^{(b-a)t}\tilde{d}_i > 0$ respectively $e^{(a-b)t}d_i + \tilde{d}_i > 0$ for all t and taking the limit $t \to \infty$ respectively $t \to -\infty$ shows that $d_i, \tilde{d}_i \geq 0$ for all $i = 1, \ldots, n$. Since $d_i + \tilde{d}_i = 1$, we conclude that

$$0 \leq d_i \leq 1 \quad \text{and} \quad 0 \leq \tilde{d}_i \leq 1 \qquad \text{for all } i = 1, \ldots, n.$$

Now consider the $(1,1)$-tensor field $D_\alpha{}^\beta = g_{\alpha\gamma}\eta^{\gamma\beta}\mathrm{vol}(g)^{-\frac{1}{n+1}}$ and its pullback $\phi_t^*(D_\alpha{}^\beta) = \phi_t^*(g_{\alpha\gamma}\eta^{\gamma\beta}\mathrm{vol}(g)^{-\frac{1}{n+1}})$. Since $g_{\alpha\beta}\mathrm{vol}(g)^{-\frac{1}{n+1}}$ is inverse to $\eta^{\alpha\beta} + \tilde{\eta}^{\alpha\beta}$, we conclude that $\phi_t^*(D_\alpha{}^\beta)$ is given by a block diagonal matrix whose i-th block is given by the following 2×2 matrix

(300)
$$\frac{e^{at}d_i}{e^{at}d_i + e^{bt}(1-d_i)}\,\mathrm{Id}_2\,.$$

By definition ϕ_t acts on the endomorphism $D_\alpha{}^\beta$ as $\phi_t^*(D_\alpha{}^\beta) = (T\phi_t)^{-1} \circ D_\alpha{}^\beta \circ T\phi_t$, which implies that the eigenvalues of $\phi_t^*(D_\alpha{}^\beta)$ at a point $x \in M$ are the same as the eigenvalues of $D_\alpha{}^\beta$ at $\phi_t(x)$. Therefore, it follows from (300) that the only possible constant eigenvalues of $D_\alpha{}^\beta$ on U are 0 and 1 (i.e. the only possible constant diagonal entries in (298) are 0 or 1). Note that $d_i = 0$ (respectively $d_i = 1$) on some open set implies that $\tilde{d}_i = 1$ (respectively $\tilde{d}_i = 0$). Hence, the only possible constant eigenvalues of A_t defined as in (299) are e^{at} and e^{bt}. Since U consists of regular points, the distinct eigenvalues of A_t (for any fixed t) are smooth real-valued functions with constant algebraic multiplicities on U. Let us write $2m$, respectively $2\tilde{m}$, for the multiplicity of the eigenvalues e^a and e^b of A_1 on U. By Lemma 5.16, the number of distinct nonconstant eigenvalues of A_1 is given by $n - m - \tilde{m}$, and $m, \tilde{m}$ are constant on the set of regular points by Corollary 5.17. We allow, of course, that m, respectively $\tilde{m}$, are zero.

LEMMA 7.7. *If at least one of the following two inequalities,*

(301) $\quad (n - \tilde{m})a + (\tilde{m} + 1)b \leq 0 \quad \text{and} \quad (m+1)a + (n-m)b \geq 0.$

is not satisfied, then the vector field Λ^α given by the gradient of $\lambda = -\frac{1}{2}D_\beta{}^\beta$ lies in the nullity space of M at x_0.

PROOF. Set $G_t := \det_{\mathbb{R}}(A_t)^{-\frac{1}{2}} A_t^{-1}$ and note that $\phi_t^* g = g(G_t \cdot, \cdot)$. We may assume that the first $2\ell := 2n - 2m - 2\tilde{m}$ elements of D are not constant (which is equivalent to assuming that $d_i(x_0) \neq 0, 1$ for $i = 1, \ldots, \ell$), the next $2m$ elements are equal to 1, and the remaining $2\tilde{m}$ elements are zero on U. Then, we deduce from (299) that G_t on U is a block diagonal matrix of block sizes $2\ell \times 2\ell$, $2m \times 2m$

and $2\tilde{m} \times 2\tilde{m}$ respectively, where the three blocks are given by
$$\Psi(t) \begin{pmatrix} \frac{1}{d_1 e^{at}+(1-d_1)e^{bt}} & & & & \\ & \frac{1}{d_1 e^{at}+(1-d_1)e^{bt}} & & & \\ & & \ddots & & \\ & & & \frac{1}{d_\ell e^{at}+(1-d_\ell)e^{bt}} & \\ & & & & \frac{1}{d_\ell e^{at}+(1-d_\ell)e^{bt}} \end{pmatrix} \tag{302}$$

$$\Psi(t)e^{-at}\,\mathrm{Id}_{2m}, \qquad \text{respectively} \qquad \Psi(t)e^{-bt}\,\mathrm{Id}_{2\tilde{m}},$$

where
$$\Psi(t) := e^{-amt}e^{-b\tilde{m}t}\prod_{i=1}^{\ell}\frac{1}{d_i e^{at}+(1-d_i)e^{bt}}.$$

Let us write $\nu_1,\ldots,\nu_\ell$, ν and $\tilde{\nu}$ for the eigenvalues of these respective diagonal matrices. Note that their asymptotic behaviour for $t \to +\infty$ respectively for $t \to -\infty$ is as follows
$$\begin{array}{llll}
t \to +\infty & \nu_i(t) \sim \frac{e^{-((n-\tilde{m}+1)a+\tilde{m}b)t}}{d_i\prod d_j} & \nu(t) \sim \frac{e^{-((n-\tilde{m}+1)a+\tilde{m}b)t}}{\prod d_j} & \tilde{\nu}(t) \sim \frac{e^{-((n-\tilde{m})a+(\tilde{m}+1)b)t}}{\prod d_j} \\
t \to -\infty & \nu_i(t) \sim \frac{e^{-(ma+(n-m+1)b)t}}{(1-d_i)\prod(1-d_j)} & \nu(t) \sim \frac{e^{-((m+1)a+(n-m)b)t}}{\prod(1-d_j)} & \tilde{\nu}(t) \sim \frac{e^{-(ma+(n-m+1)b)t}}{\prod(1-d_j)}.
\end{array} \tag{303}$$

Let us now assume that (301) is not satisfied. We can assume without loss of generality that the first inequality of (301) is not satisfied, that is to say we can assume that $(n-\tilde{m})a + (\tilde{m}+1)b > 0$, since otherwise we can change the sign of V, which interchanges the inequalities. Then it follows from (303) that all eigenvalues of G_t decay exponentially as $t \to \infty$. Consider now the sequence $(\phi_k(x_0))_{k\in\mathbb{N}}$. We claim that it is a Cauchy sequence. Indeed, note that the distance $d(\phi_k(x_0),\phi_{k+1}(x_0))$ between $\phi_k(x_0)$ and $\phi_{k+1}(x_0)$ satisfies

$$d(\phi_k(x_0),\phi_{k+1}(x_0)) \leq \int_0^1 \sqrt{g(V(\phi_{k+t}(x_0)),V(\phi_{k+t}(x_0)))}dt \tag{304}$$
$$= \int_0^1 \sqrt{(\phi_k^*g)(V(\phi_t(x_0)),V(\phi_t(x_0)))}dt.$$

Since $\phi_k^*g = g(G_k\cdot,\cdot)$ and all eigenvalues of G_t decay exponentially as $t \to \infty$, the inequality (304) shows that $d(\phi_k(x_0),\phi_{k+1}(x_0))$ decays geometrically as $k \to \infty$ (i.e. for sufficiently large k we have $d(\phi_k(x_0),\phi_{k+1}(x_0)) \leq \text{const } q^k$ for some $q < 1$). Hence, $(\phi_k(x_0))_{k\in\mathbb{N}}$ is a Cauchy sequence and completeness of M implies that $(\phi_k(x_0))_{k\in\mathbb{N}}$ converges. We denote the limit of $(\phi_k(x_0))_{k\in\mathbb{N}}$ by $p \in M$. Now consider the smooth real-valued function F on M given by

$$F = H_{\alpha\beta}{}^\gamma{}_\delta H_{\epsilon\zeta}{}^\theta{}_\eta g_{\gamma\theta} g^{\alpha\epsilon} g^{\beta\zeta} g^{\delta\eta},$$

where $H_{\alpha\beta}{}^\gamma{}_\delta$ denotes the harmonic curvature of $(M,J,[\nabla^g])$ defined as in Proposition 4.4, respectively Proposition 2.13. Since $H_{\alpha\beta}{}^\gamma{}_\delta$ is c-projectively invariant, we deduce that $(\phi_k^*F)(x_0) = F(\phi_k(x_0))$ equals

$$F(\phi_k(x_0)) = (H_{\alpha\beta}{}^\gamma{}_\delta H_{\epsilon\zeta}{}^\theta{}_\eta \,\phi_k^*g_{\gamma\theta}\,\phi_k^*g^{\alpha\epsilon}\,\phi_k^*g^{\beta\zeta}\,\phi_k^*g^{\delta\eta})(x_0). \tag{305}$$

Moreover, since F is continuous, we have $\lim_{k\to\infty} F(\phi_k(x_0)) = F(p)$.

Since in the frame we are working the matrices corresponding to g and G_t are diagonal, we see that the function $F(\phi_t(x_0))$ is a sum of the form

$$\sum_{1\leq i,j,k,\ell\leq 2n} C(ijk\ell;t)\left(H_{\alpha_i\alpha_j}{}^{\alpha_k}{}_{\alpha_\ell}(x_0)\right)^2, \tag{306}$$

where the coefficient $C(ijk\ell;t)$ is the product of the k-th diagonal entry and the reciprocals of the i-th, j-th and ℓ-th diagonal entry of the diagonal matrix that corresponds to G_t (in our chosen frame). The coefficients $C(ijk\ell;t)$ depend on t and their asymptotic behaviour for $t \to \pm\infty$ can be read off from (303). Note moreover that all coefficients $C(ijk\ell;t)$ are positive.

We claim that, if at least one of the indices i,j or ℓ is less or equal than $2n-2m-2\tilde{m}$, then $H_{\alpha_i\alpha_j}{}^{\alpha_k}{}_{\alpha_\ell}(x_0)$ vanishes. Indeed, note that, by (303), ϕ_t^*g decays exponentially at least as $e^{-((n-\tilde{m}+1)a+\tilde{m}b)t}$, which is up to a constant the smallest eigenvalue of G_t, and $\phi_t^*g^{-1}$ goes exponentially to infinity at least as $e^{((n-\tilde{m})a+(\tilde{m}+1)b)t}$ as $t \to \infty$. Suppose now that at least one of the indices i,j or ℓ is less or equal than $2n-2m-2\tilde{m}$. Then we deduce that up to multiplication by a positive constant $C(ijk\ell;t)$ behaves asymptotically as $t \to \infty$ at least as

$$e^{((n-\tilde{m})a+(\tilde{m}+1)b)t}e^{((n-\tilde{m})a+(\tilde{m}+1)b)t}e^{((n-\tilde{m}+1)a+\tilde{m}b)t}e^{-((n-\tilde{m}+1)a+\tilde{m}b)t}$$
$$=e^{2((n-\tilde{m})a+(\tilde{m}+1)b)t}.$$

Since $(n-\tilde{m})a + (\tilde{m}+1)b > 0$ by assumption, we therefore conclude that the coefficient

$$C(ijk\ell;t) \to \infty \qquad \text{as } t \to \infty.$$

Since all terms in the sum (306) are nonnegative and the sequence $F(\phi_\ell(x_0))$ converges, we therefore deduce that $H_{\alpha_i\alpha_j}{}^{\alpha_k}{}_{\alpha_\ell}(x_0) = 0$ provided that at least one of the indices i,j or ℓ is less or equal than $2n-2m-2\tilde{m}$.

Observe now that Λ^α equals the negative of the sum of the gradients of the distinct nonconstant eigenvalues $d_1,\ldots,d_{n-m-\tilde{m}}$ of D. We therefore conclude that at x_0

$$H_{\alpha\beta}{}^\gamma{}_\delta \Lambda^\alpha = 0, \quad H_{\alpha\beta}{}^\gamma{}_\delta \Lambda^\beta = 0 \quad \text{and} \quad H_{\alpha\beta}{}^\gamma{}_\delta \Lambda^\delta = 0.$$

It follows that (3) of Remark 6.3 is satisfied, which implies that the vector field Λ^α lies in the nullity space of M at x_0. $\square$

Since x_0 is an arbitrary regular point and the conditions (301) do not depend on the choice of regular point, Lemma 7.7 shows that, if the inequalities (301) are not satisfied, then (M,J,g) has nullity on the dense open subset of regular points, since Λ^α does not vanish at regular points (otherwise V is necessarily affine, which contradicts our assumption). Hence, by Theorem 7.2, we have established Theorem 7.6, except when the inequalities (301) are satisfied.

Therefore, assume now that the inequalities (301) are satisfied. Subtracting the second inequality from the first shows that

$$(n-m-\tilde{m}-1)(a-b) \leq 0.$$

Since $a-b > 0$ by assumption, we must have $n-m-\tilde{m} = 0$ or $n-m-\tilde{m} = 1$. In the first case ϕ_t^*g is parallel for the Levi-Civita connection of g for all t and hence V is affine, which contradicts our assumption. Therefore, we must have $n-m-\tilde{m} = 1$. Now substituting this identity back into (301) shows that $(m+1)a + (\tilde{m}+1)b \leq 0$

and $(m+1)a + (\tilde{m}+1)b \geq 0$, which implies $(m+1)a = -(\tilde{m}+1)b$. Hence, we conclude that we must have

(307) $\qquad n - m - \tilde{m} = 1 \quad \text{and} \quad (m+1)a = -(\tilde{m}+1)b.$

Therefore, locally around any regular point D has precisely one nonconstant eigenvalue, which we denote by ρ, and the constant eigenvalues 1 and 0 with multiplicity $2m$, respectively $2\tilde{m}$. Hence, for any regular point $x_0 \in M$ we can find an open neighbourhood U of the curve $\phi_t(x_0)$ in the set of regular points and a frame of TU such that D corresponds to a matrix of the form

(308) $$D = \begin{pmatrix} \rho & & & \\ & \rho & & \\ & & \mathrm{Id}_{2m} & \\ & & & 0_{2\tilde{m}} \end{pmatrix},$$

where ρ is a smooth function on U with $\rho(x) \neq 0, 1$ for $x \in U$. Note that we have $\lambda = -\frac{1}{2} D_\alpha{}^\alpha = -(\rho + m)$ and consequently

$$\Lambda_\alpha = -\nabla_\alpha \rho,$$

which is nowhere vanishing on U. Recall also that the pair (D, Λ) satisfies the mobility equation (131), and that by Corollary 5.17 Λ^α is an eigenvector of $D_\alpha{}^\beta$ with eigenvalue ρ. Hence, Λ^α and $J_\beta{}^\alpha \Lambda^\beta$ form a basis for the eigenspace of $D_\alpha{}^\beta$ corresponding to the eigenvalue ρ.

For later use, let us also remark that by (300) the action of the flow ϕ_t on D preserves its block structure (308). This implies, in particular, that $\mathcal{L}_V \Lambda = [V, \Lambda]$ is an eigenvector of D with eigenvalue ρ. Since furthermore $\nabla_\alpha \rho = -\Lambda_\alpha$ vanishes in direction of all vector fields orthogonal to Λ^α and $[\Lambda, J\Lambda] = 0$, we therefore conclude that $[V, J\Lambda] \cdot \rho = 0$ implying that the vector field $[V, J\Lambda] = J[V, \Lambda]$ is actually a proportional to $J\Lambda$. Hence, the c-projective vector field V preserves the orthogonal projection from TM to the 1-dimensional subspace spanned by Λ^α, which is defined on the dense open subset on which Λ^α is not vanishing (hence in particular on U).

LEMMA 7.8. *Assume* $n - m - \tilde{m} = 1$ *and* $(m+1)a = -(\tilde{m}+1)b$. *Then in a neighbourhood of any regular point there exists a real positive constant B and a smooth real-valued function μ such that*

(309) $\qquad \nabla_\alpha \Lambda_\beta = -\mu g_{\alpha\beta} + 2B D_{\alpha\beta}$

(310) $\qquad \nabla_\alpha \mu = 2B \Lambda_\alpha.$

PROOF. Fix a regular point $x_0 \in M$, an open neighbourhood U of $\phi_t(x_0)$ in the set of regular points, and a frame of TU with respect to which g corresponds to the identity and D is of the form (308). Since Λ^α is an eigenvector of $D_\alpha{}^\beta$ with eigenvalue ρ, we can furthermore assume without loss of generality that Λ^α is proportional to the first vector of our fixed local frame. We restrict our considerations from now on to U.

Differentiating the equation $D_\alpha{}^\beta \Lambda^\alpha = \rho \Lambda^\beta$, and substituting (131) and $\nabla_\alpha \rho = -\Lambda_\alpha$, we obtain

(311) $\qquad (\rho \delta_\gamma{}^\beta - D_\gamma{}^\beta) \nabla_\alpha \Lambda_\beta = -\tfrac{1}{2}(g_{\alpha\gamma} \Lambda_\beta \Lambda^\beta - \Lambda_\alpha \Lambda_\gamma - J_\alpha{}^\beta \Lambda_\beta J_\gamma{}^\epsilon \Lambda_\epsilon).$

Since $D_\alpha{}^\beta$ commutes with $\nabla_\alpha \Lambda^\beta$ by Proposition 5.13, the block diagonal form of $D_\alpha{}^\beta$ implies that $\nabla_\alpha \Lambda^\beta$ has the same block diagonal form. Equation (311) therefore shows that the second and the third block of $\nabla_\alpha \Lambda^\alpha$ are proportional to the identity

7.3. THE PROOF OF THE YANO–OBATA CONJECTURE

with coefficient of proportionality $-\frac{\Lambda_\beta \Lambda^\beta}{2(\rho-1)}$ and $-\frac{\Lambda_\beta \Lambda^\beta}{2\rho}$ respectively. The condition (309) therefore reduces to the following three equations

(312) $\quad (\nabla_\alpha \Lambda_\beta)\Lambda^\beta = (-\mu + 2B\rho)\Lambda_\alpha \quad \Lambda_\beta \Lambda^\beta = 2(\mu - 2B)(\rho - 1) \quad \Lambda_\beta \Lambda^\beta = 2\rho\mu,$

which can be equivalently rewritten as

(313) $\quad (\nabla_\alpha \Lambda_\beta)\Lambda^\beta = 2B(2\rho - 1)\Lambda_\alpha \quad \Lambda_\alpha \Lambda^\alpha = -4B(\rho - 1)\rho \quad \mu = -2B(\rho - 1).$

First note that,

(314) $\quad \Lambda_\alpha V^\alpha = \frac{d}{dt}\big|_{t=0} \phi_t^* \lambda = -\frac{d}{dt}\big|_{t=0} \frac{\rho e^{at}}{\rho e^{at} + (1-\rho)e^{bt}} = (b-a)\rho(1-\rho),$

which is nowhere vanishing on U. It follows that

(315) $\quad \phi_t^*(\Lambda_\alpha V^\alpha) = \frac{(b-a)\rho(1-\rho)e^{at}e^{bt}}{(\rho e^{at} + (1-\rho)e^{bt})^2}.$

Let us write P for the orthogonal projection of TU to the line subbundle spanned by Λ^α. We have already noticed that V preserves this projection, which implies in particular that $\mathcal{L}_V P(V) = 0$. Note also that by (314) the vector field $P(V)$ is nowhere vanishing on U. Hence, we deduce from (302) that

(316) $\quad \phi_t^*(g(P(V), P(V))) = g(G_t P(V), P(V)) = \frac{g(P(V), P(V))}{(\rho e^{at} + (1-\rho)e^{bt})^2 e^{(am+b\tilde{m})t}}.$

Since by assumption $(m+1)a + (\tilde{m}+1)b = 0$, equations (315) and (316) show that the function

(317) $\quad t \mapsto \phi_t^*\left(\frac{g(\Lambda, V)}{g(P(V), P(V))}\right) = \frac{(b-a)\rho(1-\rho)}{g(P(V), P(V))}$

is constant. Since $g(\Lambda, V) = g(\Lambda, P(V))$, we therefore obtain

$$\phi_t^*(\Lambda) = \frac{\rho(1-\rho)(b-a)}{g(P(V), P(V))} \phi_t^*(P(V)),$$

and hence (315) implies

(318) $\quad \phi_t^*(\Lambda_\alpha \Lambda^\alpha) = \frac{(b-a)^2 \rho^2 (1-\rho)^2 e^{at} e^{bt}}{(\rho e^{at} + (1-\rho)e^{bt})^2 g(P(V), P(V))}.$

Differentiating identity (318) gives

(319) $\quad (\nabla_\alpha \Lambda_\beta \Lambda^\beta)V^\alpha = \frac{d}{dt}\big|_{t=0} \Phi_t^*(\Lambda_\beta \Lambda^\beta) = \frac{(b-a)^3 \rho^2 (1-\rho)^2 (2\rho - 1)}{g(P(V), P(V))}.$

Since $(\nabla_\alpha \Lambda_\beta)\Lambda^\alpha = \frac{1}{2}\nabla_\alpha(\Lambda_\beta \Lambda^\beta)$, we can rewrite the first condition of (313) as

$$\nabla_\alpha(\Lambda_\beta \Lambda^\beta) = 4B(2\rho - 1)\Lambda_\alpha,$$

and we conclude from (319) and (314) that contracting the above identity with V^α yields

(320) $\quad B = \frac{1}{4}\frac{(b-a)^2 \rho(1-\rho)}{g(P(V), P(V))}.$

By (318) the second condition of (313) reads

(321) $\quad \frac{(b-a)^2 \rho^2 (1-\rho)^2}{g(P(V), P(V))} = 4B(1-\rho)\rho,$

which is of course equivalent to (320). Since the third condition of (313), given by $\mu = -2B(\rho-1)$, simply defines μ in terms of ρ and B, we conclude that there exist functions μ and B on U such that (309) is satisfied. Note that by (320) the function B is also positive as required. It remains to show that B is constant (in a sufficiently small neighbourhood of x_0), which implies in particular that $\mu = -2B(\rho - 1)$ is smooth and that (310) is satisfied.

The formula (320) for B shows that B is proportional with a constant coefficient to (317). Hence, we have $\nabla_V B = 0$. To show that the derivative of B also vanishes along vector field transversal to V consider the $2n - 1$ dimensional submanifold of U given by the level set of ρ

$$(322) \qquad M_y := \{x \in U : \rho(x) = \rho(y)\},$$

where y is some arbitrary point in U. Since the derivative of ρ is nontrivial along the c-projective vector field V by (314), V is transversal to M_y. We claim that the derivative of B at y vanishes along all vectors in $T_y M_y$. Indeed, note that in view of (320) this is equivalent to the vanishing of the derivative of $g(P(V), P(V))$ at y along tangent vectors of M_y. Since Λ and $P(V)$ are proportional to each other by definition, we have

$$g(P(V), P(V)) = \frac{(g(P(V), \Lambda))^2}{g(\Lambda, \Lambda)}.$$

It is thus sufficient to show that at y the derivative of $g(P(V), \Lambda)$, respectively $g(\Lambda, \Lambda)$, vanishes along tangent vectors of M_y. By (314) this follows immediately for the derivative of $g(P(V), \Lambda) = g(V, \Lambda)$. Now consider $g(\Lambda, \Lambda)$ and let $W \in T_y M_y$. Then we compute

$$W^\alpha \nabla_\alpha (\Lambda_\beta \Lambda^\beta) = 2W^\alpha (\nabla_\alpha \Lambda_\beta) \Lambda^\beta$$
$$= -2\mu W^\alpha \Lambda_\alpha + 4BW^\alpha D_{\alpha\beta} \Lambda^\beta = -2\mu W^\alpha \Lambda_\alpha + 4B\rho W^\alpha \Lambda_\alpha.$$

Since $\Lambda_\alpha = -\nabla_\alpha \rho$, we see that $W^\alpha \Lambda_\alpha$ vanishes and consequently so does $W^\alpha \nabla_\alpha (\Lambda_\beta \Lambda^\beta)$. Hence, the derivative of B vanishes at $y \in U$. Since $y \in U$ was an arbitrary point, we conclude that ∇B is identically zero on U, which completes the proof. □

We can now complete the proof of Theorem 7.6. Under the assumption that (307) holds, Lemma 7.8 shows that in an open neighbourhood of any regular point there exists a positive constant B (which a priori may depend on the neighbourhood) and a function μ such that the triple (D, Λ, μ) satisfies, in addition to the mobility equation, the equations (309) and (310). Since the set of regular points is open and dense, Theorem 6.7 implies that B is actually the same constant at all regular points and that the equations (309) and (310) hold on M for some smooth function μ. Theorem 6.7 also implies that the function $\lambda = -\frac{1}{2} D_\alpha{}^\alpha$ satisfies the equivalent conditions of Proposition 6.22 on M for a positive constant B. Since $\Lambda_\alpha = \nabla_\alpha \lambda$ is not identically zero, Tanno's result [98] completes the proof.

CHAPTER 8

Outlook

There has been considerable activity in c-projective geometry since we began work on this article in 2013. Despite this, there remain many open questions and opportunity for further work. In this final section, we survey some of the developments and opportunities which we have not discussed already in the article.

8.1. Metrisability and symmetry

One of the main focuses of this article has been metrisable c-projective structures and their mobility. However, in later sections, we restricted attention to integrable complex structures (the torsion-free case). It would be interesting to extend more of the theory to non-integrable structures with a view to applications in quasi-Kähler geometry, including 4-dimensional almost Kähler geometry—here some partial results have been obtained in [**1**].

Even in the integrable case, however, a basic question remains wide open: when is a c-projective structure metrisable? One would like to provide a complete c-projectively invariant obstruction, analogous to the obstruction found for the 2-dimensional real projective case in [**21**].

Additional questions concern the symmetry algebra $\mathfrak{cproj}(J,[\nabla])$ of infinitesimal automorphisms of an almost c-projective $2n$-manifold $(M, J, [\nabla])$. As with any parabolic geometry [**29, 36**], the prolongation of the infinitesimal automorphism equation (see Chapter 3.4, Proposition 3.9) shows that $\mathfrak{cproj}(J,[\nabla])$ is finite dimensional, with its dimension bounded above—in this case, by $2(n+1)^2 - 2$. This bound is attained only in the c-projectively flat case, and, as shown in [**65**], in the non-flat case, the dimension of $\mathfrak{cproj}(J,[\nabla])$ is at most $2n^2 - 2n + 4 + 2\delta_{3,n}$ (the so-called "submaximal dimension"). The determination of the possible dimensions of $\mathfrak{cproj}(J,[\nabla])$ remains an open question.

The symmetry algebra $\mathfrak{cproj}(J,[\nabla])$ acts (by Lie derivative) on the space of solutions to the metrisability equation, and for any nondegenerate solution, corresponding to a compatible metric g, it has subalgebras $\mathfrak{isom}(J,g) \subseteq \mathfrak{aff}(J,g)$ of holomorphic Killing fields and infinitesimal complex affine transformations. Thus the presence of compatible metrics constrains $\mathfrak{cproj}(J,[\nabla])$ further, the Yano–Obata theorems being global examples of this. Even locally, if (M, J, g) admits an *essential* c-projective vector field, i.e., an element $X \in \mathfrak{cproj}(J,[\nabla]) \setminus \mathfrak{isom}(J,g)$, then g must have mobility ≥ 2, and if $X \notin \mathfrak{aff}(J,g)$, then there are metrics c-projectively, but not affinely, equivalent to g. For example, nontrivial c-projective vector fields with higher order zeros are essential (because a Killing vector field is determined locally uniquely by its 1-jet at a point), and such *strongly essential local flows* exist only on c-projectively flat geometries [**80**].

Constraints on the possible dimensions of $\mathfrak{cproj}(J,[\nabla])/\mathfrak{isom}(J,g)$ for Kähler manifolds are given in [**79**], and an explicit classification of 4-dimensional (pseudo-)Kähler metrics admitting essential c-projective vector fields is given in [**14**].

An open question here is whether locally nonlinearizable c-projective vector fields exist on nonflat c-projective geometries.

8.2. Applications in Kähler geometry

The original motivation for c-projective geometry [**91**] was to extend methods of projective geometry to Kähler metrics (and this is one reason why we have concentrated so much on the metrisability equation). Thus one expects ideas from c-projective geometry to be useful in Kähler geometry, and indeed important concepts in Kähler geometry have c-projective origins: for instance, Hamiltonians for Killing vector fields form the kernel of the c-projective Hessian.

Apostolov et al. [**2–5**] use c-projectively equivalent metrics (in the guise of Hamiltonian 2-forms) to study *extremal Kähler metrics*, where the scalar curvature of a Kähler metric lies in the kernel of its c-projective Hessian, and it would be natural to consider extremal quasi-Kähler metrics in the same light. Kähler–Ricci solitons (and generalisations) admitting c-projectively equivalent metrics have also been studied in special cases [**5, 67, 71**], but the picture is far from complete.

A more recent development is the work of Čap and Gover [**32**], which extends previous work on projective compactification of Einstein metrics [**31**] to Kähler (and quasi-Kähler) metrics using c-projective geometry.

Let us now touch on prospective applications in the theory of finite dimensional integrable systems. As we have seen in Chapter 5, the Killing equations for Hermitian symmetric Killing tensors are c-projectively invariant. This suggests to study these equations from a c-projective viewpoint. We expect that in this way one may find interesting new examples of integrable systems, in particular on closed Kähler or Hermitian manifolds (note that only few such examples are known). Moreover, the analogy between metric projective and c-projective geometries suggests that ideas and constructions from the theory of integrable geodesic flows on n-dimensional Riemannian manifolds could be used in the construction and description of integrable geodesics flows on $2n$-dimensional Kähler manifolds. For Killing tensors of valence two this approach is very close to the one in [**59**], and we expect similar applications for Killing tensors of higher valence.

8.3. Projective parabolic geometries

We noted in the introduction that there are many analogies between methods and results in projective and c-projective geometry, and we have followed the literature in exploiting this observation. We have already noted a partial explanation for the similarities: both are Cartan geometries modelled on flag varieties G/P, one a complex version of the other. However, the c-projective metrisability equation for compatible (pseudo-)Kähler metrics is not the complexification of the corresponding projective metrisability equation. Instead, both are first BGG operators for a G-representation with a 1-dimensional P-subrepresentation. These representations determine projective embeddings of the model G/P, namely, the Veronese embedding of $\mathbb{RP}^n$ as rank one symmetric matrices in the projective space of $S^2\mathbb{R}^{n+1}$, and the analogous projective embedding of $\mathbb{CP}^n$ using rank one Hermitian matrices.

Symmetric and Hermitian matrices are examples of Jordan algebras, and this relation with projective geometry is well known (see e.g. [11]), which suggests to define a *projective parabolic geometry* as one in which the model has a projective embedding into a suitable Jordan algebra. Apart from projective and c-projective geometry, the examples include quaternionic geometry, conformal geometry, and the geometry associated to the Cayley plane over the octonions. In his PhD thesis [47], G. Frost has shown the much of the metrisability theory of these geometries can be developed in a unified framework. Further, in addition to being analogous, projective parabolic geometries are closely interrelated. For instance, S. Armstrong [6] uses cone constructions to realise quaternionic and c-projective geometry as holonomy reductions of projective Cartan connections, while the *generalised Feix–Kaledin construction* [17] shows how to build quaternionic structures from c-projective structures, modelled on the standard embedding of $\mathbb{CP}^n$ in $\mathbb{HP}^n$.

Bibliography

[1] V. Apostolov, D. M. J. Calderbank, P. Gauduchon: *The geometry of weakly selfdual Kähler surfaces*, Comp. Math. **73** (2006), 359–412.

[2] V. Apostolov, D. M. J. Calderbank, and P. Gauduchon, *Hamiltonian 2-forms in Kähler geometry. I. General theory*, J. Differential Geom. **73** (2006), no. 3, 359–412. MR2228318

[3] V. Apostolov, D. M. J. Calderbank, P. Gauduchon, and C. W. Tønnesen-Friedman, *Hamiltonian 2-forms in Kähler geometry. II. Global classification*, J. Differential Geom. **68** (2004), no. 2, 277–345. MR2144249

[4] V. Apostolov, D. M. J. Calderbank, P. Gauduchon, and C. W. Tønnesen-Friedman, *Hamiltonian 2-forms in Kähler geometry. III. Extremal metrics and stability*, Invent. Math. **173** (2008), no. 3, 547–601, DOI 10.1007/s00222-008-0126-x. MR2425136

[5] V. Apostolov, D. M. J. Calderbank, P. Gauduchon, and C. W. Tønnesen-Friedman, *Hamiltonian 2-forms in Kähler geometry. IV. Weakly Bochner-flat Kähler manifolds*, Comm. Anal. Geom. **16** (2008), no. 1, 91–126. MR2411469

[6] S. Armstrong, *Projective holonomy. I. Principles and Properties, II. Cones and complete classifications*, Ann. Global Anal. Geom. **33** (2008), no. 2, 47–69, 137–160, DOI 10.1007/s10455-007-9075-7. MR2379941

[7] T. N. Bailey, M. G. Eastwood, and A. R. Gover, *Thomas's structure bundle for conformal, projective and related structures*, Rocky Mountain J. Math. **24** (1994), no. 4, 1191–1217, DOI 10.1216/rmjm/1181072333. MR1322223

[8] R. J. Baston and M. G. Eastwood, *The Penrose transform*, Oxford Mathematical Monographs, The Clarendon Press, Oxford University Press, New York, 1989. Its interaction with representation theory; Oxford Science Publications. MR1038279

[9] S. Benenti, *Special symmetric two-tensors, equivalent dynamical systems, cofactor and bicofactor systems*, Acta Appl. Math. **87** (2005), no. 1-3, 33–91, DOI 10.1007/s10440-005-1138-9. MR2151124

[10] I. N. Bernšteĭn, I. M. Gel'fand, and S. I. Gel'fand, *Differential operators on the base affine space and a study of $\mathfrak{g}$-modules*, Lie groups and their representations (Proc. Summer School, Bolyai János Math. Soc., Budapest, 1971), Halsted, New York, 1975, pp. 21–64. MR0578996

[11] W. Bertram, *Generalized projective geometries: general theory and equivalence with Jordan structures*, Adv. Geom. **2** (2002), no. 4, 329–369, DOI 10.1515/advg.2002.016. MR1940443

[12] S. Bochner, *Curvature in Hermitian metric*, Bull. Amer. Math. Soc. **53** (1947), 179–195, DOI 10.1090/S0002-9904-1947-08778-4. MR0019983

[13] A. V. Bolsinov and V. S. Matveev, *Geometrical interpretation of Benenti systems*, J. Geom. Phys. **44** (2003), no. 4, 489–506, DOI 10.1016/S0393-0440(02)00054-2. MR1943174

[14] A. V. Bolsinov, V. S. Matveev, T. Mettler, and S. Rosemann, *Four-dimensional Kähler metrics admitting c-projective vector fields*, J. Math. Pures Appl. (9) **103** (2015), no. 3, 619–657, DOI 10.1016/j.matpur.2014.07.005. MR3310270

[15] A. V. Bolsinov and V. S. Matveev, *Local normal forms for geodesically equivalent pseudo-Riemannian metrics*, Trans. Amer. Math. Soc. **367** (2015), no. 9, 6719–6749, DOI 10.1090/S0002-9947-2014-06416-7. MR3356952

[16] A. Bolsinov, V. S. Matveev, S. Rosemann: *Local normal forms for c-projectively equivalent metrics and proof of the Yano-Obata conjecture in arbitrary signature. Proof of the projective Lichnerowicz conjecture for Lorentzian metrics*, arXiv:1510.00275. To appear, Ann. Sci. Éc. Supér.

[17] A. W. Borówka and D. M. J. Calderbank, *Projective geometry and the quaternionic Feix-Kaledin construction*, Trans. Amer. Math. Soc. **372** (2019), no. 7, 4729–4760, DOI 10.1090/tran/7719. MR4009396

[18] C. Boubel, *On the algebra of parallel endomorphisms of a pseudo-Riemannian metric*, J. Differential Geom. **99** (2015), no. 1, 77–123. MR3299823

[19] C. P. Boyer and K. Galicki, *Sasakian geometry*, Oxford Mathematical Monographs, Oxford University Press, Oxford, 2008. MR2382957

[20] T. Branson, A. Čap, M. Eastwood, and A. R. Gover, *Prolongations of geometric overdetermined systems*, Internat. J. Math. **17** (2006), no. 6, 641–664, DOI 10.1142/S0129167X06003655. MR2246885

[21] R. Bryant, M. Dunajski, and M. Eastwood, *Metrisability of two-dimensional projective structures*, J. Differential Geom. **83** (2009), no. 3, 465–499. MR2581355

[22] R. L. Bryant, G. Manno, and V. S. Matveev, *A solution of a problem of Sophus Lie: normal forms of two-dimensional metrics admitting two projective vector fields*, Math. Ann. **340** (2008), no. 2, 437–463, DOI 10.1007/s00208-007-0158-3. MR2368987

[23] D. M. J. Calderbank, *Möbius structures and two-dimensional Einstein-Weyl geometry*, J. Reine Angew. Math. **504** (1998), 37–53, DOI 10.1515/crll.1998.111. MR1656822

[24] D. M. J. Calderbank, *Two dimensional Einstein-Weyl structures*, Glasg. Math. J. **43** (2001), no. 3, 419–424, DOI 10.1017/S0017089501030051. MR1878586

[25] D. M. J. Calderbank and T. Diemer, *Differential invariants and curved Bernstein-Gelfand-Gelfand sequences*, J. Reine Angew. Math. **537** (2001), 67–103, DOI 10.1515/crll.2001.059. MR1856258

[26] D. M. J. Calderbank, V. S. Matveev, and S. Rosemann, *Curvature and the c-projective mobility of Kähler metrics with hamiltonian 2-forms*, Compos. Math. **152** (2016), no. 8, 1555–1575, DOI 10.1112/S0010437X16007302. MR3542486

[27] A. Cannas da Silva, *Lectures on symplectic geometry*, Lecture Notes in Mathematics, vol. 1764, Springer-Verlag, Berlin, 2001. MR1853077

[28] A. Čap, *Correspondence spaces and twistor spaces for parabolic geometries*, J. Reine Angew. Math. **582** (2005), 143–172, DOI 10.1515/crll.2005.2005.582.143. MR2139714

[29] A. Čap, *Infinitesimal automorphisms and deformations of parabolic geometries*, J. Eur. Math. Soc. (JEMS) **10** (2008), no. 2, 415–437, DOI 10.4171/JEMS/116. MR2390330

[30] A. Čap and A. R. Gover, *Tractor calculi for parabolic geometries*, Trans. Amer. Math. Soc. **354** (2002), no. 4, 1511–1548, DOI 10.1090/S0002-9947-01-02909-9. MR1873017

[31] A. Čap and A. R. Gover, *Projective compactifications and Einstein metrics*, J. Reine Angew. Math. **717** (2016), 47–75, DOI 10.1515/crelle-2014-0036. MR3530534

[32] A. Čap and A. R. Gover, *C-Projective compactification; (quasi-)Kähler metrics and CR boundaries*, Amer. J. Math. **141** (2019), no. 3, 813–856, DOI 10.1353/ajm.2019.0017. MR3956522

[33] A. Čap, A. R. Gover, and M. Hammerl, *Normal BGG solutions and polynomials*, Internat. J. Math. **23** (2012), no. 11, 1250117, 29, DOI 10.1142/S0129167X12501170. MR3005570

[34] A. Čap, A. R. Gover, and M. Hammerl, *Projective BGG equations, algebraic sets, and compactifications of Einstein geometries*, J. Lond. Math. Soc. (2) **86** (2012), no. 2, 433–454, DOI 10.1112/jlms/jds002. MR2980919

[35] A. Čap and H. Schichl, *Parabolic geometries and canonical Cartan connections*, Hokkaido Math. J. **29** (2000), no. 3, 453–505, DOI 10.14492/hokmj/1350912986. MR1795487

[36] A. Čap and J. Slovák, *Parabolic geometries. I*, Mathematical Surveys and Monographs, vol. 154, American Mathematical Society, Providence, RI, 2009. Background and general theory. MR2532439

[37] A. Čap and J. Slovák, *Weyl structures for parabolic geometries*, Math. Scand. **93** (2003), no. 1, 53–90, DOI 10.7146/math.scand.a-14413. MR1997873

[38] A. Čap, J. Slovák, and V. Souček, *Bernstein-Gelfand-Gelfand sequences*, Ann. of Math. (2) **154** (2001), no. 1, 97–113, DOI 10.2307/3062111. MR1847589

[39] E. Cartan, *Sur les variétés à connexion projective* (French), Bull. Soc. Math. France **52** (1924), 205–241. MR1504846

[40] U. Dini: *Sopra un problema che si presenta nella teoria generale delle rappresentazioni geografiche di una superficie su un'altra*, Ann. Mat., ser.2 **3** (1869), 269–293.

[41] V. V. Domašev and Ĭ. Mikeš, *On the theory of holomorphically projective mappings of Kählerian spaces* (Russian), Mat. Zametki **23** (1978), no. 2, 297–303. MR492674

[42] M. Eastwood, *Notes on projective differential geometry*, Symmetries and overdetermined systems of partial differential equations, IMA Vol. Math. Appl., vol. 144, Springer, New York, 2008, pp. 41–60, DOI 10.1007/978-0-387-73831-4_3. MR2384705

BIBLIOGRAPHY

[43] M. Eastwood and V. Matveev, *Metric connections in projective differential geometry*, Symmetries and overdetermined systems of partial differential equations, IMA Vol. Math. Appl., vol. 144, Springer, New York, 2008, pp. 339–350, DOI 10.1007/978-0-387-73831-4_16. MR2384718

[44] A. Fedorova, V. Kiosak, V. S. Matveev, and S. Rosemann, *The only Kähler manifold with degree of mobility at least 3 is* ($\mathbb{C}P(n), g_{Fubini\text{-}Study}$), Proc. Lond. Math. Soc. (3) **105** (2012), no. 1, 153–188, DOI 10.1112/plms/pdr053. MR2948791

[45] A. Fedorova and V. S. Matveev, *Degree of mobility for metrics of Lorentzian signature and parallel $(0,2)$-tensor fields on cone manifolds*, Proc. Lond. Math. Soc. (3) **108** (2014), no. 5, 1277–1312, DOI 10.1112/plms/pdt054. MR3214680

[46] D. J. F. Fox, *Contact projective structures*, Indiana Univ. Math. J. **54** (2005), no. 6, 1547–1598, DOI 10.1512/iumj.2005.54.2603. MR2189678

[47] G. E. Frost: *The Projective Parabolic Geometry of Riemannian, Kähler and Quaternion-Kähler Metrics*, PhD Thesis, University of Bath (2016), arXiv:1605.04406.

[48] P. Gauduchon, *Hermitian connections and Dirac operators* (English, with Italian summary), Boll. Un. Mat. Ital. B (7) **11** (1997), no. 2, suppl., 257–288. MR1456265

[49] A. R. Gover and V. S. Matveev, *Projectively related metrics, Weyl nullity and metric projectively invariant equations*, Proc. Lond. Math. Soc. (3) **114** (2017), no. 2, 242–292, DOI 10.1112/plms.12002. MR3653230

[50] A. Gray, *Spaces of constancy of curvature operators*, Proc. Amer. Math. Soc. **17** (1966), 897–902, DOI 10.2307/2036279. MR0198392

[51] A. Gray and L. M. Hervella, *The sixteen classes of almost Hermitian manifolds and their linear invariants*, Ann. Mat. Pura Appl. (4) **123** (1980), 35–58, DOI 10.1007/BF01796539. MR581924

[52] M. Hammerl, P. Somberg, V. Souček, and J. Šilhan, *On a new normalization for tractor covariant derivatives*, J. Eur. Math. Soc. (JEMS) **14** (2012), no. 6, 1859–1883, DOI 10.4171/JEMS/349. MR2984590

[53] H. Hiramatu, *Riemannian manifolds admitting a projective vector field*, Kodai Math. J. **3** (1980), no. 3, 397–406. MR604484

[54] H. Hiramatu, *Integral inequalities in Kählerian manifolds and their applications*, Period. Math. Hungar. **12** (1981), no. 1, 37–47, DOI 10.1007/BF01848170. MR607627

[55] J. Hrdina, *Almost complex projective structures and their morphisms*, Arch. Math. (Brno) **45** (2009), no. 4, 255–264. MR2591680

[56] S. Ishihara and S.-i. Tachibana, *A note on holomorphically projective transformations of a Kählerian space with parallel Ricci tensor*, Tôhoku Math. J. (2) **13** (1961), 193–200, DOI 10.2748/tmj/1178244296. MR0139128

[57] V. Kiosak and V. S. Matveev, *Proof of the projective Lichnerowicz conjecture for pseudo-Riemannian metrics with degree of mobility greater than two*, Comm. Math. Phys. **297** (2010), no. 2, 401–426, DOI 10.1007/s00220-010-1037-4. MR2651904

[58] V. F. Kiričenko, *K-spaces of maximal rank* (Russian), Mat. Zametki **22** (1977), no. 4, 465–476. MR0474103

[59] K. Kiyohara, *Two classes of Riemannian manifolds whose geodesic flows are integrable*, Mem. Amer. Math. Soc. **130** (1997), no. 619, viii+143, DOI 10.1090/memo/0619. MR1396959

[60] K. Kiyohara, *C-projective equivalence and integrability of the geodesic flow*, J. Geom. Phys. **87** (2015), 286–295, DOI 10.1016/j.geomphys.2014.07.023. MR3282374

[61] K. Kiyohara and P. Topalov, *On Liouville integrability of h-projectively equivalent Kähler metrics*, Proc. Amer. Math. Soc. **139** (2011), no. 1, 231–242, DOI 10.1090/S0002-9939-2010-10576-2. MR2729086

[62] S. Kobayashi and K. Nomizu, *Foundations of differential geometry. Vol. II*, Interscience Tracts in Pure and Applied Mathematics, No. 15 Vol. II, Interscience Publishers John Wiley & Sons, Inc., New York-London-Sydney, 1969. MR0238225

[63] S. Kobayashi, *Natural connections in almost complex manifolds*, Explorations in complex and Riemannian geometry, Contemp. Math., vol. 332, Amer. Math. Soc., Providence, RI, 2003, pp. 153–169, DOI 10.1090/conm/332/05935. MR2018338

[64] B. Kostant, *Lie algebra cohomology and the generalized Borel-Weil theorem*, Ann. of Math. (2) **74** (1961), 329–387, DOI 10.2307/1970237. MR0142696

[65] B. Kruglikov, V. Matveev, and D. The, *Submaximally symmetric c-projective structures*, Internat. J. Math. **27** (2016), no. 3, 1650022, 34, DOI 10.1142/S0129167X16500221. MR3474062

[66] J.-L. Lagrange: *Sur la construction des cartes géographiques*, Nouveaux mémoires de l'Académie royale des sciences et belles-lettres de Berlin (1779).

[67] E. Legendre and C. W. Tønnesen-Friedman, *Toric generalized Kähler-Ricci solitons with Hamiltonian 2-form*, Math. Z. **274** (2013), no. 3-4, 1177–1209, DOI 10.1007/s00209-012-1112-y. MR3078263

[68] A. Lichnerowicz, *Geometry of groups of transformations*, Noordhoff International Publishing, Leyden, 1977. Translated from the French and edited by Michael Cole. MR0438364

[69] A. Lichnerowicz, *Théorie globale des connexions et des groupes d'holonomie* (French), Edizioni Cremonese, Roma, 1957. MR0088015

[70] R. Liouville: *Sur les invariants de certaines équations différentielles et sur leurs applications*, J. Ecole Polytechnique **59** (1889), 7–76.

[71] G. Maschler and C. W. Tønnesen-Friedman, *Generalizations of Kähler-Ricci solitons on projective bundles*, Math. Scand. **108** (2011), no. 2, 161–176, DOI 10.7146/math.scand.a-15165. MR2805600

[72] V. S. Matveev, *Hyperbolic manifolds are geodesically rigid*, Invent. Math. **151** (2003), no. 3, 579–609, DOI 10.1007/s00222-002-0263-6. MR1961339

[73] V. S. Matveev, *Lichnerowicz-Obata conjecture in dimension two*, Comment. Math. Helv. **80** (2005), no. 3, 541–570, DOI 10.4171/CMH/25. MR2165202

[74] V. S. Matveev, *Proof of the projective Lichnerowicz-Obata conjecture*, J. Differential Geom. **75** (2007), no. 3, 459–502. MR2301453

[75] V. S. Matveev, *Two-dimensional metrics admitting precisely one projective vector field*, Math. Ann. **352** (2012), no. 4, 865–909, DOI 10.1007/s00208-011-0659-y. MR2892455

[76] V. S. Matveev, *Projectively invariant objects and the index of the group of affine transformations in the group of projective transformations*, Bull. Iranian Math. Soc. **44** (2018), no. 2, 341–375, DOI 10.1007/s41980-018-0024-y. MR3820550

[77] V. S. Matveev and K. Neusser, *On the groups of c-projective transformations of complete Kähler manifolds*, Ann. Global Anal. Geom. **54** (2018), no. 3, 329–352, DOI 10.1007/s10455-018-9604-6. MR3867648

[78] V. S. Matveev and S. Rosemann, *Proof of the Yano-Obata conjecture for h-projective transformations*, J. Differential Geom. **92** (2012), no. 2, 221–261. MR2998672

[79] V. S. Matveev and S. Rosemann, *Conification construction for Kähler manifolds and its application in c-projective geometry*, Adv. Math. **274** (2015), 1–38, DOI 10.1016/j.aim.2015.01.006. MR3318143

[80] K. Melnick and K. Neusser, *Strongly essential flows on irreducible parabolic geometries*, Trans. Amer. Math. Soc. **368** (2016), no. 11, 8079–8110, DOI 10.1090/tran/6814. MR3546794

[81] T. Mettler, *On Kähler metrisability of two dimensional complex projective structures*, Monatsh. Math. **174** (2014), no. 4, 599–616, DOI 10.1007/s00605-014-0636-0. MR3233113

[82] Ĭ. Mikeš, *Holomorphically projective mappings of Kähler spaces* (Russian), Ukrain. Geom. Sb. **23** (1980), 90–98, iii. MR614278

[83] J. Mikeš, *Geodesic mappings of affine-connected and Riemannian spaces*, J. Math. Sci. **78** (1996), no. 3, 311–333, DOI 10.1007/BF02365193. Geometry, 2. MR1384327

[84] J. Mikeš, *Holomorphically projective mappings and their generalizations*, J. Math. Sci. (New York) **89** (1998), no. 3, 1334–1353, DOI 10.1007/BF02414875. Geometry, 3. MR1619720

[85] J. Mikeš, N. S. Sinyukov: *Quasiplanar mappings of spaces with affine connection*, Soviet Math. **27** (1983), 63–70.

[86] T. Morimoto, *Geometric structures on filtered manifolds*, Hokkaido Math. J. **22** (1993), no. 3, 263–347, DOI 10.14492/hokmj/1381413178. MR1245130

[87] A. Moroianu and U. Semmelmann, *Twistor forms on Kähler manifolds*, Ann. Sc. Norm. Super. Pisa Cl. Sci. (5) **2** (2003), no. 4, 823–845. MR2040645

[88] P.-A. Nagy, *On nearly-Kähler geometry*, Ann. Global Anal. Geom. **22** (2002), no. 2, 167–178, DOI 10.1023/A:1019506730571. MR1923275

[89] K. Neusser, *Prolongation on regular infinitesimal flag manifolds*, Internat. J. Math. **23** (2012), no. 4, 1250007, 41, DOI 10.1142/S0129167X11007501. MR2903185

[90] A. Newlander and L. Nirenberg, *Complex analytic coordinates in almost complex manifolds*, Ann. of Math. (2) **65** (1957), 391–404, DOI 10.2307/1970051. MR0088770

[91] T. Ōtsuki and Y. Tashiro, *On curves in Kaehlerian spaces*, Math. J. Okayama Univ. **4** (1954), 57–78. MR0066024

[92] R. Penrose and W. Rindler, *Spinors and space-time. Vol. 1*, Cambridge Monographs on Mathematical Physics, Cambridge University Press, Cambridge, 1984. Two-spinor calculus and relativistic fields. MR776784

[93] U. Semmelmann, *Conformal Killing forms on Riemannian manifolds*, Math. Z. **245** (2003), no. 3, 503–527, DOI 10.1007/s00209-003-0549-4. MR2021568

[94] U. Semmelmann, *Conformal Killing forms on Riemannian manifolds*, Math. Z. **245** (2003), no. 3, 503–527, DOI 10.1007/s00209-003-0549-4. MR2021568

[95] N. S. Sinjukov, *Geodezicheskie otobrazheniya rimanovykh prostranstv* (Russian), "Nauka", Moscow, 1979. MR552022

[96] N. Tanaka, *On the equivalence problems associated with simple graded Lie algebras*, Hokkaido Math. J. **8** (1979), no. 1, 23–84, DOI 10.14492/hokmj/1381758416. MR533089

[97] S. Tanno, *4-dimensional conformally flat Kähler manifolds*, Tōhoku Math. J. (2) **24** (1972), 501–504, DOI 10.2748/tmj/1178241491. MR0317252

[98] S. Tanno, *Some differential equations on Riemannian manifolds*, J. Math. Soc. Japan **30** (1978), no. 3, 509–531, DOI 10.2969/jmsj/03030509. MR500721

[99] Y. Tashiro, *On a holomorphically projective correspondence in an almost complex space*, Math. J. Okayama Univ. **6** (1957), 147–152. MR0087181

[100] Thales: *The sphere is projectively flat*, preprint, Miletus, circa 600 BC.

[101] P. Topalov, *Geodesic compatibility and integrability of geodesic flows*, J. Math. Phys. **44** (2003), no. 2, 913–929, DOI 10.1063/1.1526939. MR1953103

[102] P. Topalov and V. S. Matveev, *Geodesic equivalence via integrability*, Geom. Dedicata **96** (2003), 91–115, DOI 10.1023/A:1022166218282. MR1956835

[103] K. Yano, *On harmonic and Killing vector fields*, Ann. of Math. (2) **55** (1952), 38–45, DOI 10.2307/1969418. MR0046122

[104] Y. Yoshimatsu, *H-projective connections and H-projective transformations*, Osaka J. Math. **15** (1978), no. 2, 435–459. MR0500679

[105] N. Woodhouse, *Geometric quantization*, The Clarendon Press, Oxford University Press, New York, 1980. Oxford Mathematical Monographs. MR605306

[106] A. Zeghib, *On discrete projective transformation groups of Riemannian manifolds* (English, with English and French summaries), Adv. Math. **297** (2016), 26–53, DOI 10.1016/j.aim.2016.04.002. MR3498793

Editorial Information

To be published in the *Memoirs*, a paper must be correct, new, nontrivial, and significant. Further, it must be well written and of interest to a substantial number of mathematicians. Piecemeal results, such as an inconclusive step toward an unproved major theorem or a minor variation on a known result, are in general not acceptable for publication.

Papers appearing in *Memoirs* are generally at least 80 and not more than 200 published pages in length. Papers less than 80 or more than 200 published pages require the approval of the Managing Editor of the Transactions/Memoirs Editorial Board. Published pages are the same size as those generated in the style files provided for $\mathcal{AMS}$-LaTeX or $\mathcal{AMS}$-TeX.

Information on the backlog for this journal can be found on the AMS website starting from http://www.ams.org/memo.

A Consent to Publish is required before we can begin processing your paper. After a paper is accepted for publication, the Providence office will send a Consent to Publish and Copyright Agreement to all authors of the paper. By submitting a paper to the *Memoirs*, authors certify that the results have not been submitted to nor are they under consideration for publication by another journal, conference proceedings, or similar publication.

Information for Authors

Memoirs is an author-prepared publication. Once formatted for print and on-line publication, articles will be published as is with the addition of AMS-prepared frontmatter and backmatter. Articles are not copyedited; however, confirmation copy will be sent to the authors.

Initial submission. The AMS uses Centralized Manuscript Processing for initial submissions. Authors should submit a PDF file using the Initial Manuscript Submission form found at www.ams.org/submission/memo, or send one copy of the manuscript to the following address: Centralized Manuscript Processing, MEMOIRS OF THE AMS, 201 Charles Street, Providence, RI 02904-2294 USA. If a paper copy is being forwarded to the AMS, indicate that it is for *Memoirs* and include the name of the corresponding author, contact information such as email address or mailing address, and the name of an appropriate Editor to review the paper (see the list of Editors below).

The paper must contain a *descriptive title* and an *abstract* that summarizes the article in language suitable for workers in the general field (algebra, analysis, etc.). The *descriptive title* should be short, but informative; useless or vague phrases such as "some remarks about" or "concerning" should be avoided. The *abstract* should be at least one complete sentence, and at most 300 words. Included with the footnotes to the paper should be the 2010 *Mathematics Subject Classification* representing the primary and secondary subjects of the article. The classifications are accessible from www.ams.org/msc/. The Mathematics Subject Classification footnote may be followed by a list of *key words and phrases* describing the subject matter of the article and taken from it. Journal abbreviations used in bibliographies are listed in the latest *Mathematical Reviews* annual index. The series abbreviations are also accessible from www.ams.org/msnhtml/serials.pdf. To help in preparing and verifying references, the AMS offers MR Lookup, a Reference Tool for Linking, at www.ams.org/mrlookup/.

Electronically prepared manuscripts. The AMS encourages electronically prepared manuscripts, with a strong preference for $\mathcal{AMS}$-LaTeX. To this end, the Society has prepared $\mathcal{AMS}$-LaTeX author packages for each AMS publication. Author packages include instructions for preparing electronic manuscripts, samples, and a style file that generates the particular design specifications of that publication series. Though $\mathcal{AMS}$-LaTeX is the highly preferred format of TeX, author packages are also available in $\mathcal{AMS}$-TeX.

Authors may retrieve an author package for *Memoirs of the AMS* from www.ams.org/journals/memo/memoauthorpac.html. The *AMS Author Handbook* is available in PDF format from the author package link. The author package can also be obtained free of charge by sending email to tech-support@ams.org or from the Publication Division, American Mathematical Society, 201 Charles St., Providence, RI 02904-2294, USA. When requesting an author package, please specify $\mathcal{AMS}$-LaTeX or $\mathcal{AMS}$-TeX and the publication in which your paper will appear. Please be sure to include your complete mailing address.

After acceptance. The source files for the final version of the electronic manuscript should be sent to the Providence office immediately after the paper has been accepted for publication. The author should also submit a PDF of the final version of the paper to the editor, who will forward a copy to the Providence office.

Accepted electronically prepared files can be submitted via the web at www.ams.org/submit-book-journal/, sent via FTP, or sent on CD to the Electronic Prepress Department, American Mathematical Society, 201 Charles Street, Providence, RI 02904-2294 USA. TeX source files and graphic files can be transferred over the Internet by FTP to the Internet node ftp.ams.org (130.44.1.100). When sending a manuscript electronically via CD, please be sure to include a message indicating that the paper is for the *Memoirs*.

Electronic graphics. Comprehensive instructions on preparing graphics are available at www.ams.org/authors/journals.html. A few of the major requirements are given here.

Submit files for graphics as EPS (Encapsulated PostScript) files. This includes graphics originated via a graphics application as well as scanned photographs or other computer-generated images. If this is not possible, TIFF files are acceptable as long as they can be opened in Adobe Photoshop or Illustrator.

Authors using graphics packages for the creation of electronic art should also avoid the use of any lines thinner than 0.5 points in width. Many graphics packages allow the user to specify a "hairline" for a very thin line. Hairlines often look acceptable when proofed on a typical laser printer. However, when produced on a high-resolution laser imagesetter, hairlines become nearly invisible and will be lost entirely in the final printing process.

Screens should be set to values between 15% and 85%. Screens which fall outside of this range are too light or too dark to print correctly. Variations of screens within a graphic should be no less than 10%.

Any graphics created in color will be rendered in grayscale for the printed version unless color printing is authorized by the Managing Editor and the Publisher. In general, color graphics will appear in color in the online version.

Inquiries. Any inquiries concerning a paper that has been accepted for publication should be sent to memo-query@ams.org or directly to the Electronic Prepress Department, American Mathematical Society, 201 Charles St., Providence, RI 02904-2294 USA.

Editors

This journal is designed particularly for long research papers, normally at least 80 pages in length, and groups of cognate papers in pure and applied mathematics. Papers intended for publication in the *Memoirs* should be addressed to one of the following editors. The AMS uses Centralized Manuscript Processing for initial submissions to AMS journals. Authors should follow instructions listed on the Initial Submission page found at www.ams.org/memo/memosubmit.html.

Managing Editor: Henri Darmon, Department of Mathematics, McGill University, Montreal, Quebec H3A 0G4, Canada; e-mail: darmon@math.mcgill.ca

1. GEOMETRY, TOPOLOGY & LOGIC

    Coordinating Editor: Richard Canary, Department of Mathematics, University of Michigan, Ann Arbor, MI 48109-1043 USA; e-mail: canary@umich.edu

    Algebraic topology, Michael Hill, Department of Mathematics, University of California Los Angeles, Los Angeles, CA 90095 USA; e-mail: mikehill@math.ucla.edu

    Logic, Mariya Ivanova Soskova, Department of Mathematics, University of Wisconsin–Madison, Madison, WI 53706 USA; e-mail: msoskova@math.wisc.edu

    Low-dimensional topology and geometric structures, Richard Canary

2. ALGEBRA AND NUMBER THEORY

    Coordinating Editor: Henri Darmon, Department of Mathematics, McGill University, Montreal, Quebec H3A 0G4, Canada; e-mail: darmon@math.mcgill.ca

    Algebra, Radha Kessar, Department of Mathematics, City, University of London, London EC1V 0HB, United Kingdom; e-mail: radha.kessar.1@city.ac.uk

    Algebraic geometry, Lucia Caporaso, Department of Mathematics and Physics, Roma Tre University, Largo San Leonardo Murialdo, I-00146 Rome, Italy; e-mail: LCedit@mat.uniroma3.it

    Analytic number theory, Lillian B. Pierce, Department of Mathematics, Duke University, 120 Science Drive Box 90320, Durham, NC 27708 USA; e-mail: pierce@math.duke.edu

    Arithmetic geometry, Ted C. Chinburg, Department of Mathematics, University of Pennsylvania, Philadelphia, PA 19104-6395 USA; e-mail: ted@math.upenn.edu

    Commutative algebra, Irena Peeva, Department of Mathematics, Cornell University, Ithaca, NY 14853 USA; e-mail: irena@math.cornell.edu

    Number theory, Henri Darmon

3. GEOMETRIC ANALYSIS & PDE

    Coordinating Editor: Alexander A. Kiselev, Department of Mathematics, Duke University, 120 Science Drive, Rm 117 Physics Bldg, Durham, NC 27708 USA; e-mail: kiselev@math.duke.edu

    Differential geometry and geometric analysis, Ailana M. Fraser, Department of Mathematics, University of British Columbia, 1984 Mathematics Road, Room 121, Vancouver BC V6T 1Z2, Canada; e-mail: afraser@math.ubc.ca

    Harmonic analysis and partial differential equations, Monica Visan, Department of Mathematics, University of California Los Angeles, 520 Portola Plaza, Los Angeles, CA 90095 USA; e-mail: visan@math.ucla.edu

    Partial differential equations and functional analysis, Alexander A. Kiselev

    Real analysis and partial differential equations, Joachim Krieger, Bâtiment de Mathématiques, École Polytechnique Fédérale de Lausanne, Station 8, 1015 Lausanne Vaud, Switzerland; e-mail: joachim.krieger@epfl.ch

4. ERGODIC THEORY, DYNAMICAL SYSTEMS & COMBINATORICS

    Coordinating Editor: Vitaly Bergelson, Department of Mathematics, Ohio State University, 231 W. 18th Avenue, Columbus, OH 43210 USA; e-mail: vitaly@math.ohio-state.edu

    Algebraic and enumerative combinatorics, Jim Haglund, Department of Mathematics, University of Pennsylvania, Philadelphia, PA 19104 USA; e-mail: jhaglund@math.upenn.edu

    Probability theory, Robin Pemantle, Department of Mathematics, University of Pennsylvania, 209 S. 33rd Street, Philadelphia, PA 19104 USA; e-mail: pemantle@math.upenn.edu

    Dynamical systems and ergodic theory, Ian Melbourne, Mathematics Institute, University of Warwick, Coventry CV4 7AL, United Kingdom; e-mail: I.Melbourne@warwick.ac.uk

    Ergodic theory and combinatorics, Vitaly Bergelson

5. ANALYSIS, LIE THEORY & PROBABILITY

    Coordinating Editor: Stefaan Vaes, Department of Mathematics, Katholieke Universiteit Leuven, Celestijnenlaan 200B, B-3001 Leuven, Belgium; e-mail: stefaan.vaes@wis.kuleuven.be

    Functional analysis and operator algebras, Stefaan Vaes

    Harmonic analysis, PDEs, and geometric measure theory, Svitlana Mayboroda, School of Mathematics, University of Minnesota, 206 Church Street SE, 127 Vincent Hall, Minneapolis, MN 55455 USA; e-mail: svitlana@math.umn.edu

    Probability theory and stochastic analysis, Davar Khoshnevisan, Department of Mathematics, The University of Utah, Salt Lake City, UT 84112 USA; e-mail: davar@math.utah.edu

SELECTED PUBLISHED TITLES IN THIS SERIES

1296 **Christophe Cornut,** Filtrations and Buildings, 2020

1295 **Lisa Berger, Chris Hall, René Pannekoek, Jennifer Park, Rachel Pries, Shahed Sharif, Alice Silverberg, and Douglas Ulmer,** Explicit Arithmetic of Jacobians of Generalized Legendre Curves Over Global Function Fields, 2020

1294 **Jacob Bedrossian, Pierre Germain, and Nader Masmoudi,** Dynamics Near the Subcritical Transition of the 3D Couette Flow I: Below Threshold Case, 2020

1293 **Benjamin Jaye, Fedor Nazarov, Maria Carmen Reguera, and Xavier Tolsa,** The Riesz Transform of Codimension Smaller Than One and the Wolff Energy, 2020

1292 **Angel Castro, Diego Córdoba, and Javier Gómez-Serrano,** Global Smooth Solutions for the Inviscid SQG Equation, 2020

1291 **Vasilis Chousionis, Jeremy Tyson, and Mariusz Urbanski,** Conformal Graph Directed Markov Systems on Carnot Groups, 2020

1290 **Harold Rosenberg and Graham Smith,** Degree Theory of Immersed Hypersurfaces, 2020

1289 **Pavel M. Bleher and Guilherme L. F. Silva,** The Mother Body Phase Transition in the Normal Matrix Model, 2020

1288 **Alexander Blokh, Lex Oversteegen, Ross Ptacek, and Vladlen Timorin,** Laminational Models for Some Spaces of Polynomials of Any Degree, 2020

1287 **Mike Hochman,** On Self-Similar Sets with Overlaps and Inverse Theorems for Entropy in $\mathbb{R}^d$, 2020

1286 **Andrew J. Blumberg and Michael A. Mandell,** Localization for $THH(ku)$ and the Topological Hochschild and Cyclic Homology of Waldhausen Categories, 2020

1285 **Zhaobing Fan, Chun-Ju Lai, Yiqiang Li, Li Luo, and Weiqiang Wang,** Affine Flag Varieties and Quantum Symmetric Pairs, 2020

1284 **Rodney G. Downey, Keng Meng Ng, and Reed Solomon,** Minimal Weak Truth Table Degrees and Computably Enumerable Turing Degrees, 2020

1283 **Antonio Alarcón, Franc Forstnerič, and Francisco J. López,** New Complex Analytic Methods in the Study of Non-Orientable Minimal Surfaces in $\mathbb{R}^n$, 2020

1282 **David Carchedi,** Higher Orbifolds and Deligne-Mumford Stacks as Structured Infinity-Topoi, 2020

1281 **S. V. Ivanov,** The Bounded and Precise Word Problems for Presentations of Groups, 2020

1280 **Michael Handel and Lee Mosher,** Subgroup Decomposition in $\text{Out}(F_n)$, 2020

1279 **Cristian Gavrus and Sung-Jin Oh,** Global Well-Posedness of High Dimensional Maxwell–Dirac for Small Critical Data, 2020

1278 **Peter Poláčik,** Propagating Terraces and the Dynamics of Front Like Solutions of Reaction-Diffusion Equations on $\mathbb{R}$, 2020

1277 **Henri Lombardi, Daniel Perrucci, and Marie-Françoise Roy,** An Elementary Recursive Bound for Effective Positivstellensatz and Hilbert's 17th Problem, 2020

1276 **Victor Beresnevich, Alan Haynes, and Sanju Velani,** Sums of Reciprocals of Fractional Parts and Multiplicative Diophantine Approximation, 2020

1275 **Laurent Berger, Peter Schneider, and Bingyong Xie,** Rigid Character Groups, Lubin-Tate Theory, and (φ, Γ)-Modules, 2020

1274 **Gonzalo Fiz Pontiveros, Simon Griffiths, and Robert Morris,** The Triangle-Free Process and the Ramsey Number $R(3, k)$, 2020

1273 **Massimiliano Berti and Riccardo Montalto,** Quasi-periodic Standing Wave Solutions of Gravity-Capillary Water Waves, 2020

For a complete list of titles in this series, visit the
AMS Bookstore at **www.ams.org/bookstore/memoseries/**.